Recent Developments in Gauge Theories

NATO ADVANCED STUDY INSTITUTES SERIES

A series of edited volumes comprising multifaceted studies of contemporary scientific issues by some of the best scientific minds in the world, assembled in cooperation with NATO Scientific Affairs Division.

Series B. Physics

Recent Volumes in this Series

Volume 52 – Physics of Nonlinear Transport in Semiconductors
edited by David K. Ferry, J. R. Barker, and C. Jacoboni

Volume 53 – Atomic and Molecular Processes in Controlled Thermonuclear Fusion
edited by M. R. C. McDowell and A. M. Ferendeci

Volume 54 – Quantum Flavordynamics, Quantum Chromodynamics, and Unified Theories
edited by K. T. Mahanthappa and James Randa

Volume 55 – Field Theoretical Methods in Particle Physics
edited by Werner Rühl

Volume 56 – Vibrational Spectroscopy of Molecular Liquids and Solids
edited by S. Bratos and R. M. Pick

Volume 57 – Quantum Dynamics of Molecules: The New Experimental Challenge to Theorists
edited by R. G. Woolley

Volume 58 – Cosmology and Gravitation: Spin, Torsion, Rotation, and Supergravity
edited by Peter G. Bergmann and Venzo De Sabbata

Volume 59 – Recent Developments in Gauge Theories
edited by G. 't Hooft, C. Itzykson, A. Jaffe, H. Lehmann, P. K. Mitter, I. M. Singer, and R. Stora

Volume 60 – Theoretical Aspects and New Developments in Magneto-Optics
Edited by Jozef T. Devreese

Volume 61 – Quarks and Leptons: *Cargèse 1979*
edited by Maurice Lévy, Jean-Louis Basdevant, David Speiser, Jacques Weyers, Raymond Gastmans, and Maurice Jacob

Volume 62 – Radiationless Processes
edited by Baldassare Di Bartolo

This series is published by an international board of publishers in conjunction with NATO Scientific Affairs Division

A Life Sciences B Physics	Plenum Publishing Corporation London and New York
C Mathematical and Physical Sciences	D. Reidel Publishing Company Dordrecht, Boston and London
D Behavioral and Social Sciences	Sijthoff & Noordhoff International Publishers
E Applied Sciences	Alphen aan den Rijn, The Netherlands, and Germantown U.S.A.

Recent Developments in Gauge Theories

Edited by

G. 't Hooft
Institute for Theoretical Physics
Utrecht, The Netherlands

C. Itzykson
CEN Saclay
Gif-sur-Yvette, France

A. Jaffe
Harvard University
Cambridge, Massachusetts

H. Lehmann
University of Hamburg
Hamburg, Federal Republic of Germany

P. K. Mitter
University of Paris
Paris, France

I. M. Singer
University of California
Berkeley, California

and

R. Stora
Center of Theoretical Physics, CNRS
Marseille, France

PLENUM PRESS • NEW YORK AND LONDON
Published in cooperation with NATO Scientific Affairs Division

Library of Congress Cataloging in Publication Data

Main entry under title:
Recent developments in gauge theories.

(NATO advanced study institutes series: Series B, Physics; v. 59)
"Published in cooperation with NATO Scientific Affairs Division."
Includes index.
1. Gauge fields (Physics)–Addresses, essays, lectures. I. 't Hooft, G. II. Series.
QC793.3F5R42 530.1'43 80-18528
ISBN 978-1-4684-7573-9 ISBN 978-1-4684-7571-5 (eBook)
DOI 10.1007/978-1-4684-7571-5

Proceedings of the NATO Advanced Study Institute on
Recent Developments in Gauge Theories, held in Cargèse, Corsica,
August 26–September 8, 1979.

Softcover reprint of the hardcover 1st edition 1980
A Division of Plenum Publishing Corporation
227 West 17th Street, New York, N.Y. 10011

PREFACE

Almost all theories of fundamental interactions are nowadays based on the gauge concept. Starting with the historical example of quantum electrodynamics, we have been led to the successful unified gauge theory of weak and electromagnetic interactions, and finally to a non abelian gauge theory of strong interactions with the notion of permanently confined quarks. The early theoretical work on gauge theories was devoted to proofs of renormalizability, investigation of short distance behaviour, the discovery of asymptotic freedom, etc.., aspects which were accessible to tools extrapolated from renormalised perturbation theory. The second phase of the subject is concerned with the problem of quark confinement which necessitates a non-perturbative understanding of gauge theories. This phase has so far been marked by the introduction of ideas from geometry, topology and statistical mechanics in particular the theory of phase transitions. The 1979 Cargèse Institute on "Recent Developments on Gauge Theories" was devoted to a thorough discussion of these non-perturbative, global aspects of non-abelian gauge theories. In the lectures and seminars reproduced in this volume the reader will find detailed reports on most of the important developments of recent times on non perturbative gauge fields by some of the leading experts and innovators in this field. Aside from lectures on gauge fields proper, there were lectures on gauge field concepts in condensed matter physics and lectures by mathematicians on global aspects of the calculus of variations, its relation to geometry and topology, and related topics. The presence of mathematicians as enthusiastic participants and masterful lecturers in this school deserves special mention. We hope this trend will continue in the future and that, in the last quarter of this century, common concerns about the fundamental interactions will bring ever closer the physical and mathematical communities as in the days of yore.

We wish to express our gratitude to NATO whose generous financial contribution made it possible to organise this school. We also thank the Centre National de la Recherche Scientifique, the Délégation à la Recherche Scientifique et Technique, the

C.E.N. de Saclay, as well as the University of Hamburg for financial help. We thank the University of Nice for making available to us the facilities of the Institut d'Etudes Scientifiques de Cargèse. Grateful thanks are due to Marie-France Hanseler for much help with the material aspects of the organisation. Last but not least we thank the lecturers and participants for their enthusiastic involvement which contributed much to the scientific atmosphere of the school.

G. 't Hooft
C. Itzykson
A. Jaffe
H. Lehmann
P.K. Mitter
I.M. Singer
R. Stora

CONTENTS

REMARKS ON MORSE THEORY

M. F. Atiyah

University of Oxford
Mathematical Institute
24-29 St. Giles, Oxford

§1. INTRODUCTION

Morse theory is a topological approach to the Calculus of Variations. It aims to relate the critical points of a functional to the topology of the function space on which the functional is defined. It is only directly applicable in special rather restrictive conditions, notably for problems involving one independent variable. However I will discuss a number of special examples, in some of which the Morse theory really works, and others in which it clearly fails but where nevertheless some aspects appear still to survive. These examples include those of physical interest and it would be interesting to investigate these further. One can make a number of speculations in this direction.

§2. GEODESICS ON A GROUP

The classical example for which Morse developed his theory is that of geodesics. This can be formulated as follows. Given a compact Riemannian manifold M we consider closed paths in M described by a periodic function $f(t)$ with values in M or more formally a map

$$f: S^1 \to M$$

where S^1 is the unit circle. As our functional we take

$$E(f) = \int |f'(t)|^2 dt$$

where $f'(t)$ is a tangent vector to M and $|f'(t)|$ denotes its length in the Riemannian metric of M. The critical points of

E , that is the solutions of the corresponding Euler-Lagrange equations, correspond to closed geodesics on M parametrized by arc length.

As a special example we may take $M = G$ a compact Lie group with bi-invariant metric. The closed geodesics through the identity of G are just the closed 1-parameter subgroups, and so are completely known in terms of the group structure of G . Using the Morse theory one can then derive information about the topology of the loop space $\Omega(G)$, i.e. the space of closed paths based at the identity. This programme was carried through many years ago by Bott [3]. As a simple example if $G = SU(2)$, the 3-sphere, the closed geodesics are the great circles (including their n-fold iterates and for $n = 0$ the degenerate point-map). From this one finds that the homology of ΩG is infinite cyclic in every even dimension and zero in odd dimensions.

§3. YANG-MILLS IN TWO DIMENSIONS

We consider Yang-Mills theory over a compact two-dimensional surface, for example the 2-sphere S^2 . The Yang-Mills equations are rather trivial in 2-dimensions since they assert that the field $F_{\mu\nu}$ (or curvature) is covariant constant. From this one can easily describe all solutions and one finds that they are precisely the <u>homogeneous</u> connections, i.e. connections which admit the action of SU(2) (the double cover of the rotation group SO(3)). By a general principle homogeneous connections with group G are given by homomorphisms of the isotropy group, in this case the circle subgroup of SU(2), into G . Thus we see that the critical points of the Yang-Mills action correspond to the critical points of §2.

If we look at the function space situation, the space of all connections is contractible but since our action is invariant under gauge transformations it is appropriate to factor out by these (actually for technical reasons it is best to use only gauge transformations which are fixed to be the identity at some base point of S^2). The resulting space turns out to have the same homotopy type as ΩG . Thus we see that our example is, in its essential features, like that of §2. This should not be too surprising to physicists since, if we put the Yang-Mills theory in Hamiltonian form (treating 1-dimensional space as compact) we get the motion of a free particle on G , which is precisely the Hamiltonian version of §2.

Mathematically there is interest in replacing S^2 by a general surface of genus g and this case has been extensively studied by Bott and myself. For a preliminary account see [1].

It is perhaps worth pointing out at this stage that solutions of Yang-Mills over S^2 occur naturally as asymptotic data of stationary (hypothetical) magnetic monopoles with group G . Unstable critical points (i.e. those which are not minima of the action) correspond to unstable monopoles, according to recent results of S. Coleman and others.

§4. CP_n-MODELS

Our next example is also two-dimensional but is much more interesting physically than that of §3. We consider maps $f: S^2 \to CP_n$ where S^2 is the standard 2-sphere and CP_n is complex projective n-space with its standard metric (arising from its description as a homogeneous space $U(n+1)/U(1)\times U(n)$). As our functional we take

$$E(f) = \int_{S^2} \|df\|^2$$

where $\|df\|^2$ is the Hilbert-Schmidt norm of the linear map df (Recall that this norm for any linear map $T: V \to W$ of Euclidean spaces is defined by $\|T\|^2 = \text{Trace } T^*T$).

The map f has an integer topological invariant k (its "degree") and the absolute minimum of E(f) , for given k , is achieved by holomorphic (in fact rational) maps f . For $n = 1$, when $CP_n = S^2$, it is known that there are no other critical points. This shows that the Morse theory cannot apply here, otherwise we could deduce that the whole function space (of all smooth maps f) could be retracted, along paths of steepest descent for E , to the space of rational maps. But this would give a contradiction since the space of rational maps of degree k is itself finite dimensional (in fact of dimension $2k + 1$) whereas the function space in question is known by topologists to have homology in arbitrarily large dimensions.

On the other hand if we look at the space M(k,n) of rational maps $S^2 \to CP_n$, this increases in dimension with k and one can wonder whether it approximates the whole function space as $k \to \infty$. The answer is affirmative, according to a remarkable recent theorem of G. B. Segal [4] . This result of Segal's would be a consequence of Morse theory if this applied and if, in addition, it were true that a critical map f of degree k is either an absolute minimum or else has large Morse index (in fact index $>k$). This has suggested that one should try to modify the Morse theory in some appropriate way so that it will apply to this case and explain Segal's result. One possibility is to try to identify and define some kind of "ideal critical point", namely a limit of a suitable sequence of maps f which is not itself an admissible map, but which plays the role of critical point for the Morse theory. Candidates have been suggested for such ideal maps, and for these the Lagrange density would typically consist of a number of delta-functions located at different points of S^2

§5. YANG-MILLS IN FOUR DIMENSIONS

For the physically most interesting case of Yang-Mills over the 4-sphere S^4 there is considerable similarity with the model in §4. In particular there is a topological invariant k and absolute minima of the action, for given k, are given by (suitable) rational functions. The situation is naturally a good deal more complicated than in the simpler CP_n-model but the main features appear similar. For $SU(2)$ no solutions are known which are not absolute minima and it is widely believed that no such solutions exist. Moreover it was shown in [2], that for large k all the homology of the function space (as in §3) is contained in the space of instantons.

One can therefore speculate, as in §4, on ways to modify the Morse theory so that it will apply to this situation and explain the topology.

One approach which has both mathematical and physical interest would be to alter the Yang-Mills Lagrangian by addition of terms involving other (linear) fields. This would not affect the topology of the function space (linear spaces being contractible) but more critical points, for the coupled equations, might exist. The original minima of the Yang-Mills Lagrangian would still be there (with all other fields put to zero), but new critical points might require non-zero values for the other fields. Topological considerations (see [2]) do indeed suggest that something on these lines is reasonable with the linear fields being taken as spinors (i.e. fermions).

In conclusion one might ask what relevance this discussion of Morse theory has for the physics. A plausible answer runs as follows. We would like to understand the Euclidean functional integral approach to a quantized Yang-Mills theory. Integration over a non-linear manifold (namely the space of connections modulo gauge transformations) requires a good hold of the topological nature of the manifold. It seems reasonable that the topological complications should reflect themselves at various points in the analysis. Studying the classical solutions is one aspect of the problem and may provide some guidance.

REFERENCES

[1] M. F. Atiyah and R. Bott, Yang-Mills and bundles over algebraic curves, Volume dedicated to V. K. Patodi

to be published by the Indian Academy of Science.

[2] M. F. Atiyah and J.D.S. Jones, Topological aspects of Yang-Mills theory, Comm. Math. Phys. 61 (1978), 97-118.

[3] R. Bott, An application of the Morse theory to the topology of Lie groups, Bull. Soc. Math. France, 84 (1956), 251-281.

[4] G. B. Segal, The topology of spaces of rational functions. Acta Math. 143 (1979), 39-72.

MORSE THEORETIC ASPECTS OF YANG-MILLS THEORY *

Raoul Bott

Department of Mathematics
Harvard University
Cambridge, Massachusetts 02138

Let me start in the manner I have learned of late from all you Physicists, with a modest list of topics to be covered in these two lectures. My topics are:

(i) Algebraic topology
(ii) Morse theory
(iii) Equivariant Morse theory
(iv) Pertinence of (i), (ii), and (iii) to the solutions of the classical Yang-Mills Equations.

Unfortunately I have no direct applications to Quantum theory of this framework to boast of. Still these ideas are sufficiently close to other topological considerations that I have heard here, that I hope they could be of use to you at some future time.

(i) Algebraic topology. The simplest topological invariant of a space X, is the number of connected pieces into which it falls. This number, $\#(X)$, can also be defined as the dimension of the vector-space $H^0(X)$, of locally constant functions on the space:

$$\#(X) = \dim H^0(X) . \tag{1.1}$$

* This work was supported in part through funds provided by the National Science Foundation under grant 33-966-7566-2.

In case X is a smooth manifold M we may restrict ourselves to smooth functions and then the space of locally constant functions is simply the space of solutions of the equation

$$df = 0 \quad ,$$

where d is the operator assigning to a smooth function its gradient

$$df = \frac{\partial f}{\partial x^i} dx^i \quad .$$

Now the first miraculous fact of differential topology is that this operator d, admits a natural extension to a whole sequence of tensorial differential operators - also written d, - going from the space of smooth antisymmetric covariant tensorfields with k indices, to those with k + 1 indices.

Writing $\Omega^k(M)$ or simply Ω^k for the vector space of k-indexed tensors, one thus obtains a sequence:

$$(1.2) \qquad \Omega^0 \xrightarrow{d} \Omega^1 \xrightarrow{d} \cdots \xrightarrow{d} \Omega^n \qquad n = \dim M \quad .$$

Furthermore it follows from the symmetry

$$\frac{\partial^2 f}{\partial x^i \partial x^j} = \frac{\partial^2 f}{\partial x^j \partial x^i}$$

that $d^2 = 0$ in (1.2). This in turn leads one to define

$$H^i(M) = \{\text{space of solution } d\omega = 0 \\ \omega \text{ in } \Omega^i/d\eta, \ \eta \text{ in } \Omega^{i-1}\}$$

and the second miraculous fact of differential topology is now that this vector space is finite dimensional provided for instance, that M can be covered by a finite number of open sets each diffeomorphic to $\mathbb{R}^n$.

The dimensions of these spaces therefore give rise to diffeomorphism invariants of M, and can be conveniently combined into a polynomial

$$(1.3) \qquad P(M;t) = \Sigma t^k \dim H^k(M) \quad .$$

This "Poincaré Polynomial" of M then turns out to be invariant under homeomorphisms and even under homotopy equivalences. In many instances the number $P_1(M)$ can be thought of as a measure of the topological complexity of M.

Examples.

$$P_t(\mathbb{R}^1) = 1 = P_t(\mathbb{R}^n)$$

$$P_t(S^1) = 1+t \qquad (P_t(S^1) \equiv P(S^1;t))$$

$$P_t(S^n) = 1+t^n$$

$$P_t(\mathbb{C}P^n) = 1+t^2+t^4+\cdots+t^{2n} \quad .$$

In this table, $\mathbb{R}^n$ denotes Euclidean n-space; S^1 denotes the circle, S^n the n-sphere: $x_1^2+\cdots+x_{n+1}^2 = 1$ in $\mathbb{R}^{n+1}$; and $\mathbb{C}P^n$

complex projective n-space. Thus $\mathbb{C}P^n = \mathbb{C}^{n+1} - 0/\sim$ where $(x_1, \cdots, x_{n+1}) \sim (\lambda x_1, \cdots, \lambda x_{n+1})$ for any nonzero complex number λ.

Note that as $n \to \infty$, this formula gives

$$P_t(\mathbb{C}P_\infty) = \frac{1}{1-t^2} \tag{1.4}$$

for the "Poincaré series" of the infinite complex projective space of rays in Hilbert space.

Under the disjoint union, and cartesian product the Poincaré polynomial has the following simple behavior:

$$\begin{aligned} P_t(M \amalg N) &= P_t(M) + P_t(N) \\ P_t(M \times N) &= P_t(M) \cdot P_t(N) \quad . \end{aligned} \tag{1.5}$$

Thus in particular for the torus $T = S^1 \times S^1$,

$$P_t(S^1 \times S^1) = 1+2t+t^2 \quad . \tag{1.6}$$

So much for the first topic - at least for the time being. I turn now to the second one:

(ii) Morse theory. Suppose then that on M we are given a smooth function f. A point p on M is then called a critical point of f, iff : $df|_p = 0$ i.e. iff

$$\left.\frac{\partial f}{\partial x^i}\right|_p = 0 \;;\quad x^1, \cdots, x^n \text{ local coordinates at } p. \tag{2.1}$$

At such a point two aspects of the matrix $\left.\dfrac{\partial^2 f}{\partial x^i \partial x^j}\right|_p$ make invariant sense. These are its rank and the number of negative (or positive) Eigenvalues.

Morse called n-rank the nullity of p as a critical point of f, and the number of negative Eigenvalues the index of p as a critical point of f. Finally he called a function nondegenerate iff all its critical points have nullity $= 0$, and to such a function he assigned a polynomial:

$$\mathfrak{m}_t(f) = \sum_{df|_p = 0} t^{\lambda_p} \qquad \lambda_p = \text{index}(p)$$

which we will call the "Morse Polynomial of f."

Examples: In the figures below I have sketched some surfaces in $\mathbb{R}^3$. The z-coordinate restricts to a function on each of these and I have indicated the critical points of this height-function - together with their indices - on each figure.

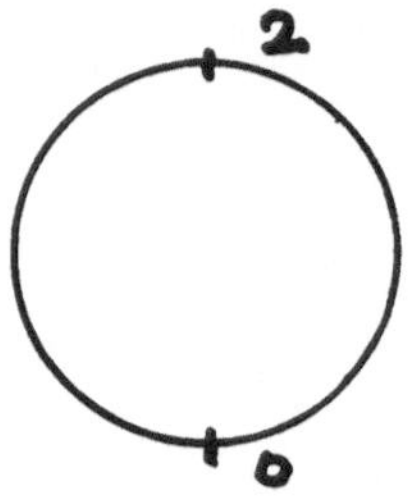

$$\mathfrak{m}_t(z) = 1 + t^2 \qquad M = S^2$$

figure 1

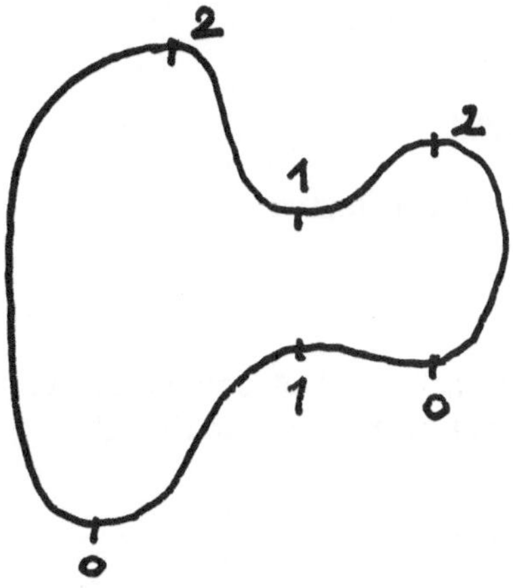

$\mathfrak{m}_t(z) = 2 + 2t + 2t^2$; $M = S^2$

figure 2

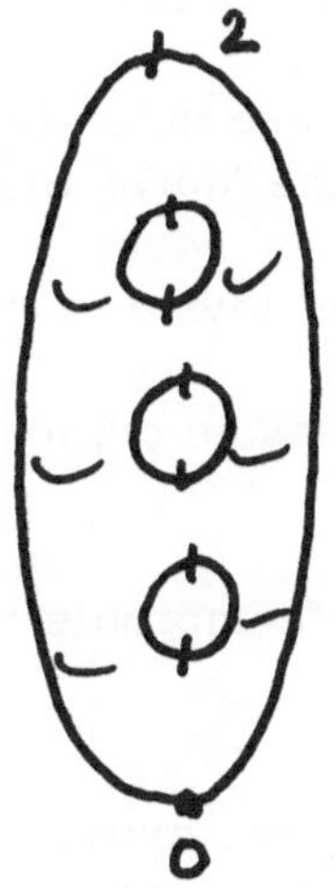

$\mathfrak{m}_t(z) = 1 + 6t + t^2$

M = Surface of genus 3

figure 3

Morse's beautiful observation - going back to the 1920's - is now the following one. He found first of all, that on compact manifolds M the polynomial $P(M;t)$ <u>plays the role of a lower bound for all Morse polynomials</u> $\mathfrak{m}(f;t)$ <u>with</u> f <u>a nondegenerate function on</u> M. <u>Thus</u>

(2.1) $$\mathfrak{m}(f;t) \geq P(M;t) \quad \text{for all} \quad t > 0 \quad .$$

Acually more is true - the coefficients of $\mathfrak{m}(f;t)$ and $P(M;t)$ satisfy the so-called "Morse inequalities." Precisely, this amounts to the following:

<u>If we define the polynomial</u> $Q(t)$ <u>by</u>

(2.2) $$P(M;t) - \mathfrak{m}(f;t) = (1+t)\,Q(t)$$

<u>then all coefficients of</u> $Q(t)$ <u>are nonnegative.</u>

Note that an immediate corollary is the relation

(2.3) $$P(M;-1) = \mathfrak{m}(f;-1)$$

for all nondegenerate f. In general $\mathfrak{m}(f;t)$ will of course be much larger than $P(M;t)$ but if f is nondegenerate and satifies the equation

$$\mathfrak{m}(f;t) = P(M;t)$$

we speak of a perfect <u>Morse function.</u> Such functions then have the minimal extremal behavior tolerated by the underlying manifold.

Let me say a few words concerning the proof of these inequalities. They are based on the following two basic theorems of the Morse theory.

Let f be given and let us write M_a for the set where $f \leq a$.

THEOREM A. Suppose M is compact and that f has no critical points in the range $[a,b]$.

Then $M_a \simeq M_b$, in the sense that there is a diffeomorphism between these two spaces.

THEOREM B. Suppose that there is a single nondegenerate critical point p, $a < f(p) < b$ in the range $[a,b]$, and that its index is λ. Then

(2.2) $$M_b \cong M_a \underset{\alpha}{\cup} (e^{\lambda} \times e^{n-\lambda})$$

The meaning of (2.2) is that M_b is obtained from M_a by attaching a "thickened handle of dim λ". Technically, e_{λ} is a cell of dimension λ, i.e. the set:

$$x_1^2 + \cdots + x_{\lambda}^2 \leq 1 \quad \text{in} \quad \mathbb{R}^{\lambda},$$

and the precise meaning of the formula

$$M_b = M_a \underset{\alpha}{\cup} e_\lambda \times e_{n-\lambda}$$

is that M_b is obtained from the disjoint union

$$M_a \amalg e_\lambda \times e_{n-\lambda}$$

by glueing the set $(\partial e_\lambda) \times e_{n-\lambda} = S^{\lambda-1} \times e_{n-\lambda}$ into ∂M_a via α.

The following diagram should make this theorem intuitively clear.

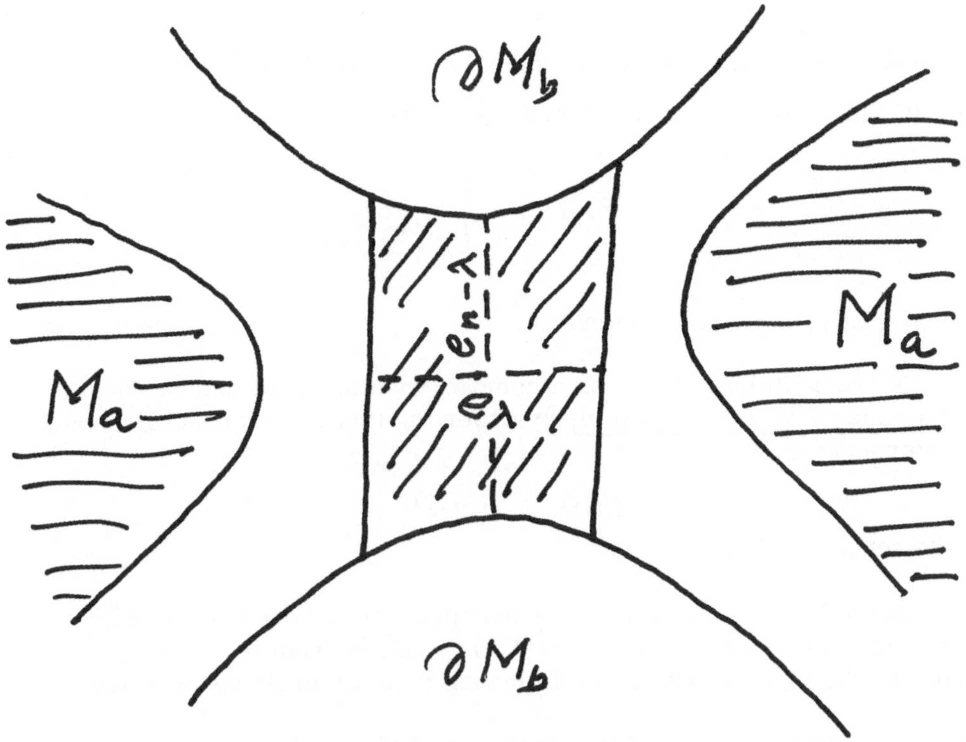

figure 4

Here we have drawn the sets M_b and M_a as they appear near a critical point p of a function f, which near p is of the form

$$f = y^2 - x^2 \quad .$$

Clearly if we remove the "thickened handle" $e_\lambda \times e_{n-\lambda}$ from M_b, the rest is diffeomorphic to M_a. Q. E. D.

In any case, equipped with these two theorems the Morse inequalities follow from the - by now - standard properties of $P(M;t)$. Namely these imply that a priori the change in $P(M;t)$ under the attaching of a thickened λ-handle is either t^λ or $-t^{(\lambda-1)}$:

$$\Delta P(M;t) = \begin{cases} t^\lambda \\ -t^{\lambda-1} \end{cases} .$$

On the other hand the change i. e. the contribution of the new critical point in M_b to $\mathfrak{m}(f;t)$ is clearly t^λ. Thus

$$\Delta P(M;t) - \Delta \mathfrak{m}(f;t) = \begin{cases} 0 \\ t^{\lambda-1} \overset{\text{or}}{(1+t)} \end{cases} ,$$

and so (2. 1) is proved inducitvely.

To recapitulate, then, on a compact manifold, the <u>Morse theory refines the maximum principle</u>, by asserting that for all nondegenerate functions f:

$$\mathfrak{m}(f;t) \geq P(M;t)$$

in the sense of (2. 1).

Actually there are fortuitous instances where one can use the Morse inequalities to compute $P(M;t)$. Indeed from the $(1+t)$ factor on the right of (2. 1) the following is an elementary exercise:

<u>The Lacunary Principle: Suppose that in</u> $\mathfrak{m}(f;t)$ <u>the product of any two consecutive coefficients</u> $m_i \cdot m_{i+1}$ <u>is zero. Then</u> f <u>is a perfect Morse function for</u> M, <u>i. e.</u>

$$\mathfrak{m}(f;t) = P(M;t) .$$

Thus for example, just knowing that the height function x_{n+1} on the n-sphere

$$x_1^2 + \cdots + x_{n+1}^2 = 1$$

has as Morse series $1+t^n$, implies - for $n > 1$ - that $P_t(S^n) = 1+t^n$! The formula is of course true for $n = 1$, also, but there does not follow from the inequalities. As a more interesting example of this principle, let us compute $P(\mathbb{C}P_n; t)$. For this purpose let A be a hermitian matrix on $\mathbb{C}^{n+1}$ and consider the expression

$$f(x) = \frac{\langle Ax|x\rangle}{\langle x|x\rangle} \qquad x \in \mathbb{C}^{n+1}$$

we clearly have $f(\lambda x) = f(x)$ for $\lambda \neq 0$, so that f is a function on $\mathbb{C}P_n$. The stationary points of this f, are now - using your beloved Lagrange multipliers - easily seen to be the Eigen-rays of A:

$$A x_i = \lambda_i x_i \quad .$$

If we now order these: $\lambda_1 < \lambda_2 < \cdots < \lambda_{n+1}$ (assuming no two coincide - as we may -) then λ_1 is the minimum of f, and has index 0, while one computes with just a little effort that the index at λ_i is $t^{2(i-1)}$. (Moving in the earlier directions $x_1, \ldots, x_{i-1}$ decreases f to second order and as we are in complex space the total number independent descent directions is $2(i-1)$.)

Thus

$$\mathfrak{m}(f;t) = 1 + t^2 + \cdots + t^{2n} \quad ,$$

and hence by the Lacunary Principle:

$$P(M;t) = 1 + t^2 + \cdots + t^{2n} \quad .$$

Q. E. D.

So far we have been discussing what I like to call the Baby Morse Theory. The Morse theory proper deals with much more general situations. To explain these it will be convenient to use the following terminology:

I will say that the Morse theory works (for a certain class of functions on a certain class of spaces) if a priori the Morse inequalities hold in this framework. Thus for instance the Morse theory doe not work for nondegenerate functions on noncompact manifolds. Indeed such a manifold admits functions without any critical points! On the other hand,

and this is the Morse theory proper, it does work in a large class of Calculus of Variation problems. In particular Morse showed that if Ω_a^b denotes the space of piecewise smooth curves μ, on a complete Riemann manifold M (compact or not):

$$\mu : [0,1] \to M \ , \quad u(0) = a \ , \ u(1) = b \ ,$$

parametrized by arc-length, and if $E(\mu)$ is the energy functional on Ω_a^b :

$$E(u) = \int_0^1 |\dot{u}|^2 \, dt$$

then the Morse theory works.

This infinite dimensional setting of the Morse theory is beautifully geometric and in every way satisfying. First of all one has the following interpretation of the pertinent objects:

Critical point of E	<—>	geodesic segment joining a to b
the geodesic s is a nondegenerate cricial point,	<—>	b is not conjuage to a along s
index of the critical points	<—>	number of conjugate points of a in the interior of s .

To illustrate the power of this dictionary let us compute the Morse "polynomial" $\mathfrak{m}(E;t)$ when M is the n-sphere S^n, in its in its usual metric, a is the North-pole and b is any point - other than one of the poles.

The diagram below then indicates the first few geodesics on S^n joining a to b with their indices.

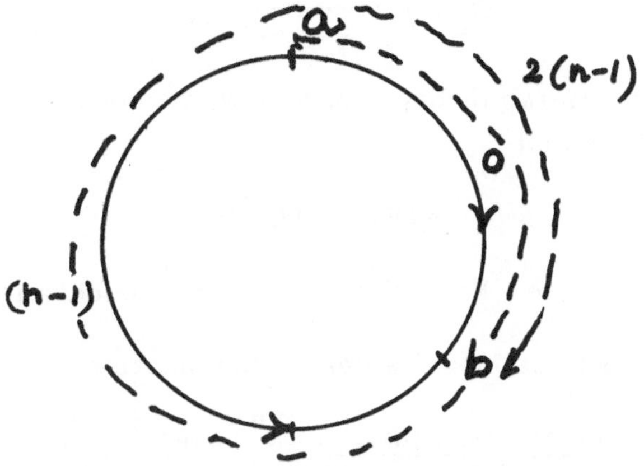

figure 5

The indices are here very easy to compute as the only conjugate point of a is its antipode and it has multiplicity is $(n - 1)$ as a conjugate point.

With this understood, it is clear that the Morse "polynomial" of E is given by

$$\mathfrak{m}(E;t) = 1 + t^{n-1} + t^{2(n-1)} + \cdots \quad . \tag{2.4}$$

In short then $\mathfrak{m}(E;t)$ has become a series rather than a polynomial - which seems reasonable enough as Ω_a^b is an infinite dimensional manifold.

Because the Morse theory works in this instance, we would expect $P_t(\Omega_a^b)$ also to become a formal series, and indeed, for $n > 1$, we would expect by the Lacunary Principle that

$$\mathfrak{m}(E;t) = P(\Omega_a^b;t) \tag{2.5}$$

This turns out to be the case so that - as was already found by Morse in the twenties -

$$P(\Omega_a^b; t) = \frac{1}{1 - t^{2(n-1)}} \quad . \tag{2.6}$$

Turning the Morse theory around now, we conclude from the Morse inequalities that

$$\mathfrak{M}_t(E) > P(\Omega_a^b; t)$$

for any nondegenerate E of the type under consideration.

Thus one obtains Morse's celebrated theorem:

For any Riemann structure on S^n, there must be an infinite number of geodesics joining two points a, b which are not conjugate along any geodesic segment joining them.

Actually this turns out to be a generic condition on a and b, and so we can also assert the same result for "almost all pairs" a, b. If one then pushes a bit more the assertion can be established for all pairs a, b.

In this framework it is now natural to ask whether the Morse theory works for the Yang-Mills functional, and if not, by how much it fails. Now recall that the Yang-Mills functional

$$A \longrightarrow S(A)$$

is defined on the space $\mathfrak{a}(P)$ of connections of some principal bundle P over some manifold M, and furthermore that $\mathfrak{a}(P)$ is naturally an affine space. In particular then $\mathfrak{a}(P)$ is contractible to a point and hence has Poincaré series 1! Thus the extrema of Yang-Mills are certainly not forced by the topology of $\mathfrak{a}(P)$.

What does force extrema is the fact that S has a large group of symmetries - namely the whole group of local gauge transformations - and, as I will try to explain in my next lecture, symmetries also force critical points although great care has to be taken to measure their effect properly.

But let me close here with an extension of the concept of non-degeneracy which I noticed a long time ago, which has many uses and which will be essential in the program just outlined. Returning to the Baby Morse theory for the moment, we will from now on call f nondegenerate iff df vanishes along smooth manifolds $\overline{\{N\}}$, with the property that

$$\det \left. \frac{\partial^2 f}{\partial y^i \partial y^j} \right|_{y=0} \neq 0$$

along N, when the y's are local coordinates normal to N, - i.e. in terms of which

$$N = \{y_1 = 0, y_2 = 0, \cdots, y_k = 0\} \quad .$$

On each connected N of this type the "index"

$$(2.7) \qquad \lambda_N = \# \text{ of negative eigenvalues of } \frac{\partial^2 f}{\partial y^i \partial y^j}$$

is then well defined, that is a constant along N.

For functions f, nondegenerate in this sense one now defines the Morse Series by

$$\mathfrak{m}(f;t) = \sum_N t^{\lambda_N} P(N;t)$$

where N runs over the critical manifolds of f. Thus in this situation each N contributes $t^{\lambda_N} P(N;t)$. With this understood one then has the following theorem.

Provided f satisfies a mild orientability along each N, the Morse ineqaulities:

$$\mathfrak{m}(f;t) \geq P(M;t) \quad ,$$

persist in this new setting. (That is for any nondegenerate f on a compact M.)

The orientability condition here is simply that the negative bundle ν_N along N, formed by the directions of steepest descent be orientable. (If this fails one must replace P(M;t) by a slightly modified Poincaré polynomial to keep the Morse inequalities.)

As a very simple verification consider a doughnut placed on the table in the usual manner - i. e. stably. Its boundary is then a torus, and the height function has as extrema two nondegenerate circles. The bottom one is a minimum, while the top one clearly has index 1. Thus

$$\mathfrak{M}(f;t) = (1+t) + t(1+t) \tag{2.8}$$

yielding once again $1 + 2t + t^2$, so that this f is again a perfect Morse function.

My first contribution to Morse theory was really the observation that for a compact Lie group G, in its left and right invariant metric the energy function E, was a nondegenerate perfect Morse function on Ω_a^b for any two points a and b. Later H. Samelson and I extended this result to all symmetric spaces.

For instance on the spheres, say on S^2, taking a and b as antipodes, the critical sets are now all circles, and the indices - computed as before - are 0, 2, 4, etc. Thus

$$\begin{aligned} \mathfrak{M}_t(E) &= (1+t) + t^2(1+t) + \cdots \\ &= \frac{1+t}{1-t^2} = \frac{1}{1+t}, \end{aligned}$$

as it should be. The implications of this result on $\Omega_a^b G$ actually turned out to be rather greater than I at first expected, for instance they led me a few years later to the periodicity phenomena in the homotopy of the classical groups. All that is too far afield to explain here.

However these notions do seem pertinent in our present context.

For, as M. Atiyah and I discovered recently the "Equivariant Yang-Mills Theory" over a Riemann Surface does lead to new perfect nondegenerate function.

(iii) Equivariant Morse Theory. The problem I want to explore now is the following one. Suppose that f is a nondegenerate function on M, which in addition is invariant under the action of a Lie group G on M. What is then the minimal critical behavior of f? Or put differently are there any additional critical points forced on f by the symmetry condition? Two simple examples may serve to set the stage.

Example 1. Suppose f is defined on $\mathbb{R}^1$ and is invariant under the subgroup (δ) of integers, $\mathbb{Z}$, acting on $\mathbb{R}^1$ by translation. It is surely clear how to proceed here. Our symmetry condition on f implies that

$$f(x) = f(x+n) \qquad n \in \mathbb{Z} \quad ,$$

i.e. that f is periodic - and hence really a function, f/, on $\mathbb{R}^1/\mathbb{Z}$ - that is - the circle. Thus

$$\mathfrak{m}(f/ \ ;t) > P(S^1;t)) = 1 + t \quad .$$

In particular then f/ must have at least two critical points, while an f on $\mathbb{R}^1$ without the symmetry condition need have no critical points at all. Encouraged, we proceed to

Example 2. Let f be defined on S^2 and be invariant under the circle action of S^1 given by rotations about the z-axis.

The principle of Example 1, now leads us to consider the space S^2/S^1 and the function f/ which f induces on this quotient space by virtue of its symmetry. On the other hand S^2/S^1 is clearly just an interval. Thus this space is not anymore a manifold in the proper sense of the word. Furthermore $P(S^2/S^1;t) = 1$ and so a Morse principle relative to this polynomial would predict only one critical point for f/. On the other hand f must have at least two, all symmetry considerations aside, by virtue of the straight Morse theory on S^2!

The essential difference between these two situations is that in the first the action of G on M is free - that is the orbit of each point in M is G and this happens uniformly in a certain sense - while in the second it fails to be free. In fact the north and south poles are now fixed under all of G. Thus what seems to be happening, is, that the quotient construction $f \to f/G$; $M \to M/G$ is fine if G acts freely - and wrong if it does not.

Now this phenomenon has long been understood in homotopy theory and in this discipline they invented the concept of a homotopy quotient M//G, which turns out to be equivalent to M/G for free actions but which produces a quite different and very interesting space - and even a manifold in some sense - when G fails to be free. The idea of this construction is as follows. Observe first that if G acts freely on a manifold W, then the diagonal action of G on $M \times W$ will be free, whatever the action of G on M was. The homotopy theorist then argued that to define the homotopy quotient one should first simply pass from M to $M \times W$, with W some space which does not change the homotopy of M, and on which G acts freely. That is, we first seek a space W which has two properties:

(3.1) (1) G acts freely on W,

(3.2) (2) W is contractible to a point,

and equipped with such a W we define the homotopy quotient of an action of G on M by setting

$$M//G = M \times W/G \quad . \tag{3.3}$$

Now this direction of inquiry has been one of the most fundamental in modern topology, for, as you would discover upon a little reflection, the two conditions (1) and (2) on W at first seem well neigh incompatible, but then finally turn out to have - as far as homotopy is concerned - an essentially unique, even if surprising, solution. In the following table I have listed the W's satisfying (3.1) and (3.2) for some standard groups. I have also listed with them the quotient space W/G - which in view (3.3) may be thought of as the homotopy quotient of the most nonfree action imaginable, namely the action of G on a point:

$$\text{point}//G = W/G \quad . \tag{3.4}$$

Examples

G	W	W/G
$\mathbb{Z}$	$\mathbb{R}$	$\mathbb{R}/\mathbb{Z} = S^1$
$\mathbb{Z}^n$	$\mathbb{R}^n$	$\mathbb{R}^n/\mathbb{Z}^n = \underbrace{S^1 \times \cdots \times S^1}_{(n)}$
$\mathbb{Z}_2$	$S(H)$	$\mathbb{R}P_\infty$
$U(1) = S^1$	$S(H)$	$\mathbb{C}P_\infty$
$U(2)$	2-frames in H	$G_2(H)$
$U(n)$	n frames in H	$G_n(H)$.

Here $\mathbb{Z}$ denotes the integers, $\mathbb{Z}^n$ the direct product of $\mathbb{Z}$ with itself n times, $\mathbb{Z}_2$ the group $\{\pm 1\}$, and U(n) of course the unitary group. The first two W's of course come to mind immediately, on the other hand the rest should strike you as way out. In all of these think of H as a complex infinite-dimensional Hilbert space, and of S(H) as its unit sphere. The space of n-frames on H is then the space of n-tuples $\{x_1, \cdots, x_n\}$ of elements in S(H) which are mutually orthogonal. U(n) clearly acts on these

$$\{x_i\} \longrightarrow \{U_{ij} x_j\}$$

and the quotient gives precisely the Grassmannian $G_n(H)$ of all n-dimensional subspaces of H. For $n = 1$, this is simply the projective space.

All these examples then rely on the beautiful fact that the unit sphere in an infinite dimensional Hilbert Space is contractible!

I do not have time here to explain to you how absolutely fundamental the above table is in modern topology. The spaces W/G are often denoted by BG and are in some sense the topological embodiment of G - as both a space and a group.

They are also referred to as "classifying spaces" because of the following theorem:

Every principal G-bundle, P, over a reasonable space X is the pull-back by a map

$$f_P : X \longrightarrow BG \tag{3.5}$$

from the bundle $W \to W/G$. Furthermore homotopic maps give rise to isomorphic bundles over X.

Here let me just try and convince you that these constructions also fit the bill for an equivariant Morse theory. The principle - which I am embarrassed to say occurred to Atiyah and me only quite recently - is simply to pass from an invariant function f on M, to the function $f/\!/G$ on $M/\!/G$ which is induced in the obvious manner by f, and then to do ordinary Morse theory for $f/\!/G$.

Precisely, if f is invariant on M then its pull-back to $M \times W$ is clearly invariant under the diagonal action and hence descends to

$M \times W/G$ giving rise to $f/\!\!/G$! Q. E. D.

Now this passage from f to $f/\!\!/G$ would of course be a cheat unless at least the Morse Series of $f/\!\!/G$ could be read off directly from the critical sets of f, and the behavior of these sets under G. Luckily this turns out to be the case as the following dictionary clearly indicates:

There is a one-to-one correspondence $N \mapsto N/\!\!/G$ between the critical sets of f and those of $f/\!\!/G$, with the following properties:

(3.5) f nondegenerate $\longrightarrow$ $f/\!\!/G$ nondegenerate

(3.6) index of critical set N = index of $N/\!\!/G$

(3.7) if N is an orbit G/H, then $N/\!\!/G$ is BH.

The first two of these are very plausible but of course essentially use nondegeneracy in the extended sense. The third may seem surprising at first. But consider the extreme case when the critical manifold N of f reduces to a point $p \in M$. Then, in view of invariance, p must be fixed under G, so that then $H = G$. Now, the inverse image of p in $M \times W$ is simply $p \times W$ and hence in $M/\!\!/G$ is given by $W/G = BG$. Thus p in M goes over to BG in $M/\!\!/G$. In any case equipped with these formulae $\mathfrak{m}(f/\!\!/G;t)$ is essentially computed from f on M. For example when f is the z-coordinate on S^2, in our second example,

$$\mathfrak{m}(f/\!\!/G;t) = \frac{1}{1-t^2} + \frac{t^2}{1-t^2} \tag{3.8}$$

corresponding to the minimum and maximum each of which contributes a $BS^1 = \mathbb{C}P_\infty$ in $M/\!\!/G$. In view of (3.7) our formula follows.

The right-hand side $P(M/\!\!/G;t)$ is also easily computed in this case and as one would expect yields the same answer:

$$P(M/\!\!/G;t) = 1 + t^2/1 - t^2 , \tag{3.9}$$

so that this f is a perfect Morse-function in the equivariant sense.

Finally two remarks. a) I should put your mind at rest - if the action of G on M is free to start with then $M/\!\!/G \approx M/G$ in the sense of homotopy, so that in that case these two situations give isomorphic results on both sides.

b) We will at times call the Morse Series of $f/\!\!/G$, the equivariant Morse Series of M, and the Poincaré series of $M/\!\!/G$ the equivariant Poincaré series of M. One may think of these as the old objects now suitably modified to incorporate the G-action.

(iv) Yang-Mills over Riemann Surfaces. At this stage it should be clear that if the Morse theory works for the Yang-Mills functional, it is only in the equivariant sense that we could expect interesting results. Thus if we are studying the Yang-Mills functional for a principal bundle P over a manifold M, with structure - that is gauge - group G, then the pertinent space is $\mathcal{A}(P)/\!\!/\mathcal{G}(P)$ where $\mathcal{A} = \mathcal{A}(P)$ is the space of connections over P, and $\mathcal{G}(P)$ is the group of local gauge transformations of P.

Now because $\mathcal{A}$ is contractible this space is just another copy of $B\mathcal{G}$ - the classifying space of the group $\mathcal{G}$!

$$\mathcal{A}(P)/\!\!/\mathcal{G} = B\mathcal{G} \quad .$$

A first result is then that up to homotopy the following holds:

$$B\mathcal{G} = \mathrm{Map}_P(M; BG) \quad . \tag{3.10}$$

Here Map denotes the space of all maps of the first space into the second, and the BG is of course the classifying space of the gauge group G. Finally the subscript P denotes the component of this space in which the classifying map f_P of P is found.

This last formula makes the equivariant Poincaré series of $\mathcal{A}(P)$ computable by standard techniques provided M is reasonably simple, e.g. if $M = S^4$. For instance when M is a Riemann Surface of Genus g, and $G = U(n)$ the answer is:

$$P(B\mathcal{G};t) = \frac{\left\{(1+t)(1+t^3)\dots(1+t^{2n-1})\right\}^{2g}}{\left\{(1-t^2)(1-t^4)\dots(1-t^{2n-2})\right\}^2(1-t^{2n})} .$$

It is for these Yang-Mills problems that M. Atiyah and I recently noticed a quite remarkable fact. Namely we first conjectured that in this Riemann Surface case the Yang-Mills functional should yield a perfect equivariant function on $\mathfrak{a}(P)/\!/\mathcal{G} = B\mathcal{G}$! Now for $G = U(2)$ (and $c_1(P) = 1$!) we then looked at the solutions of Yang-Mills and with the aid of our dictionary found that the equivariant Morse series of Yang-Mills took the form:

$$\text{(3.11)} \qquad \mathfrak{m}(Y\text{-}M/\!/\mathcal{G};t) = \frac{P(\mathrm{Min};t)}{(1-t^2)} + \frac{t^{2g}(1+t)^4}{(1-t^2)^2(1-t^4)}$$

where Min denotes the variety of stable solutions of Y-M over M. (These are the ones where Y-M attains its absolute minimum.) The space Min is in a natural way an algebraic variety but has a very difficult and imperfectly known topological structure. In view of (3.11) the equivariant perfection of Y-M in this situation therefore leads to the equation:

$$(*) \qquad \frac{P(\mathrm{Min};t)}{1-t^2} + \frac{t^{2g}(1+t^4)}{(1-t^2)^2(1-t^4)} = \frac{(1+t)^{2g}(1+t^3)^{2g}}{(1-t^2)^2(1-t^4)} ,$$

for $P(\mathrm{Min};t)$.

We then found first of all, that this equation agreed with explicit computations of Newsteads of $P(\mathrm{Min};t)$ for low g, and then much more surprisingly that this equation (*) was already to be found in the literature. Namely that Harder, [2], using number theoretic methods and the Weil conjectures - which at that time were not proved yet - produced this identical formula from deep results in number theory over finite fields.

At this stage we are pretty certain that A - the Morse theory works in the equivariant sense for M of dimensions (guess what!) less than 4 - and B - that the Y-M functional gives rise to a perfect equivariant theory over Riemann surfaces if the Gauge Group is $U(n)$, although the details of all of this remain formidable. Eventually we therefore hope to bring quite new proofs of Harder's formulae, and in any case, using both Harder's and our approaches, we can at least refine our topological insight into the structure of these varieties Min to show that <u>their homology is torsion-free.</u> Their Poincaré Series are formidable though! For instance for $U(3)$ one finds

$$P_t(\mathrm{Min}) = \frac{(t^5+1)^{2g}(t^3+1)^{2g}-(t^2+1)^2 t^{4g-2}(1+t)^{2g}(1+t^3)^{2g}+(1+t^2+t^4)t^{6g-2}(1+t)^{4g}}{(t^2-1)(t^4-1)^2(t^6-1)} .$$

Let me conclude by summarizing what I hope you will remember from this whole story. The first is of course that the topology of a space forces extrema on certain functions on it and this behavior can be made quantitative. Of course here all critical sets have to be considered - not only the minima - for the theory to be complete. Second, that symmetry conditions also force critical points and that a quantitative theory exists here too, but that in the theory the singularities of the action of the symmetry group should not be ignored. I hope you will also remember that the extrema of the Y-M functional in the simplest instance - i.e. over Riemann Surfaces - are the minimal ones forced on that functional by its Symmetry Group. Finally I hope you will keep the concept of the classifying space of a group in reserve somewhere, and that in this topological theory, the quotient $\mathfrak{a}/\mathcal{G}$ - which I have often heard discussed here - is replaced by $B\mathcal{G}$.

I include here the bibliography of a paper on this subject by M. Atiyah and myself which we hope to publish in the near future.

Bibliography

1. R. Bott and H. Samelson, Applications of the theory of Morse to symmetric spaces, Amer. J. of Math. vol. 80 (1968), pp. 964-1029.
2. G. Harder, Eine Bemerkung Zu einer Arbeit von P. E. Newstead, Jour. für Math. 242 (1970), 16-25.
3. G. Harder and M. S. Narasimhan, On the cohomology groups of moduli spaces of vector bundles over curves, Math. Ann. 212 (1975), 215-248.
4. D. Mumford and P. E. Newstead, Periods of a moduli space of bundles on curves, Amer. J. Math. 90 (1968), 1201-1208.
5. M. S. Narasimhan and S. Ramanan, Moduli of vector bundles on a compact Riemann surface, Ann. of Math. 89 (1969), 19-51.

6. M. S. Narasimhan and S. Ramanan, Vector bundles on curves, Proceedings of the Bombay Colloquium of Algebraic Geometry, 335-346, Oxford University Press, 1969.

7. M. S. Narasimhan and C. S. Seshadri, Stable and Unitary vector bundles on a compact Riemann surface, Ann. of Math. 82 (1965), 540-567.

8. P. E. Newstead, Topological properties of some spaces of stable bundles, Topology 6 (1967), 241-262.

9. ______, Stable bundles of rank 2 and odd degree over a curve of genus 2, Topology 7 (1968), 205-215.

10. ______, Characteristic classes of stable bundles of rank 2 over an algebraic curve, Trans. Amer. Math. Soc. 169 (1972), 337-345.

11. ______, Rationality of moduli spaces of stable bundles, to appear.

12. C. S. Seshadri, Space of unitary vector bundles on a compact Riemann surface, Ann. of Math. 85 (1967), 303-336.

13. ______, Moduli of π-vector bundles over an algebraic curve, Questions on Algebraic Varieties, Roma, 1970.

A SEMICLASSICAL APPROACH TO THE STRONG COUPLING PHYSICS OF QCD

Curtis G. Callan, Jr.

Physics Department
Princeton University
Princeton, N. J. 08544

INTRODUCTION

Over the past few years, a reasonably simple picture of the workings of QCD has emerged: It assigns different physics to different distance scales via an effective coupling, g(d), governing the quantum fluctuations of scale size d. At short distances, g(d) is small, and the theory resembles a simple weak-coupling perturbation theory. The effective coupling grows with increasing d and eventually becomes large enough that the fluctuations resemble those of a strong coupling lattice theory (the value, d_c, of d at which this happens is presumably roughly equal to the size of a typical hadron). The strong coupling limit of lattice QCD is almost as simple as perturbation theory: confinement is automatic and the physics is completely characterized by the energy per unit length, σ, of the flux tube which connects static charges.

This picture of the limiting behaviors of QCD in turn poses some difficult physics problems. The strong coupling limit is characterized by the dimensional quantity σ. The weak coupling limit is characterized by a totally different dimensional quantity, Λ (defined by the asymptotic freedom behavior of g(d) at small distance: $g^{-2}(d) \propto \ln \Lambda d$). It must be possible to establish a relation of the form $\sigma = \phi \Lambda^2$, where ϕ is a pure number, and so to make a precise numerical connection between the physics of weak and strong coupling. In more general terms, we want to know how the scales of dimensional physical quantities (hadron masses, radii, etc.) are set in terms of Λ. In order to make progress on any of these problems it is obviously necessary to understand the basic physics of the transition between weak and strong coupling.

In these lectures we shall develop the notion that instantons are the driving force behind this transition. The work on which I shall report[1] is the result of a collaboration with Roger Dashen and David Gross and I will accordingly make use of the pronoun "we". The reader is assumed to be familiar with the basic mechanics of instantons and the context in which they arise: the semiclassical approximation to the QCD functional integral. In particular, he is assumed to be familiar with the particular version of this context which has been developed by the above-mentioned collaborators and myself.[2] (The reader should refer to this reference for a more general list of contributions to this subject than we shall present here.) The basic point is that instantons represent a new weak-coupling but non-perturbative effect which must be added to the usual perturbative treatment of g(d). In practice, the new physics turns on at a well-defined distance scale, d_c, and causes g(d) to increase very sharply. At the same time, d_c is so small that any conventional estimate would indicate that standard perturbation theory methods should be quite accurate. Within a narrow range in d, g(d) becomes so large that weak coupling methods of any kind, perturbative or non-perturbative, cease to make sense. What is not clear is whether at this point g(d) is large enough to put the theory in the strong coupling limit. If so, we have solved the dimensional transmutation problem:[3] On the one hand, the scale, d_c, at which the theory passes to strong coupling is essentially the same as the physical size of typical hadrons; on the other hand, since d_c has been determined by weak coupling methods, it is known in terms of asymptotic freedom Λ. If not, we have to call on further non-weak-coupling and a fortiori non-perturbative effects to carry the system into the strong coupling limit. In this case there would be little hope of gaining any useful qualitative understanding of the ultimate passage to the strong coupling limit.

In what follows we will present evidence that instantons alone suffice to bridge the gap between weak and strong coupling and will draw quantitative conclusions about dimensional transmutation. For simplicity we shall, for the most part, restrict our discussion to the unrealistic, yet instructive case of pure gauge theory with at most static external color sources. Massless fermions and chiral symmetry breaking pose special problems which we do not yet know how to deal with in this context, although we shall see that they can and must have a large effect on dimensional transmutation.

Our basic tool for studying the weak to strong coupling transition is an effective lattice theory, of variable scale, in which all degrees of freedom of continuum QCD, except those of the surviving lattice links, have been integrated out. We of course do not actually do the "integrating out", but make use of our knowledge of the behavior of the continuum theory on various short distance scales to make reasonable statements about the behavior of various terms in the effective action. It turns out that the effective lattice action notion very efficiently organizes what little information we have and allows us to make surprisingly concrete predictions.

THE EFFECTIVE LATTICE THEORY

In this section we propose to explore the general qualitative aspects of replacing a confining continuum theory by an effective lattice theory of variable scale. We will draw freely on standard lore about flux tubes, strong coupling lattice theory and so on.

The simplest way to proceed is to imagine that the continuum theory is actually defined as a lattice theory on a very fine lattice of spacing a_o (the Planck length, say!) and that the lattice simply acts as a particular sort of ultraviolet cutoff. The conventional choice for the lattice action is the Wilson action[4]

$$S_W = \frac{1}{g^2(a_o)} \mathcal{L}_W(\{U_i\}) = \frac{1}{g^2(a_o)} \sum_p (\mathrm{tr}(\prod_{\ell\varepsilon p} U_\ell) + \mathrm{h.c.}),$$

where U_i are the link variables and p stands for an elementary plaquette. If we set $g^{-2}(a_o) = -(11/12\pi^2)\,\ell n\,(a_o\Lambda)$(for an SU_2 gauge group), then by usual asymptotic freedom/renormalization group arguments, we can construct the continuum theory "simply" by taking the limit $a_o \to 0$. Once a_o is well within the weak-coupling regime, no significant change in the physics occurs as a_o decreases. For our purposes then, it will suffice to replace the continuum theory by its lattice approximation on a lattice finer than the hadron scale.

There is then no conceptual problem with integrating over the basic link variables while keeping compound link variables $\{\bar{U}_I\}$ associated with a larger sublattice fixed:

$$e^{-S_a(\{\bar{U}\})} = \int \Pi dU_i \; e^{-S_W(\{U\})} \prod_{\bar{I}} \delta(\bar{U}_I - \prod_{i\varepsilon\bar{I}} U_i)$$

The result, as indicated, is an effective action for the variables $\bar{U}$ defined on a lattice of scale a. Since a is any integer multiple of a_o, it is for all practical purposes a continuous variable. We will, for reasons soon to be explained, single out the Wilson term in S_a,

$$S_a(\{\bar{U}\}) = \frac{1}{g^2(a)} \mathcal{L}_W(\{\bar{U}\}) + \text{"higher terms"},$$

thus defining a running coupling constant, g(a), associated with the Wilson action at any scale. The full effective action S_a contains all the information needed to answer any question which can be formulated on the lattice of spacing a, and of course gives the same answer as the original action would have if used directly.

The usual expectations about confinement lead one to believe that for a large enough to put the system well within the strong coupling regime: (i) the individual terms in S_a become small enough that the strong-coupling expansion, $e^{-S} = 1 - S + \frac{1}{2}S^2 \ldots$, can be used to compute arbitrary expectation values (ii) the expectation value of any planar Wilson loop is proportional to C^{-N}, where N is the number of basic plaquettes in the loop (i.e. it obeys the lattice version of Wilson's area law). The only way to meet these expectations is for the "higher terms" to be such that, at least to leading order in $g^{-2}(a)$, the Wilson term dominates the expectation value of any planar loop. There is no particular reason why non-planar loops should be dominated in the same way by the Wilson term. Fortunately, there is at least one physical quantity which is expressible in terms of planar loop properties— the tension of the string between a widely separated quark antiquark pair— so that interesting physics can be extracted from the simple planar loop problem.

The fact that for strong coupling, planar loops are governed by the Wilson term means that g(a) satisfies a closed renormalization group equation: The expectation of the Wilson loop operator for a planar loop of area $Na^2=A$, as computed from the Wilson action term alone, is

$$\langle W \rangle = (1/g^2(a))^N \; (1 + 0(1/g^2(a))$$

(for an SU_2 gauge group). The standard physical interpretation of the Wilson loop implies that $\langle W \rangle$ should also equal $\exp(-A\sigma)$ where σ is the tension of the physical string connecting widely separated heavy quarks. Thus the value of σ determined by the effective lattice of spacing a is

$$\sigma = a^{-2} \; \ell n \; g^2(a) + 0(g^{-2}(a)).$$

Since σ is a quantity of direct physical significance, it cannot depend on which strong coupling effective lattice is used to determine it, and σ must be independent of a. This gives a renormalization group equation

$$\frac{d \; \ell n g(a)}{d \; \ell n a} = \hat{\beta} \; (g(a)) = \ell n \; g^2(a)$$

which must govern the evolution of the coefficient of the Wilson term in the large a limit. At the opposite extreme of small a (weak coupling) one also obtains a closed renormalization group equation for g(a). This is of course just the usual asymptotic freedom result (for SU_2, for example) that

$$\frac{d\,\ell n\, g(a)}{d\,\ell n\, a} = \hat{\beta}\,(g) = \frac{11}{24\pi^2}\, g^2(a) + O(g^4) \;.$$

Because the coupling constant associated with the Wilson term in the effective action renormalizes in a simple way for both weak and strong coupling (small and large lattice spacing), it will be particularly useful to study the transition between weak and strong coupling by following the evolution of the single coupling g(a) and its associated β function, $\hat{\beta}(g(c))$.

An interesting aspect of this problem is displayed in Fig. 1. There we plot both the strong and weak coupling limits of $\hat{\beta}(g)$ (the solid curves) as well as the curves obtained by including the first correction in the corresponding perturbation expansions of $\hat{\beta}$ (the dashed curves). In the region $1 < g < 4$ both expansions appear to be valid but give different values of $\hat{\beta}$. We eventually resolve this paradox by showing that non-perturbative weak coupling effects cause a rapid but smooth transition between the two asymptotic forms to occur right in the middle of this region.

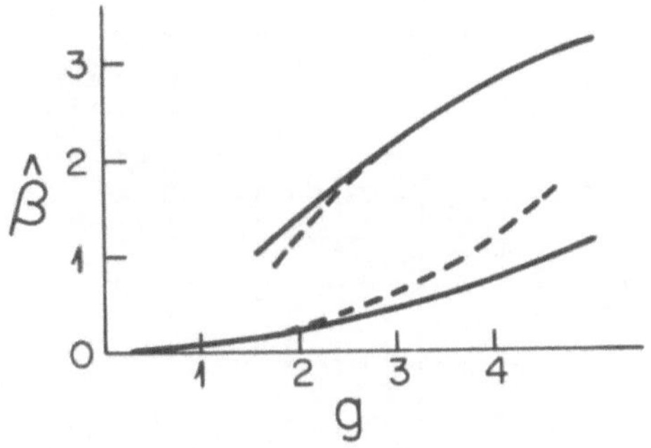

Fig. 1. Expected strong and weak coupling limits of $\hat{\beta}$.

Let us return to our qualitative discussion of the effective lattice action in a confining theory. An important physical manifestation of confinement is the existence of tubes (of some fixed finite radius, R) of trapped color flux connecting heavy quark pairs. This fact correlates very neatly with the expected behavior of the effective lattice theory. Consider first a flux tube upon which a lattice of spacing 2R has been superposed, as in Fig. 2. Since, in the usual picture, there is no correlation between fields inside and outside the flux tube, we must conclude that neighboring links on the lattice of spacing 2R are uncorrelated (except of course for those that neighbor one another along the flux tube axis). But decorrelation of neighboring links is characteristic of the strong coupling limit of a lattice theory and we must conclude that the effective lattice of spacing a = 2R is in the strong coupling regime. Consider next the same flux tube overlaid with a lattice of spacing a=R (Fig.3). Now on a time slice of the lattice there are <u>five</u> chains of links within the flux tube and parallel to the axis. Since the field within a flux tube is coherent there must be strong correlations between neighboring links of the effective lattice. In short, the effective lattice of spacing a=R must be in the weak coupling regime.

In terms of the expected behavior of the β function shown in Fig. 1, this means that for a<R (a>2R) g(a) must be less than

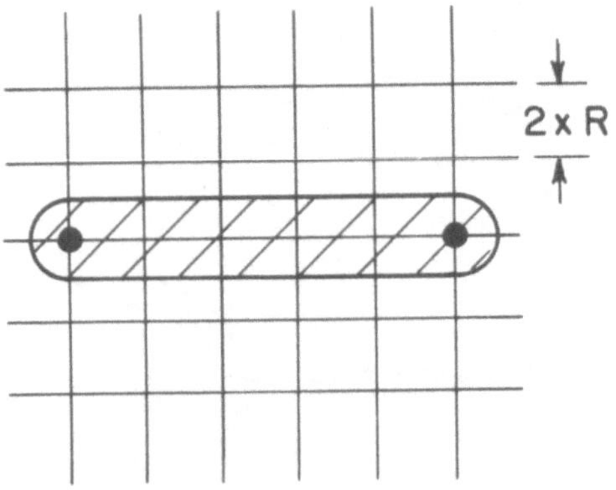

Fig. 2. A flux tube of radius R superposed on a lattice of spacing 2R.

(greater than) the range in which the transition between the weak coupling and strong coupling behaviors of β occurs. The conclusion

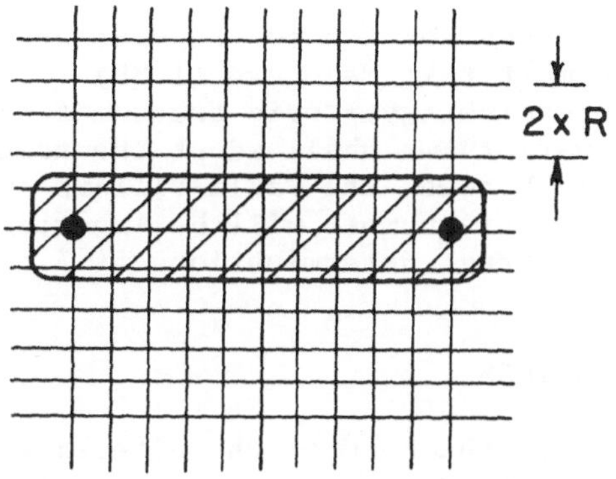

Fig. 3. A physical flux tube of radius R superposed on a lattice spacing R.

that the transition from weak to strong coupling takes place over at most a doubling of the distance scale is of course a direct consequence of our assumption that the flux tube has a well-defined radius. We do not know that this is so (although it is the most attractive of various possibilities) but our calculations will give an a posteriori justification. Finally, since our methods naturally lead us to measure a in units of Λ^{-1}, we will, by determining the value of a at which the weak-to-strong coupling transition occurs, easily obtain one of the desired dimensional transmutation results, a determination of R in units of Λ^{-1}. At the same time we will also get the string tension, σ, in units of Λ^2.

At this point we should make an important remark about Λ: It is not the same as the Λ associated with more conventional perturbation theory regulation schemes (Pauli-Villars, dimensional regulation, ...). A rather simple-minded estimate of the difference can be made by comparing the integral $I = \int d^4p/p^4$ as regulated by a lattice and by a Pauli-Villars scheme. A modest calculation shows that

$$I_{Lattice} - I_{Pauli-Villars} = \left(\frac{a}{2}\right)^4 \int_{-\pi/a}^{\pi/a} d^4p \left[\sum_{i=1}^{4} \sin^2\left(\frac{P_i a}{2}\right)\right] - \int d^4p \left[p^{-4} - (p^2 + M^2)^{-2}\right] = C\ (\ell n\ Ma - 1.89).$$

This means that in order for the two regulation methods to produce the same result we must take Ma = exp (1.89) = 6.6. Since the integrals which determine asymptotic freedom Λ are of the same massless one-loop variety we shall adopt the same number as a rough measure of the ratio between the Pauli-Villars and lattice versions of Λ: $\Lambda_{PV}/\Lambda_L = 6.6$. This number will be important to us when we wish to make use of direct determinations of Λ from scaling violation experiments.

THE EFFECT OF INSTANTONS

Roughly speaking, the form of the effective lattice action on scale a, $\mathcal{L}_a$, is determined by the quantum fluctuations on scales less than a. For small enough a, these are just the fluctuations of perturbation theory and $\mathcal{L}_a$ is just the Wilson action with a coupling constant which varies with a à la asymptotic freedom. Previous work on the semiclassical method suggests that the first new effects to appear as we increase the limiting scale size are those due to instantons. There ought then to be some range of a over which $\mathcal{L}_a$ is determined by instanton and perturbative effects only. In that range we should be able reliably to compute the diagnostic quantities σ and β and observe their deviation from perturbative behavior. If we increase a too far, our semiclassical methods of course break down, but before they do we find evidence that the system has reached the strong coupling limit (σ, in particular, settles down to a scale-independent constant). In this manner we will be able to use semiclassical methods to evaluate strong-coupling quantities.

To actually evaluate $\mathcal{L}_a$ in the manner described previously would be a monstrous undertaking which could only be done numerically. Since we are trying to develop insight we will use a shortcut whose accuracy is only imperfectly known, but which should give an accurate picture of the physics. To compute $\mathcal{L}_a$ we hold the link variables $\{U_i\}$ associated with the a-lattice fixed and integrate over all configurations of the primitive lattice consistent with the choice of $\{U_i\}$. Since the primitive lattice is just a way of cutting off the continuum theory, it will have recognizable multi-instanton configurations, and it is precisely over these configurations which we would like to integrate. Instantons of scale size significantly greater than a are clearly not relevant since they can be specified by imposing a suitable variation on the link variables of the a-lattice. Instantons of scale size less than a, if not too dense, have an effect which can be summarized by an effective paramagnetic vacuum permeability, $\mu(a)$ (the argument for this, as well as explicit formulas for μ in terms of an integral over the density of instantons of various scale sizes has been given in a number of places[2]). In a continuum theory the bare action function, $\mathcal{L}_o = \frac{1}{4}F_{\mu\nu}$, is modified by the presence of a medium of permeability μ to $\mathcal{L}_\mu = 1/4\mu\ F^2_{\mu\nu}$. Since the Wilson action is just the lattice approximation to $\frac{1}{4}F^2_{\mu\nu}$, we make the reasonable assumption that the effect of the "sublattice instantons" on the Wilson action is seen as a similar multiplicative renormalization: $g^2(a) = g^2_{AF}(a)\ \mu(a)$ (where $g^2_{AF}(a)$ accounts for the perturbative asymptotic freedom coupling constant renormalization).

In this procedure, $\mu(a)$ is determined by integrating over instantons of scale size up to some cutoff a_c roughly equal to a. The best information we have on how to choose a_c comes from our study of instanton effects on the heavy quark potential.[5] There, rather than fixing link variables, we put a Wilson loop into the functional integral and integrated over dilute instanton gas configurations. The interesting point is that if the maximum instanton scale size was taken to be less than 2/3 the Wilson loop diameter, the exact result was indistinguishable from that obtained by ignoring the instantons and just renormalizing the perturbation theory coupling by the corresponding μ. At the same time, if the Wilson loop diameter was taken to be less than one times the instanton scale size, the result is very accurately equal to the extrapolation of the zero diameter limit of the Wilson loop. We take this to mean that, in computing $\mathcal{L}_a$, instantons of scale size less than 2/3 a renormalize the Wilson action coupling constant via the permeability μ, and that instantons of scale size greater than a can be neglected since their effect will be obtained by subsequent integration over the lattice link variables. We don't know how to deal with instantons of scale size intermediate between these two limits and will assume that this ignorance can be compensated for by choosing a_c intermediate between 2/3 a and a. For definiteness we shall choose $a_c = 5/6$ a. Varying the cutoff ratio a_c/a will be seen to cause a proportional change in dimensional strong coupling quantities (σ, hadron radii, etc.).

With these preliminaries out of the way, we are ready to construct σ and β in the instanton-dominated regime. The basic formulae, culled from this discussion and earlier papers, are (we specialize to SU_2)

$$g^2(a) = g^2_{AF}(a)\ \mu(a)$$

$$\frac{8\pi^2}{g^2_{AF}(a)} = \frac{22}{3}\ \ell n\ \frac{1}{\Lambda a} = x_{AF}(a)$$

$$\mu(a) = \eta + \sqrt{1 + \eta^2}$$

$$\eta = \frac{4\pi^2}{2} \int_0^{a_c} \frac{d\rho}{\rho} D(\rho)\, x_{AF}(\rho)$$

where D (ρ) is the instanton density function and Λ is the lattice renormalization mass parameter, Λ_L. As is pointed out in the original papers on the dilute instanton gas method, these formulas make sense only so long as a_c is small enough that the integrated instanton density is quite a bit less than one.

To make an explicit calculation we must specify D (ρ). For isolated instantons in the weak coupling limit D (ρ) = C $x_{AF}^4(\rho)$ $\exp(-x_{AF}(\rho))$, where C is a constant which depends on which coupling constant definition has been adopted.[6] As we have already had occasion to point out, the method we are using clearly calls for the use of the lattice coupling constant definition. The values of C associated with two different coupling definitions are related by the values of the associated Λ's by (for SU_2) $C_1/C_2 = (\Lambda_1/\Lambda_2)$. From the work of 't Hooft, we know that $C_{PV} = .015$ and we have already determined that $\Lambda_{PV}/\Lambda_L \sim 6.6$. Therefore $C_L \sim 1.7 \times 10^4$.

Another important point is that the density of instantons in a medium is, on the average, greater than the density of isolated instantons. This is due to the presence of the same long range attractive interactions which lead to behavior of the dilute instanton gas as a permeable medium. In our original dilute instanton gas studies,[7] we showed that a reasonable representation of this effect is obtained by replacing exp(-x) by exp(-3/4x) in the expression for D(ρ). We shall again adopt that approximation.

It is now a matter of rather trivial numerical computation to determine $\hat{\beta}$ and σ. We plot $\hat{\beta}(g)$ in Fig. 4 and $\sigma(g) = (\log g^2(a))/a^2$ in Fig. 5. The dashed lines in Fig. 4 show the expected weak and strong coupling limits of β, while the solid line is our computed value. One notes a rapid (as a function of g) departure from weak coupling behavior at g $\sim$1.4 and an equally rapid settling down to strong coupling behavior at a slightly larger g. The values of integrated instanton density appropriate to the transition region are in the range of only a few percent and our dilute gas calculations make sense well into the region where the effective coupling is exhibiting strong coupling behavior. In Fig. 5 we plot $\sigma(g)$ in order to see how close we come to the true strong coupling (constant string tension) limit: After g has passed through the transition region defined by the behavior of β, σ is indeed very accurately equal to a constant. That constant, $\sigma = (210\ \Lambda)^2$, is our prediction for the string tension. In Figs. 4 and 5 we have labelled the abscissal with a as well as g so that we can read off

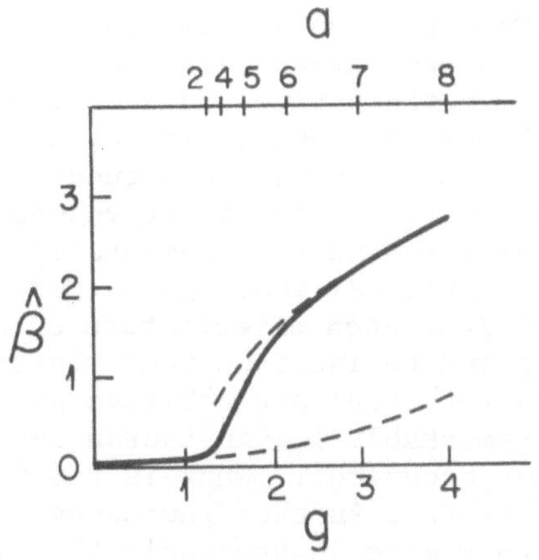

Fig. 4. A plot of our approximation to $\hat{\beta}$ as a function of g and a (measured in units of $10^{-3}\ \Lambda_L^{-1}$).

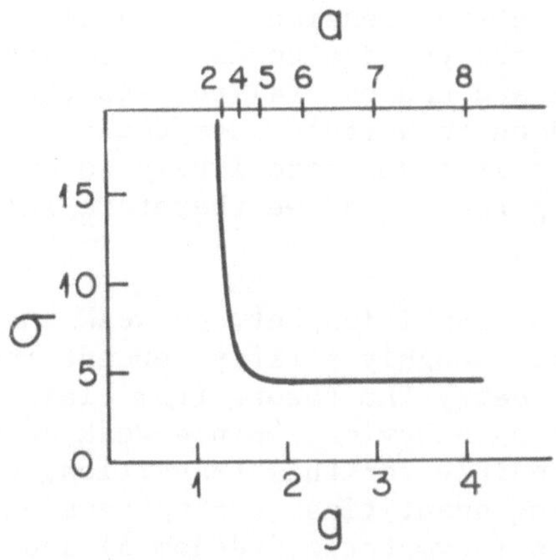

Fig. 5. A plot of σ (in units of $10^4\ \Lambda^2$) as a function of g and a (measured in units of $10^{-3}\ \Lambda_L^{-1}$).

the physical distance scale (in units of Λ^{-1}) where the transition occurs. This will eventually allow us to determine the physical size of "hadron states" by the sort of argument given in the previous section.

We can summarize the situation as follows: With the help of some reasonable assumptions we have managed to convert our understanding of the continuum perturbative and semiclassical physics of QCD into a picture of how the coefficient of the Wilson action in an effective lattice theory behaves as a function of increasing lattice spacing, a. For small a, the effective coupling behaves (by construction) according to the requirements of perturbative asymptotic freedom. At a critical scale, $a_c \sim .003\ \Lambda_L^{-1}$ and corresponding coupling $g_c \sim 1.4$, instanton effects turn on abruptly and cause the effective coupling to increase very rapidly as a function of a. This initial departure from perturbative physics is caused by configurations of a remarkably low instanton density (of order one percent of space time occupied). What is remarkable, and to some extent miraculous, is that further increases in a soon cause the effective coupling to evolve in precisely the manner characteristic of the lattice strong coupling limit. This pseudo strong-coupling behavior is generated by configurations of low instanton density and can be explicitly followed through about a factor 2 increase in lattice spacing. We assume that this instanton-generated strong-coupling behavior is the ultimate strong-coupling behavior of the theory and that the string tension one reads off from Fig. 4 is the true string tension. It is certainly possible that some totally different physics comes in at larger scales than we can directly examine and causes $\sigma(a)$ to pass through yet another transition and settle down to a truly asymptotic value which we can, of course, not compute. It seems more likely to us that there is only one strong coupling region and we therefore ignore this possibility.

In our picture, the transition between weak and strong coupling behavior is very abrupt. Roughly a fifty percent increase in distance scale suffices to carry the theory from clearly weak-coupling to clearly strong-coupling behavior. Since weak coupling semiclassical effects are responsible for this transition, we are able to calculate strong coupling quantities (string tension) in terms of weak coupling quantities (asymptotic freedom Λ) and so realize dimensional transmutation. The "paradox" posed by the existence of overlapping regions of validity of the strong and weak coupling expansion of the β function is of course resolved by the observation that weak coupling non-perturbative effects overwhelm ordinary radiative corrections precisely in this overlapping region of validity. Finally, since instanton effects can be dealt with in any SU_N gauge theory we can easily carry out the same analysis for the physically more relevant case of an SU_3 gauge theory: we find the same qualitative behavior of β and σ and virtually the same value for the string tension.

Let us now discuss the value we have found for the string tension in terms of expected phenomenology. We find $\sigma \sim (200\ \Lambda_L)^2$. Now σ is in principle to be extracted from data such as the Regge slope, while Λ must be determined from studies of scaling violations in deep inelastic scattering. In order to credibly extract Λ, the right coupling constant definition must be used. According to the analysis of Celmaster and Gonsalves,[8] a momentum space subtraction scheme is a good choice. The corresponding Λ will be called Λ_M and one can show, by a combination of their arguments and our calculation of Λ_{PV}/Λ_L that $\Lambda_M/\Lambda_L \sim 20$! Then a more useful expression for σ is $\sigma \sim (10\Lambda_M)^2$. Preliminary estimates of Λ_M suggest that it lies in the range of several hundred MeV, while σ is known quite accurately to be $(420\ \text{MeV})^2$. In the real world, then, the dimensionless number $\sqrt{\sigma}\ \Lambda_M^{-1}$ is probably in the neighborhood of one, while our calculations yield a value near 10!

Whatever the merits of our method, we happen to know, after the fact, that we have obtained the correct value of σ for a pure gauge theory — Creutz[9] has carried out a computer evaluation of quarkless QCD and also found that $\sigma \sim (200\ \Lambda_L)^2$. The real world is not a pure gauge theory of course — there are at least five quark flavors, of which three are very low in mass. This is potentially very important since light quarks have the effect of suppressing instantons[10,2] — precisely the objects we use to drive the weak to strong coupling transition. A qualitative argument can be constructed to the effect that adding more and more light flavors delays the weak to strong coupling transition to larger and larger scales (in units of Λ_L^{-1}) and correspondingly decreases σ (in units of Λ_L^2).* Whether adding these light flavors will reduce $\sqrt{\sigma}\ \Lambda_M^{-1}$ from approximately 10 to approximately 1 is not yet known, but it seems quite reasonable. In any event we should be warned that dimensional transmutation in QCD must be very sensitive to the number of light quarks.

DETAILED STRUCTURE OF FLUX TUBES

In the previous section we showed how to use semiclassical methods to calculate one strong-coupling quantity, the string tension. Our methods are such that the string tension is almost the only quantity we can hope to compute on a first-principles basis. However, if we allow ourselves to use some of the ideas of the QCD bag model, we can extract more information about the flux tube than just the string tension. At the same time, we can show that the bag model is consistent in a quantitative sense with those aspects of QCD dynamics which we can handle.

*What is involved here is the notion that the instantons drive spontaneous mass generation for the light flavors, so that the zero mass fermion suppression of instantons eventually undoes itself.[2,11]

In the bag model, widely separated static color charges are joined by a stationary flux tube of some finite radius, R. Such flux tubes correspond physically to the strings of the strong coupling limit of a lattice theory. Our general picture of the behavior of the effective lattice action also leads us to expect that the flux tube has a finite radius, about equal to the lattice spacing at which instanton effects first turn on strongly. The bag model is such that, given the radius of the flux tube, its energy per unit length, which is identical to the string tension, is determined. We will use our determination of the flux tube radius in units of Λ^{-1} to determine the string tension according to the bag model and will see that it is equal to the string tension directly determined from our arguments. We take this as a piece of evidence that the instanton-induced strong coupling behavior of the effective coupling is the true strong-coupling behavior.

The arguments given in the preceding sections imply that the radius, R, of the flux tube lies somewhere between the value of the lattice spacing at which g(a) takes leave of weak coupling ($R<$) and that at which it settles down to strong coupling behavior ($R>$). $R<$ is most easily read off from Fig. 2 while $R>$ is most easily inferred from Fig. 3: $R< = .003\ \Lambda$, $R> = .0045\ \Lambda_L^{-1}$. That these two values are so close to each other implies that the flux tube "wall" is thin compared to the flux tube diameter. This must be true for the bag model, which assumes an infinitely thin wall, to make sense. We simply take the mean of $R<$ and $R>$ as an estimate of the flux tube diameter: $R = .0038\ \Lambda^{-1}$

The bag model equations for the flux tube are

$$\sigma = \frac{1}{g_o^2} E_c^2 \quad (\pi R^2)$$

$$E_c \quad (\pi R^2) = Q\ g_o^2$$

where E_c is the electric field strength in the tube, g_o is the perturbation theory coupling constant describing the coupling of a static charge to the static color field inside the tube and Q is the intrinsic strength of the charge producing the flux tube. If the source is in the fundamental representation, $Q^2 = 3/4$ for SU_2, $Q^2 = 4/3$ for SU_3, etc. Putting this together we have

$$\sqrt{\sigma}\ R \;=\; \frac{g_o Q}{\sqrt{\pi}} \;=\; \sqrt{\frac{Q^2\ 8\pi}{x_o}}$$

The coupling strength, g_o, depends on the physics of linearized gluon propagation inside the geometry of the flux tube.

Crudely speaking we should take it equal to $g_M(p_o)$ where g_M is the coupling defined by momentum subtraction[8] and p_o is a typical momentum encountered on inverting the Laplacian in a flux tube of radius R. For a flux tube the eigenvalues of the Laplacian can be written $p_r^2 + p_\ell^2$ (r and ℓ standing for radial and longitudinal momenta). The radial eigenvalues satisfy $J_1(p_r R) = 0$ and the lowest relevant one is $p_r = 2.85/R$. To estimate p_ℓ we note that g_o enters the calculation through the solution of Gauss' law at the end of a flux tube, a roughly spherical calculation. This suggests that we take $p_\ell^2 = \frac{1}{2} p_r^2$ and gives a rough estimate for p_o of $4.72\ R^{-1}$. Consequently

$$\frac{8\pi^2}{g_o^2} = x_o = \frac{22}{3} \ell n \frac{4.72}{R\Lambda_M} = \frac{22}{3} \ell n \left(\frac{.236}{R\Lambda_L}\right)$$

where we have chosen an SU_2 gauge group and used $\Lambda_M \simeq 20\ \Lambda_L$ to convert back to the lattice scale parameter.

For a fundamental flux tube of radius in an SU_2 gauge theory we have the bag model prediction

$$\sqrt{\sigma} = \frac{1}{R} \sqrt{\frac{9\pi}{11\ \ell n\ (.236/R\ \Lambda_L)}}$$

Upon inserting our estimate, $R = .0038\ \Lambda_L^{-1}$ for the flux tube diameter, we obtain $\sqrt{\sigma} \sim 200\ \Lambda_L$, which is gratifyingly close to the string tension which we obtained directly from our semiclassical, effective lattice method. For the reasons already cited, we take this as additional support for our method and for the bag picture of the physical structure of the flux tube. The same analysis could be carried through for other gauge groups of course. The agreement is not as spectacular for SU_3 as for SU_2 but it is good enough to support the same qualitative conclusions.

SUMMARY AND OUTLOOK

Our aim in this discussion has been relatively modest. We have not tried to prove confinement but rather, assuming confinement, to show that everything we know about QCD is consistent in a precise numerical sense with the notion that weak-coupling semiclassical effects (i.e. instantons) bridge the gap between weak coupling (short distance) and strong coupling (large distance) physics. Our surprising finding is that, within our framework, weak coupling but nonperturbative effects produce behavior indistinguishable from strong coupling behavior and allow us to identify certain strong coupling quantities in terms of weak coupling parameters (i.e. to solve the dimensional transmutation problem).

Since we have not done a real calculation but rather indulged in physically motivated guesses about the results of certain calculations, our quantitative results are subject to many uncertainties. If we have identified the right physical mechanism underlying the transition from weak to strong coupling, however, our crude arguments cannot lead us too far astray, even on quantitative questions. In this respect it is worth remarking that Kogut, Pearson and Shigemitsu,[12] working with Padé approximants to strong coupling expansions, also find a very rapid transition between weak coupling and strong coupling behavior with the transition occurring for a value of g very near what we find.

The real value of our method is that since it relies on rather simple physics, it should in principle be possible to see how the overall physics of QCD is quantitatively changed by varying parameters such as: the gauge group, the number of light fermions, the value of the vacuum angle. We have already pointed out that there is good reason to believe that light fermions have a remarkably large effect on quantitative questions. This is a subject which will be actively pursued.

REFERENCES

1. Curtis G. Callan, Jr., Roger Dashen, and David J. Gross, Instantons as a Bridge Between Weak and Strong Coupling in QCD, Princeton preprint (1980).
2. Curtis G. Callan, Jr., Roger Dashen, and David J. Gross, Semiclassical Methods in Quantum Chromodynamics, in "Quantum Chromodynamics," W. Frazer and F. Henyey, eds., American Institute of Physics, New York (1978).
3. Sidney Coleman and Erick Weinberg, Phys. Rev. D7:1888 (1972).
4. K.G. Wilson, Phys. Rev. D10: 2445 (1975).
5. Curtis G. Callan, Jr., R. Dashen, and D.J. Gross, Phys. Rev. A. Zee, Phys. Rev. D18:4684 (1978).
6. G. 't Hooft, Phys. Rev. D14:3432 (1976).
7. Curtis G. Callan, Jr., R. Dashen, and D. J. Gross, Phys. Rev. D20 (1979)
8. W. Celmaster and R. Gonsalves, Phys. Rev. 42:1435 (1979).
9. M. Creutz, BNL Preprint (1979).
10. G. 't Hooft, Phys. Rev. Lett. 37:8 (1976).
11. D. Caldi, Phys. Rev. Lett. 39:121 (1977); R. Carlitz, Phys. Rev. D17:3225 (1978).
12. J. Kogut, R. Pearson, and J. Shigemitsu, Phys. Rev. Lett. 43:484 (1979).

LATTICE GAUGE THEORIES

James Glimm[1]

The Rockefeller University

New York, N. Y. 10021

ABSTRACT

Recent mathematical results concerning lattice gauge theories are surveyed. The distinction between color screening and color confinement is explained in terms of the center of the color gauge group, for heavy quarks. The role of the center in terms of non-abelian vortices and confinement is also discussed.

1. INTRODUCTION AND DEFINITIONS

The lattice gauge theories are highly successful - at least qualitatively - in that the strong coupling regime is under control for these models. For the lattice Z^d, let $L(L_0)$ be the set of (oriented) links and P the set of plaquettes. For G a compact gauge group, we define G^{L_0} as the space of functions γ_ℓ from L_0 to G. Then G^{L_0} serves as a space of gauge potentials. The group of local gauge transformations G^{Z^d} is the space of functions γ_i from Z^d to G, which take the value $\gamma_i = e =$ identity, except for a finite number of lattice sites i. This group is a symmetry

1. Supported in part by the National Science Foundation under Grant PHY-78-08066.

group of the theory. If $i_{\ell,-}$ and $i_{\ell,+}$ are the initial and final sites of the link ℓ, then the group action is

$$\gamma_\ell \to \gamma_{i_{\ell,-}} \gamma_\ell \gamma_{i_{\ell,+}}^{-1}$$

The number of degrees of freedom is reduced from $G^{\ell 0}$ to G^{ℓ} by the requirement

$$\gamma_\ell = \gamma_{\ell'}^{-1}$$

if ℓ' is the link ℓ with orientation reversed.

The problem of lattice gauge theory is to integrate over the space G^{L}. The basic measure is $\Pi_{\ell \in L} d\mu(\gamma_\ell)$ where $d\mu(\gamma_\ell)$ is Haar measure on G. This measure corresponds to an infinite gauge coupling constant. Since it is completely trivial, it is a good starting point for perturbations, and it is the feature which makes the lattice gauge fields tractable at strong coupling.

For a plaquette $p \in P$ and a real character χ of G, define

$$V(\gamma) = \sum_{p \in P} \chi(\gamma_p)$$

where $\gamma_p = \gamma_{\ell_1} \gamma_{\ell_2} \gamma_{\ell_3} \gamma_{\ell_4}$ if $\gamma_{\ell_1}, \ldots, \gamma_{\ell_4}$ are the links which border p. Note that $\chi(\gamma_p)$ is independent of the starting point and direction in which the border ∂_p is traversed. Then

$$Z^{-1} e^{\beta V(\gamma)} d\mu(\gamma) \quad d\mu_\beta(\gamma) \ ,$$

where

$$Z = \int e^{\beta V(\gamma)} d\mu(\gamma) \ ,$$

defines the Gibbs measure $d\mu_\beta$ on G^{L} with coupling $1/\beta$.

2. BASIC PROPERTIES

The lattice is an ultraviolet cutoff. As the lattice spacing tends to zero, the ordinary continuum gauge theory is obtained, at least on a formal level. Good cutoffs preserve as many as possible of the properties of the exact solution. They never preserve all properties of the exact solution, and for this reason, it is usually desirable to have a variety of cutoffs available. Properties of the lattice cutoff include

(a) Local gauge invariance

(b) Lattice translation and lattice rotation invariance

(c) Positivity of the quantum mechanical Hilbert space inner product (Osterwalder-Schrader positivity)

The significance of (b)-(c) is most easily explained if we set the time-direction lattice spacing to zero, in the infinite volume theory. Then (b)-(c) allows the reconstruction of quantum mechanics and analytic continuation to a Minkowski metric, via the Osterwalder-Schrader reconstruction theorem.

The limits lattice volume $\to \infty$ and time lattice spacing $\to 0$ can be studied by existing methods for β small, using the cluster expansion, see below. For arbitrary β, this analysis is not complete, but the infinite volume limit follows from correlation inequalities, at least for a restricted class of groups G [1,2].

To give a technical statement of (c), let θ be the reflection about the time $t = 0$ hyperplane. For f a gauge invariant function supported in the half space $t \geq 0$, (c) is the statement [15]

$$0 \leq \int (\theta f)^{-} f \, d\mu_\beta \quad .$$

In case the time direction is a lattice, and quark fields are included in the model, the lattice is chosen so that $t = 0$ falls midway between lattice points.

The proof of (a) can be found in [3]. We sketch the proof, since the ideas are very simple. Let F be an observable depending only on links ℓ in a finite volume Λ. By the local Markov property,

$$\int F \, d\mu_\beta = \int F \, f_{\partial\Lambda} \, \Pi_{p \, \subset \, \Lambda} \, e^{\beta X(\gamma_p)} \, \Pi_{\ell \, \subset \, \Lambda} \, d\mu_\ell$$

where $f_{\partial\Lambda}$ is some function depending only on links γ which cross $\partial\Lambda$. For a gauge transformation γ_i with $\gamma_i = e$ for $i \notin \Lambda$, the right side is invariant, which completes the proof. Lüscher's statement [12] is stronger, but for a more restricted class of groups G: Any non-gauge invariant state has energy strictly above the energy of the vacuum.

3. CONFINEMENT AND SCREENING

The basic tool is the cluster expansion [9]. The basic expansion step is

$$Z^{-1} \prod_{p \varepsilon P} e^{\beta\chi(\gamma_p)} = Z^{-1} \prod_{p \varepsilon P} [1 + (e^{\beta\chi(\gamma_p)}-1)]$$

$$= Z^{-1} \sum_{P_0 \subset P} \Pi_{p \varepsilon P_0} (e^{\beta\chi(\gamma_p)}-1)$$

$$= \sum_{\text{connected subsets } P_0} (Z(P_0)/Z) \prod_{p \varepsilon P_0} (e^{\beta\chi(\gamma_p)}-1) ,$$

where $Z(P_0)$ is a partition function, with contributions from P_0 excluded. These methods do not require a lattice and are valid for continuum fields (near Gaussian) whenever the existence of the continuum fields can be controlled. In the present case, the expansions converge for small β [15,3].

For infinitely heavy quarks, confinement can be formulated in terms of the Wilson loop (parallel transport) operator

$$(P)\prod_{\gamma \varepsilon C} U_\gamma = W_C$$

where $(P)\Pi$ is a path ordered product. Let $\sigma \varepsilon G^\wedge$ be an irreducible representation of G. Then σ labels a gauge color particle species, and confinement vs. nonconfinement of this species is distinguished by the asymptotic behavior

$$\text{confinement:} \quad \langle Tr_\sigma W_C \rangle \sim e^{-0(\text{area enclosed by } C)}$$

$$\text{nonconfinement:} \quad \langle Tr_\sigma W_C \rangle \sim e^{-0(\text{Length of } C)} .$$

(To avoid trivial complications, we suppose that the representation defined by χ (in the measure $d\mu_\beta$) has kernel $= \{e\}$.)

A basic class of representation of G is

$$\Sigma = \{\sigma \varepsilon \hat{G} : \sigma \subset \Sigma_{\tau \varepsilon G^\wedge} \tau^* \otimes \tau\}.$$

By inspection, all representations in Σ annihilate the center of G. We believe that Σ is exactly the representations in $G^\wedge$ which annihilate the center. This is true for G abelian, for $G = SU(2)$ or $SU(3)$ by a special computation, and for $G = SU(n)$, making use of results of Kostant and Rallis [11] on the structure of real and complex lie groups. We now assume G is a group for which the conjecture is valid.

Theorem [15,8]. Let $d \geq 3$ and G as above. Then $\sigma \in G^\wedge$ is confined if and only if $\sigma \notin \Sigma$ (e.g. if and only if σ has non-zero triality). For $d = 2$, all $\sigma \neq \text{id}$ are confined. The basic idea is that Gauss' Theorem holds only for abelian gauge groups, or more precisely for the abelian part, i.e. the center Z, for an arbitrary gauge group G. Using the cluster expansion to represent the vacuum as products of $\chi(\gamma_p)$ times the product measure $d\mu_{\beta=\infty}$, we see that many terms give zero after integration over the group variable $d\mu(\gamma_\ell)$. In fact in the simplest case, see Figure 1,

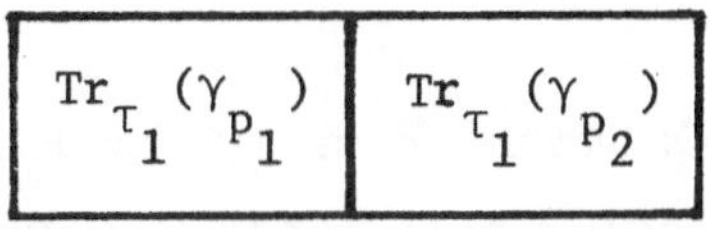

Figure 1

integration over the common bond γ_ℓ gives zero unless $\tau_1 = \tau_2$. Also for bond(s) running through C, we have (see Figure 2)

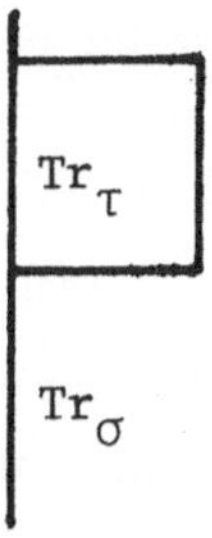

Figure 2a.

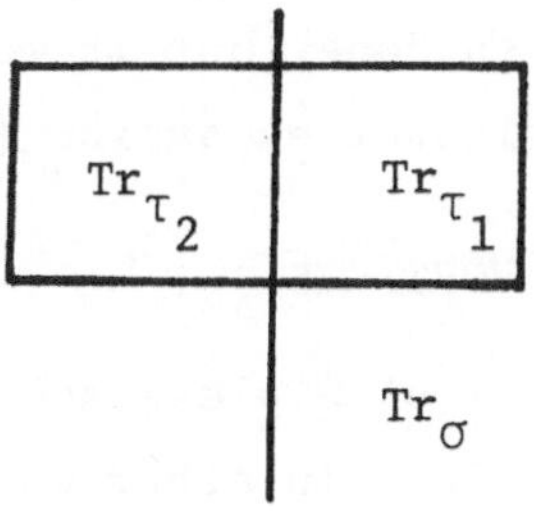

Figure 2b.

zero unles $\sigma = \tau$ (case (a)) and zero unless $\sigma \subset \tau_2^* \otimes \tau$, (case b).

For more complicated combinations, the $z^\wedge$ part of $G^\wedge$ is always preserved (conservation of abelian flux lines). Here $G^\wedge \to z^\wedge$ is defined by restriction: $\sigma \to \sigma|z$.

If $\sigma|z \neq$ identity, i.e. $\sigma \notin \Sigma$, then because the abelian flux quantum number is preserved, the entire area of C must be filled in to get a nonzero contribution. Otherwise a nonzero contribution only requires plaquettes forming a tube around C.

From this construction (for $d \geq 3$) we see that σ is confined if and only if $\sigma \in \Sigma$.

The failure of Gauss' Theorem in the nonabelian case is color screening. The Tr_σ particle, if its color charge is measure by Gauss' Theorem, is color neutral. In fact $\int En$ in such a state does not see the flux, since surface flux does not extend beyond a microscopic neighborhood of the particle. Thus as the lattice spacing goes to zero, the color information in this state disappears. Alternately we can introduce a shpere of infinite radius, to determine the color quantum number of the Tr_σ particle. Then again it is color neutral.

The final version of the confinement/screening problem is the statement:

All particle species which occur in the Hilbert space of in/out states have only color zero quantum numbers.

In this form, the confinement problem is an aspect of asymptotic completeness. Thus it could be studied in lattice gauge theories, with the help of the Bethe-Salpeter equation. This has not been done, but it would certainly be possible to do so, using the cluster expansion [16,17,7].

4. NONABELIAN VORTICES.

The center z of G plays a key role in confinement, as we have just seen. z arises in other ways as well. The t'Hooft disorder operator is a singular gauge transformation, which introduce a vortex in the gauge field. A vortex is a singular $d-2$ surface

S with $F_{\mu\nu} = \delta_S(x)$ where $\mu\nu$ span the directions normal to S and $\delta_S(x)$ is a delta function in the directions normal to the surface. The strength of the δ-function is chosen so that parallel transport around the vortex has strength $\gamma \varepsilon$ z. Because of this restriction, the singular gauge transformation which introduces the vortex, although multivalued in itself, is single valued in its action on the vector potentials. (The latter transform under G/z.)

One can check (by comparing high and low temperature expansions) that the vortices dominate the Z_2 phase transition, in the sense that they have low density at low temperature (concentrating on low density bounded closed surfaces S) but have high density at high temperature (lying on unbounded surfaces S).

Following [6,8], we write $d_G\gamma = d_z z\, d_{G/z}(\gamma z)$ and integrate out the G/z degrees of freedom in order to reduce a G-gauge theory to an effective z-gauge theory, we estimate effective G coupling constants for the vortex phase transition as follows: The SU(n) critical coupling is bounded by the Z_n critical coupling [8,4,13,14] and for $n = 2$, is estimated at about 30% lower [8]. An earlier, and mathematically rigorous implementation of these ideas can be found in [10].

REFERENCES

1. G.F. DeAngelis and D. deFalco, Correlation inequalities for lattice gauge fields. Lett Nuovo Cimento 18, 536 (1977).
2. G.F. DeAngelis, D. deFalco and F. Guerra, Scalar quantum electrodynamics as classical statistical mechanics. Commun. Math. Phys. 57, 201 (1977).
3. G.F. DeAngelis, D. de Falco, F. Guerra and R. Marra, Gauge fields on a lattice. Preprint (1978).
4. J. Fröhlich, Confinement in Z_n lattice gauge theories implies confienment in SU(n) lattice Higgs theories. Phys. Lett. 83B, 195 (1979).
5. J. Glimm and A. Jaffe, Quark trapping for lattice U(1) gauge fields. Phys. Lett. 66B, 67 (1977).
6. J. Glimm and A. Jaffe, Instantons in a U(1) lattice gauge theory: a Coulomb dipole gas. Commun. Math. Phys. 56, 195 (1977).
7. J. Glimm and A. Jaffe, The resummation of one particle lines. Commun. Math. Phys. 67, 267 (1979).

8. J. Glimm and A. Jaffe, Charges, vortices and confinement. Nuclear Phys. B149, 49 (1979).
9. J. Glimm, A. Jaffe and T. Spencer, In: Lecture notes in physics, vol. 25. Springer-Verlag, New York (1973).
10. J. Glimm, A. Jaffe and T. Spencer, A convergent expansion about mean field theory I,II. Ann. Phys. 101, 610 (1976).
11. B. Kostand and S. Rallis, Orbits and representations associated with symmetric spaces. Amer. J. Math. 93, 753 (1971).
12. M. Luscher, Absence of spontaneous gauge symmetry breaking in Hamiltonian lattice gauge theories. Preprint (1977).
13. G. Mack and V. Petkova, Sufficient condition for the confinement of static quarks by a vortex condensation mechanism. Preprint (1978).
14. G. Mack and V. Petkova, Z_2 monopoles in the standard SU(2) lattice gauge theory model. Preprint (1979).
15. K. Osterwalder and E. Seiler, Gauge field theories on a lattice. Ann. Phys. 110, 440 (1978).
16. T. Spencer, The decay of the Bethe-Salpeter kernel in $P(\phi)_2$ quantum field. Commun. Math. Phys. 44, 143 (1975).
17. T. Spencer and F. Zirilli, Scattering states and bound states in $\lambda P(\phi)_2$. Commun. Math. Phys. 49, 1 (1976).

SOME RESULTS AND COMMENTS ON QUANTIZED GAUGE FIELDS

Jürg Fröhlich

Institut des Hautes Études Scientifiques
F-91440 Bures-sur-Yvette

ABSTRACT

A few basic facts concerning the geometry of classical gauge fields are summarized; in particular, it is asserted that a principal bundle with connection can be characterized uniquely by its "Wilson loops". The quantization of gauge fields is then shown to consist of converting the Wilson loops into "random fields" on a manifold of oriented loops, a problem in "random geometry". Other examples in random geometry are briefly sketched. A general theorem permitting to reconstruct quantized Wilson loops from a sequence of Schwinger functionals is stated, the quark-antiquark potential is introduced, and "disorder fields" are discussed in general terms. The status of the construction of quantized gauge fields in the continuum limit is indicated, and some random-geometrical arguments are applied to lattice gauge theories and used to derive estimates on the expectation of the Wilson loop, resp. the disorder field.

1. Introduction

These lecture notes are organized as follows:

§2. Some elementary facts about the geometry of gauge fields.

§3. Random (or stochastic) geometry.

§4. Schwinger functionals and relativistic quantum fields.

§5. Existence of quantized gauge fields.

§6. Random geometrical methods in lattice gauge theories.

§7. Conclusions and acknowledgments.

The purpose of these notes is to introduce the reader to some basic, conceptual aspects of the problem of constructing quantized gauge fields and to summarize some rigorous results concerning the general (axiomatic) theory of quantized gauge fields, the existence of models in the continuum limit and some physical properties of lattice gauge theories, in particular the confinement of static quarks, the proof of which is based on random geometrical methods. This is the content of §6 which, due to page limitations, has come out to be too short. (That material will probably be treated in more detail elsewhere. Much of it is contained in the references quoted in the text). Unfortunately, we were forced to omit all proofs and to even state some of the main results in a somewhat cavalier way, but the necessary precision can be achieved by consulting the references given in the text. Our choice of references does not represent a value judgment. It reflects the author's taste and ignorance and a certain emphasis on developments in which he has been involved.

The main problem of quantizing gauge fields and thereby constructing a mathematically consistent and physically realistic model of the fundamental interactions is among the central problems of theoretical physics. We are still far from having complete and satisfactory solutions to that problem, and the technical barrier separating super-renormalizable from renormalizable theories is still not overcome, at all, in spite of the advent of asymptoically free theories, the Yang-Mills theories. (Some progress may be in sight, though). In view of the main problem these notes and many of the references quoted may seem naive; (they certainly are). They may however help to see some conceptual and mathematical problems through a perspective which we hope is not completely useless.

§2. Some Elementary Facts about the Geometry of Gauge Fields

In this section we briefly review some mathematical building blocks of the theory of classical matter and gauge fields, emphasizing some geometrical aspects. For mathematical details see e.g. |1| and the notes by Singer and Mitter. In |2| we have attempted to give a rather detailed, pedagogical "introduction for physicists" to this subject which might also be useful.

Let M be a manifold, the space-(imaginary) time manifold.

For particle physics one would choose $M = E^4$, but for well known, technical reasons one also chooses $M = E^\nu$ or S^ν, $\nu = 2,3,(4)$.

Let G be some compact Lie group, the gauge group. Let V be a topological space (typically a vector space carrying a representation of G or a homogeneous space) on which G acts as a homeomorphism group. Physically, V is the space of internal degrees of freedom of some matter field.

Let $\mathcal{F} = (B,M,V,G,\pi)$ be a fibre bundle with bundle space B, base space M, fibre V, group G and projection π. If V = G, the gauge group itself, we shall denote $\mathcal{F}$ by $\mathcal{P} = (P,M,G,\pi)$, and such a bundle is usually called a principal bundle. For definitions and results concerning fibre-and principal bundles see e.g. |1|. We propose to view classical matter fields as sections of a fibre bundle $\mathcal{F}$ and classical gauge fields as connections on a fibre-or principal bundle.

Let $\{\Omega_i\}_{i\varepsilon I}$ be a covering of M by open, simply connected coordinate neighborhoods such that the bundle space B restricted to Ω_i is homeomorphic to $\Omega_i \times V$, for all $i\varepsilon I$.

Let $\xi_{\Omega_i}, \xi'_{\Omega_i} : \Omega_i \times V \to \pi^{-1}(\Omega_i)$ be two coordinate functions.

Let $\xi^{(')}_{\Omega_i,x} := \xi^{(')}_{\Omega_i}(x,\cdot)$. By definition of a fibre bundle,

$\xi^{-1}_{\Omega_i,x}\xi'_{\Omega_i,x} \equiv h(x)$ is an element of the gauge group G, depending continuously on $x\varepsilon\Omega_i$, for all $i\varepsilon I$. The G-valued function h thus determines a change of coordinates and is called in physics a gauge transformation. Let $x\varepsilon\Omega_i \cap \Omega_j$. The gauge transformations

$h_{ij}(x) := \xi^{-1}_{\Omega_i,x}\xi_{\Omega_j,x}$ are called transition functions. They determine the bundle uniquely and also serve to associate to each fibre bundle a principal bundle: the one with the h_{ij}'s as its transition functions.

A connection, A, on a fibre bundle is a family of 1-forms $\{A^{(i)}\}_{i\varepsilon I}$ with values in the Lie algebra $\mathcal{G}$ of G such that $A^{(i)}$ is defined on Ω_i, $i\varepsilon I$, and for $x\varepsilon\Omega_i \cap \Omega_j \neq \emptyset$

$$A^{(i)}(x) = h^{-1}_{ji}(x)A^{(j)}(x)h_{ji}(x) - h^{-1}_{ji}(x)(dh_{ji})(x) \tag{2.1}$$

Moreover, if h is a gauge transformation defined on Ω_i, $A^{(i)}$ transforms according to

$$A^{(i)} \to A^{(i)h} = h^{-1}A^{(i)}h - hdh \tag{2.2}$$

In physics, A is called <u>gauge field potential</u>. It serves to define the notion of <u>parallel transport</u>. To explain this we choose some $\Omega = \Omega_i$, $i\varepsilon I$. Let $\gamma_{yx} \subset \Omega$ be an oriented curve connecting $x\varepsilon\,\Omega$ to some point $y\varepsilon\Omega$. We propose to construct a homeomorphism $g^{(i)}_{\gamma_{yx}} \varepsilon\, G$ from V into V in terms of A which describes the parallel transport of some $\Phi(x)\varepsilon V$ from x to y along γ_{yx}. If $y = x+dx$ is infinitely proximate to x we set

$$g^{(i)}_{x+dx,x}\ \Phi(x) := (1_V + A^{(i)}(x))\,\Phi(x), \tag{2.3}$$

with
$$A^{(i)}(x) = \sum_{j=1}^{\nu} A^{(i)}_j(x)dx^j,$$
(in local coordinates; $\nu = \dim M$).

If γ_{yx} is bounded, oriented, continuous and piecewise smooth, (2.3) can be integrated along γ_{yx}, yielding

$$g^{(i)}_{\gamma_{yx}} = P\{\exp \int_{\gamma_{yx}} A^{(i)}_j(z)dz^j\}, \tag{2.4}$$

where the r.s. is an infinite product obtained as a limit of finite products of factors
$(1+A^{(i)}_j(x_{k,m})\Delta^j_{k,m})$, with $|\Delta^j_{k,m}| \searrow 0$, as $m \to \infty$, for all j and k.

Under gauge transformations, h, $g^{(i)}_{\gamma_{yx}}$ transforms according to

$$g^{(i)}_{\gamma_{yx}} \to g^{(i)h}_{\gamma_{yx}} = h^{-1}(y)g^{(i)}_{\gamma_{yx}}h(x), \tag{2.5}$$

and if $\gamma_{yx} \subset \Omega_i \cap \Omega_j$

$$g^{(i)}_{\gamma_{yx}} = h^{-1}_{ji}(y)g^{(j)}_{\gamma_{yx}}h_{ji}(x). \tag{2.6}$$

Thanks to equations (2.5) and (2.6) one can now define parallel transport on the bundle space B of F as follows: If $\gamma_{yx} \subset \Omega_i$ parallel transport on B is given by a homeomorphism $\Gamma_{\gamma_{yx}} : V_x \to V_y$, with $V_x = \pi^{-1}(x)$ the fibre over x, which is defined by $\Gamma_{\gamma_{yx}} = \xi_{\Omega_i,y}\, g^{(i)}_{\gamma_{yx}}\, \xi^{-1}_{\Omega_i,x}$. (2.7)

If γ_{yx} is not contained in a single coordinate neighborhood Ω_i, one cuts up γ_{yx} into pieces $\gamma_{x_{m+1}x_m} \subset \Omega_{i_m}$ with $\Omega_{i_{m+1}} \cap \Omega_{i_m} \neq \emptyset$ and sets

$$\Gamma_{\gamma_{yx}} = \Gamma_{\gamma_{x_N x_{N-1}}}\, \Gamma_{\gamma_{x_{N-1}x_{N-2}}} \cdots \Gamma_{\gamma_{x_2 x_1}}, \tag{2.8}$$

with $x_N = y$ and $x_1 = x$. By equations (2.5)-(2.7), $\Gamma_{\gamma_{yx}}$ is independent of the choice of coordinate neighborhoods and coordinates. By the parallel displacement of some $\Phi(x)\varepsilon V_x$ from x to y along γ_{yx} we mean the element

$$\Gamma_{\gamma_{yx}}\, \Phi(x) \text{ in } V_y. \tag{2.9}$$

Next, we propose to characterize a fibre bundle with parallel transport in terms of a convenient family of gauge invariant functionals of the connection A. For this purpose we introduce the notion of holonomy groups.

Let x be some point in M, and let $\Omega(x)$ be the manifold of all bounded, continuous, piecewise smooth, oriented paths, ω_x, starting and ending at x, called loops. Given $\omega_x \varepsilon \Omega(x)$, let ω_x^{-1} denote the same curve as ω_x but with reversed orientation. On $\Omega(x)$ we define multiplication as the composition of paths, i.e. $(\omega_x, \omega_x') \to \omega_x o \omega_x'$, the composition of ω_x with ω_x'. With the obvious equivalence relation imposed, $\omega_x o \omega_x^{-1} = \omega_x^{-1} o \omega_x = 1_x$, the identity element in $\Omega(x)$, and $\Omega(x)$ is seen to be an infinite dimensional group.

Given a connection A and some loop $\omega_x \varepsilon \Omega(x)$ we set

$$g_\omega = \xi^{-1}_{\Omega_i,x}\, \Gamma_{\omega_x}\, \xi_{\Omega_i,x}\, \varepsilon G, \tag{2.10}$$

where Ω_i is a coordinate neighborhood containing x, and $\xi_{\Omega_i,x}$ are local coordinates on B. This defines a representation $g\colon \omega_x\, \varepsilon \Omega(x) \to g_{\omega_x}\, \varepsilon G$ of $\Omega(x)$ on V. The image of $\Omega(x)$,

$$H_x(A) = \{ g_{\omega_x} : \omega_x \in \Omega(x) \}, \tag{2.11}$$

is called the <u>holonomy group</u> of A.

For continuous $A, H_x(A)$ is a closed subgroup $\subseteq$ G. If M is connected $H_x(A)$ is independent of x, up to conjugacy, and if M is simply connected $H_x(A)$ is connected. If $H_x(A) = G$ the connection A is <u>irreducible</u>.

Under a gauge transformation, h, g_{ω_x} transforms according to

$$g_{\omega_x} \to g^h_{\omega_x} = h^{-1}(x) g_{\omega_x} h(x), \tag{2.12}$$

as follows from (2.10) and (2.5)-(2.7). Thus the elements $g_{\omega_x} \in H_x(A)$ depend on the choice of local coordinates (the gauge).

Let χ be a character of G. Then Y, given by

$$Y(\omega_x) = \chi(g_{\omega_x}) \tag{2.13}$$

is a character of $\Omega(x)$.

By (2.12), $Y(\omega_x)$ is <u>gauge-invariant</u>. We define $a(\omega_x)$, $\omega_x \in \Omega(x)$, to be the infimum of the areas of all smooth surfaces bounded by ω_x.

<u>Theorem 2.1.</u>

Assume M is simply connected. Let Y be an irreducible character of $\Omega(x)$ with the properties:

(1) Y is of positive type on $\Omega(x)$.

(2) $Y(1_x) = n$, for some natural number $n < \infty$.

(3) $|Y(\omega_x) - n| \leq O(a(\omega_x))$, as $a(\omega_x) \searrow 0$.

Then there exist an irreducible, connected Lie subgroup $H \subseteq U(n)$ and a representation $h: \omega_x \in \Omega(x) \to h_{\omega_x} \in H$ of $\Omega(x)$

such that $Y(\omega_x) = tr(h_{\omega_x})$. The representation h of $\Omega(x)$ is unique up to unitary equivalence. Moreover, there exists a connection A with values in the Lie algebra of H such that h_{ω_x}, $\omega_x \in \Omega(x)$, is the parallel transport around ω_x determined by A. If $H=U\pi(G)U^{-1}$, where π is an n-dimensional, faithful representation of some compact, connected Lie group G, and $U \in U(n)$ then $h_{\omega_x} = U\pi(g_{\omega_x})U^{-1}$, and $g_{\omega_x} \in G$ is unique up to (G-valued) gauge transformations. □

For a proof of Theorem 2.1 see |3|. This result says that a principal bundle with structure group G and a connection on it are uniquely determined by the numbers $\{ \chi(g_{\omega_x}) : \omega_x \in \Omega(x) \}$ if χ is a faithful, unitary character of G.

Let P be some principal bundle with structure group G, and A^I the space of all continuous, irreducible connections on P. Let G be the group of all gauge transformations modulo those which take values in the center of G. Clearly G acts as a transformation group on A^I. We define the orbit space O as $A^I/_G$. Given a connection $A \in A^I$, the corresponding orbit of A under G is denoted by $[A]$. Unfortunately O is generally not a linear space, but an infinite dimensional manifold with rather complicated geometrical properties, unless G is abelian. This is an expression of the intrinsic non-linearity of non-abelian gauge fields. Singer has shown that A^I is a principal bundle with base space O, fibre G and projection π given by $\pi(A) = [A]$, |4|. (If $M = S^3$ or S^4 A^I is not homeomorphic to $O \times G$, i.e. gauge fixing is impossible; see |4|. If $M = \mathbb{E}^\nu$ this conclusion is however not valid).

Since O is a manifold one can define a space $C(O)$ of continuous functions on O. The elements of $C(O)$ represent the "observables" of a classical Yang-Mills theory. "Euclidean quantization" consists, in a vague sense, in converting the elements of $C(O)$ into random variables. This procedure requires some more explicit knowledge of the structure of $C(O)$. We thus describe a convenient dense subspace of $C(O)$: Let $A \in A^I$ and let $g_{\omega_x}(A)$ denote the parallel transport around ω_x determined by A. The functions

$Y(\omega_x;A) := \chi(g_{\omega_x}(A))$, $\omega_x \in \Omega(x)$, χ a character of G, are gauge invariant, i.e. depend only on $[A]$ and are continuous in $[A]$. Therefore they belong to $C(O)$.

Theorem 2.2.

Let M be connected and suppose χ is some faithful, unitary character of G. Then the algebra of functions generated by

$$W: = \{ Y(\omega_x;A): \quad \omega_x \in \Omega(x) \}$$

is dense in $C(\mathcal{O})$.

Proof. By Theorem 2.1 the functions $\{ Y(\omega_x;A) \}$ separate points in $\mathcal{O}$. Moreover $\overline{Y(\omega_x;A)} = Y(\omega_x^{-1};A)$ also belongs W, and finally $Y(1_x;A) = \text{const.} > 0$. By the Stone-Weierstrass theorem the algebra generated by W is dense in $C(\mathcal{O})$; ($\mathcal{O}$ is supposed to be compact). ∎

Remarks.

1. Theorems 2.1 and 2.2 serve as one motivation for viewing the functions $Y(\omega_x) = \chi(g_{\omega_x})$ as the basic "observables" of a pure Yang-Mills theory.

2. Let Ω_d be a denumerable set of bounded, closed, piecewise smooth, oriented loops, e.g. the loops of some lattice on M. Let $\overline{\Omega}_d$ be the closure of Ω_d under inversion of orientation and composition of loops; ($\overline{\Omega}_d$ is a groupoid). Let $C_{\Omega_d}(\mathcal{O})$ be the algebra generated by $\{ Y(\omega): \omega \in \overline{\Omega}_d \}$. If all connections in A^I are continuous then $C_{\Omega_d}(\mathcal{O}) \nearrow C(\mathcal{O})$, (in the supremum topology), as $\overline{\Omega}_d \nearrow \overline{\Omega} \supseteq \Omega(x)$, for some $x \in M$, with M connected.

Let Ω_L denote all bounded, oriented loops in a lattice L. Approximating $C(\mathcal{O})$ by $C_{\Omega_L}(\mathcal{O})$ is the starting point of the lattice approximation to Yang-Mills theory.

It would be interesting to make a systematic study of all convenient, separable approximations to $C(\mathcal{O})$ that could serve to construct gauge-invariant regularizations of (quantized) Yang-Mills theory.

§3. Random (or Stochastic) Geometry

Throughout these notes we follow the Euclidean (time purely imaginary) approach to quantizing relativistic quantum field theory. In this approach the problem of field quantization is converted into one of constructing random fields and functional integrals, (unless there are Fermi fields in the theory which are ignored in these notes). In the Euclidean approach to quantized, pure Yang-Mills theory the basic random fields turn out to be the variables

$$Y(\omega) = \chi_Y(\omega),$$

studied in §2; χ_Y is a unitary character of the gauge group G, ω is a bounded, oriented loop, and g_ω is a "random holonomy operator" assigned to ω. (If the theory also contains a matter field Φ, assumed here to be spinless, transforming under a representation U^Φ of G then, in addition to the variables $Y(\omega)$, one must consider the variables

$$(\Phi(y), U^\Phi(g_{\gamma_{yx}})\Phi(x)),$$

where $(\cdot,\cdot)$ is an inner product on the fibre V of the bundle whose sections are given by Φ).

We thus see that in the Euclidean approach to quantized Yang-Mills theory one wants to construct random fields on spaces of geometrical objects, the oriented paths and loops in Euclidean space-time. According to Theorem 2.1, the random fields $Y(\omega)$ are in correspondence with a random connection on a random principal bundle.

The construction of such random fields can thus be viewed as a problem in a hypothetical branch of mathematics attempting to combine geometry and probability theory which one might call <u>random (or stochastic) geometry</u>.

We now give a short list of some problems in random geometry and then discuss a few of them in more detail.

1) Convert geometric objects (loops, surfaces, clusters, holonomy operators, etc.) into random variables, resp. random currents.

2) Construct stochastic processes whose state space is a space of geometrical objects.

3) Construct random fields on spaces of geometrical objects.

4) Construct random holonomy operators on a (random) fibre bundle.

5) Investigate random operators associated with a foliation |6|; etc.

In many situations random geometry is really measure theory on infinite dimensional manifolds, or manifolds modulo the action of some infinite dimensional transformation group, (e.g. a group of gauge transformations, the diffeomorphism group of the circle or a sphere, etc.).

Of concern to us are the following specific problems in random geometry:

(A) Theory of random holonomy operators on random bundles with fixed base space.

(B) Diffusion processes whose state space is a space of loops or a manifold of open paths with fixed endpoints, (modulo the action of the group of reparametrizations).

(C) Theory of random surfaces bounded by some fixed loops.

These problems are relevant for the understanding of quantized gauge fields, as we hope to explain in the remainder of these notes. We emphasize that there are numerous other branches in theoretical physics which pose their own problems in random geometry. In particular, statistical mechanics is rich in such problems.

Unfortunately, it turns out that random geometry in the continuum is very difficult and forces one to study very singular objects. For example, the holonomy operators of a Euclidean quantized Yang-Mills theory on ν-dimensional space-time, with $\nu \geq 3$ cannot be expected to be random fields in the precise sense of the word. To see this one may consider the free electromagnetic field:

Let $A(\cdot) = (A_1(\cdot), \ldots, A_\nu(\cdot))$ be the $\mathbb{R}^\nu$-valued Gaussian process with mean 0, i.e. $\langle A(\cdot) \rangle = 0$, and covariance

$$\langle A_j(0) A_m(x) \rangle = \delta_{jm} (2\pi)^{-\nu/2} \int e^{ikx} k^{-2} d^\nu k \tag{3.1}$$

One may attempt to define random holonomy operators (random phase factors), g_ω, by

$$g_\omega := \exp i \oint_\omega A_j(x) dx^j \tag{3.2}$$

Unfortunately g_ω does not exist as a random field on the space of loops: $g_\omega = 0$, almost surely. To give meaning to g_ω it needs to be "normal-ordered":

$$g_\omega \to N(g_\omega) = \text{"}\exp\left[\tfrac{1}{2} V_C(0) |\omega|\right] g_\omega\text{"} . \tag{3.3}$$

Here $|\omega|$ is the length of the loop ω, and V_C is the $(\nu-1)$-dimensional Coulomb potential. The r.s. of (3.3) can be defined rigorously as a limit of regularized objects if ω is sufficiently smooth (C^2) and $\nu \leq 4$. Since A is Gaussian, it is easy to calculate $< \overline{N}(g_\omega)N(g_{\omega'}) >$. One checks that if the relative positions and orientations of ω and ω' are suitably chosen one gets

$$< N(g_\omega)N(g_{\omega'}) > \sim \exp\left[\text{const. dist. } (\omega,\omega')^{-1}\right], (\text{for } \nu=4). \quad (3.4)$$

Thus, the objects $N(g_\omega)$ are too singular to be random fields in the usual sense of the word. For a (heuristic) theory of normal ordering of holonomy operators in three-dimensional, interacting theories see |7|.

The above discussion suggests that random geometry in the continuum may be plagued with serious difficulties. One way of regularizing the objects studied in random geometry is to pass to random combinatorial geometry by replacing continuum geometry by discrete geometry (combinatorics); see e.g. |8| for some discussion.

We conclude §3 with sketches of three examples of random combinatorial geometry.

I. Let $\mathcal{P} = (P, \mathbb{E}^\nu, G, \pi)$ be a principal bundle over $\mathbb{E}^\nu$. As discussed at the end of §2, we may approximate the space $C(\mathcal{O})$ of continuous functions over the orbit space $\mathcal{O}$ (the "observables") by $C_{\Omega_L}(\mathcal{O})$, where Ω_L is the set of all bounded, oriented loops in a lattice L which we choose to be $\varepsilon\, \mathbb{Z}^\nu = \{ x: \varepsilon^{-1} x \in \mathbb{Z}^2 \}$. We assign to each link (nearest neighbor pair) $xy \in L$ an element $g_{xy} \in G$. Given $\omega \in \Omega_L$, let

$$g_\omega = \overset{\circlearrowleft}{\prod_{xy\subset\omega}} g_{xy} \quad (3.5)$$

Let χ_Y be a character of G. We set

$$Y(\omega) = \chi_Y(g_\omega) \quad (3.6)$$

The algebra generated by the Y's is dense in $C_{\Omega_L}(\mathcal{O})$. Thus, in order to convert the elements of $C_{\Omega_L}(\mathcal{O})$ into random variables, it suffices to construct the joint distribution of the "Wilson loop variables," i.e. to construct a measure on $\{ g_{xy} \}$. The standard proposal |9|, due to Wilson, is the following: Let χ be a unitary character of G, let p denote a plaquette (2-cell) of $L = \varepsilon\, \mathbb{Z}^\nu$, ∂p its boundary.

Let $\Lambda \in \varepsilon\mathbb{Z}^\nu$ be a bounded set. Define an action, A^{YM}_Λ, by

$$A^{YM}_\Lambda = - \sum_{p \subset \Lambda} \mathrm{Re}\, \chi(g_{\partial p}) \tag{3.7}$$

Let dg denote normalized Haar measure on $G, \beta > 0$. We define

$$d\mu^{(\varepsilon)}_\Lambda(g) = Z^{-1}_\Lambda e^{-\beta A^{YM}_\Lambda} \prod_{xy} dg_{xy} \tag{3.8}$$

with Z_Λ such that $\int d\mu^{(\varepsilon)}_\Lambda(g) = 1$.

By a standard compactness argument one can choose a sequence $\{\Lambda_n\}^\infty_{n=o}$ increasing to $\varepsilon\mathbb{Z}^\nu$ such that

$$d\mu^{(\varepsilon)}(g) = w^*\text{-}\lim_{n\to\infty} d\mu^{(\varepsilon)}_\Lambda(g) \text{ exists.} \tag{3.9}$$

(Conditions for existence and uniqueness of the limit $d\mu$ are given e.g. in |9,10|). The measure $d\mu$ is now interpreted as the joint distribution of the random variables $\{ Y(\omega) : \omega \in \Omega_L \}$. Of particular interest in the discussion of the resulting theory are the <u>Schwinger functionals</u>

$$S^{(\varepsilon)}_n(Y_1(\omega_1),\ldots,Y_n(\omega_n)) = \int d\mu^{(\varepsilon)}(g) \prod_{j=1}^n \chi_{Y_j}(g_{\omega_j}) \tag{3.10}$$

What we have introduced here is the standard lattice approximation to quantized, pure Yang-Mills theory |9|.

II. Let $\Gamma_L(x,y)$ be the set of all finite, oriented curves in $L = \varepsilon\mathbb{Z}^\nu$ starting at x and ending at y. This is clearly a countable set. Let $\ell_{2,r}$ be the Hilbert space of functions F on $\Gamma_L(x,y)$ with the property that

$$\sum_{\gamma_{xy} \in \Gamma_L(x,y)} e^{r|\gamma_{xy}|} |F(\gamma_{xy})|^2 < \infty, \tag{3.11}$$

for some $r \geqq 0$. Let p be an oriented plaquette.

If $\partial p \cap \gamma_{xy} \neq \emptyset$ we define $\gamma_{xy} \circ \partial p$ by the following figure:

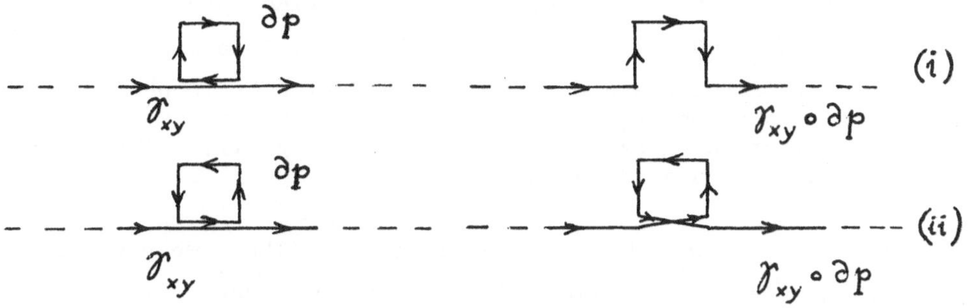

For $F \in \ell_{2,r}$ define

$$(\delta_p F)(\gamma_{xy}) = \begin{cases} F(\gamma_{xy} \circ \partial p) - F(\gamma_{xy}) \text{ if } \partial p \cap \gamma_{xy} \neq \emptyset \\ \text{and } \gamma_{xy} \circ \partial p \text{ is connected;} \\ 0, \text{ otherwise} \end{cases} \tag{3.12}$$

One may now define a functional Laplacean, $\mathcal{D}_1$, as the unique selfadjoint operator determined by the quadratic form

$$F \to \sum_p (\delta_p F, \delta_p F)_{\ell_2}, \tag{3.13}$$

defined e.g. on $\ell_{2,r}$, $r > 0$; ($\ell_2 = \ell_{2,r=0}$).
Let V_α be the multiplication operator on ℓ_2 given, for example, by

$$(V_\alpha F)(\gamma_{xy}) = \alpha |\gamma_{xy}| F(\gamma_{xy}), \tag{3.14}$$

where $|\gamma_{xy}|$ is the number of links in γ_{xy}.

The operator sum $\mathcal{D}_1 + V_\alpha$ is still selfadjoint, and it follows from a general theorem in |11| that the kernel $(\exp[-t(\mathcal{D}_1 + V_\alpha)])(\gamma_{xy}, \gamma'_{xy})$ is non-negative, for all $\gamma_{xy}, \gamma'_{xy}$ in $\Gamma_L(x,y)$. Thus $\exp -t(\mathcal{D}_1 + V_\alpha)$ is the transition function of a stochastic process on $\Gamma_L(x,y)$. This process describes the diffusion of an oriented string with fixed endpoints x and y. It has some significance in the analysis of confinement in lattice gauge theories, |7|. See also §6. If the deformation (ii) in the definition of $\gamma_{xy} \circ \partial p$ is omitted, the resulting process may be of interest in the study of selfavoiding random walks).

II'. Let Ω_L be the set of all finite, oriented loops in L; Ω_L is still countable. Therefore one may define spaces $\ell_{2,r} = \ell_{2,r}(\Omega_L)$ and operators δ_p, $\mathcal{D}_1$, V_α, etc. in a similar way as above. There results a model for the diffusion of loops in the lattice L. For some results concerning a general theory of diffusion of discrete, geometrical objects see e.g. |11|. (They have applications in statistical mechanics).

III. Let Ω_L, $\mathcal{D}_1$, V_α,... be as in (II), (II'). we propose to give an example of a random field Φ on Ω_L. To each $\omega \in \Omega_L$ we assign an n×n matrix, $\Phi(\omega)$, with a priori distribution $d\Phi(\omega)$ given by the Lebesgue measure on $\mathbb{C}^{n^2}$. There exists a random field Φ the distribution of which corresponds to the formal measure

$$d\mu(\Phi) = Z^{-1}\exp\Big[-\sum_{\omega} tr(\Phi(\omega)^*[(\mathcal{D}_1+V_\alpha)\Phi](\omega))\Big].$$

$$\cdot \prod_{\omega\in\Omega_L} e^{-2tr(\Phi(\omega)^*\Phi(\omega))^2} d\Phi(\omega).$$

The measure $d\mu$ can be constructed as a limit of cutoff measures. The field Φ is conveniently described by its "Schwinger functionals"

$$S_n(\Phi_{\alpha_1\beta_1}(\omega_1)\dots\Phi_{\alpha_n\beta_n}(\omega_n)) = \int d\mu(\Phi) \prod_{j=1}^{n} \Phi_{\alpha_j\beta_j}(\omega_j) \tag{3.15}$$

which one may interpret as Schwinger functions of a lattice string theory |12|.

If the constraints

$$\prod_{\omega,\omega'} \delta(\Phi^*(\omega\circ\omega')\Phi(\omega)\Phi(\omega')-\mathbb{1}) \tag{3.16}$$

are inserted into $d\mu$ and the couplings are suitably rescaled, the above theory becomes a lattice gauge theory with $G = U(n)$; see |11|. This example is admittedly somewhat naive. It may serve as a challenge for a serious study of more interesting random geometrical models. The most important problem is to find interesting models of this sort for which the continuum limit ($\varepsilon \searrow 0$) exists. This is the subject of the renormalization group ("block spin transformations") and non-perturbative renormalization.

§4. Schwinger Functionals and Relativistic Quantum Fields

In this paragraph we briefly discuss the question whether Schwinger functionals of the sort defined in (3.10) and (3.15) determine a relativistic quantum field theory. The answer to this question is, for conventional, local field theories, the Osterwalder-Schrader reconstruction theorem |13,14|. We quote here a generalization of that result which accounts for theories of fields defined on spaces of geometrical objects such as the "Wilson loops" of pure Yang-Mills theory. The theorem is first stated for a class of continuum theories and represents a special case of more general results of this type |15|.

The Euclidean space-time manifold is $\mathbb{E}^\nu, \nu=2,3,4$. Let $\Omega^{(d)}$ be the family of all oriented C^∞ d-dimensional surfaces in $\mathbb{E}^\nu$ without self-intersections, (i.e., topologically, d-dimensional spheres), with $d \leqq \nu-2$. For ω, ω' in $\Omega^{(d)}$, set

$$d(\omega,\omega') = \text{dist}\,(\omega,\omega') \equiv \min_{\substack{x\varepsilon\omega \\ y\varepsilon\omega'}} |x-y|.$$

Let $\Omega^{(d)n}_{\neq} = \{\omega_1,\ldots,\omega_n \text{ in } \Omega^{(d)}: d(\omega_i,\omega_j) > 0, \text{ for } i\neq j\},$ (4.1)

$$\Omega^{(d)n}_{>} = \{(\omega_1,\ldots,\omega_n) \,\varepsilon\, \Omega^{(d)n}_{\neq}: \omega_j \,\varepsilon \{x=(\vec{x},t): t>0\}, j=1,\ldots,n\} \tag{4.2}$$

We now assume that we are given a sequence of Schwinger functionals $\{S_n(Y_1(\omega_1),\ldots,Y_n(\omega_n))\}_{n=0}^\infty$ with the following properties:

(S1) $S_0=1$; $S_n(Y_1(\omega_1),\ldots,Y_n(\omega_n))$ is well-defined on $\Omega^{(d)n}_{\neq}$ and continuous under small C^∞ deformations of $\omega_1,\ldots,\omega_n$ in $\Omega^{(d)n}_{\neq}$.

Moreover, the growth of $|S_n(Y_1(\omega_1),\ldots,Y_n(\omega_n))|$, as $d_n \equiv \min_{i\neq j} d(\omega_i,\omega_j) \searrow 0$, is bounded by $O(\exp[\text{const.}\ d_n^{-\alpha}])$, for some $\alpha \geqq 0$ and constants that depend on n in a suitable way; see |13,15|.

(S2) (Osterwalder-Schrader positivity)
Let r be reflection at $\{t=0\}$ and let $Y \to Y_r$ be some reflection map (in the case of Yang-Mills theory $Y_r = \bar{Y}$, the complex conjugation of Y). The N x N matrix C with matrix elements C_{ij} given by

$$S_{n(i)+n(j)}(Y^i_{n(i),r}(\omega^i_{n(i),r}),\ldots,Y^i_{1,r}(\omega^i_{1,r}),Y^j_1(\omega^j_1),\ldots,Y^j_{n(j)}(\omega^j_{n(j)})),$$

$i,j = 1,\ldots,N$, is positive semi-definite, provided $(\omega_1^k,\ldots,\omega_{n(k)}^k) \in \Omega_>^{(d)n(k)}$, for all $k = 1,\ldots,N$; $N=1,2,3\ldots$.

(S3) (Symmetry)
$S_n(Y_1(\omega_1),\ldots,Y_n(\omega_n))$ is symmetric under arbitrary permutations of its arguments, for all n.

(S4) (Invariance)
Let β be a proper Euclidean motion. Then $S_n(Y_1(\omega_1),\ldots,Y_n(\omega_n)) = S_n(Y_1(\omega_{1,\beta}),\ldots,Y_n(\omega_{n,\beta}))$, for all n. (Here ω_β is the image of ω under β).

If we consider a lattice theory we replace (S4) by (S4'): Invariance under the symmetries of the lattice.

(S5) (Clustering; see |13|)

Heuristically, the Schwinger functionals of a Yang-Mills theory satisfy additional properties, in particular an extended version of (S2) (Osterwalder-Schrader positivity) to which we refer as $(S2^{ext.})$; see |10,15|.

The main theorem about sequences of Schwinger functionals satisfying (S1)-(S4) is

Theorem 4.1.

If $\{S_n(Y_1(\omega_1),\ldots,Y_n(\omega_n))\}_{n=0}^\infty$ satisfies (S1), (S2) and (S4) then one can reconstruct from those Schwinger functionals a separable physical Hilbert space $\mathcal{H}$, a vacuum vector $\Omega \in \mathcal{H}$, with $<\Omega,\Omega> = 1$, and a unitary representation U of the proper Poincaré group

$$P_+^\uparrow \text{ on } \mathcal{H} \text{ with } U(a,\Lambda)\,\Omega = \Omega, \tag{4.3}$$

for all $(a,\Lambda) \in P_+^\uparrow$. The spectrum of the generators $(\vec{P},H)$ of the space-time translations is contained in the forward light cone $\overline{V}_+$.

If, in addition, (S3) holds there exist "local fields"

$$y(\omega;Y),\ \omega \in \Omega^{(d)},\ \omega \subset \{x=(\vec{x},t) \in M^\nu : t = \text{const.}\},$$

with $[y(\omega;Y),y(\omega';Y')] = 0$ if ω and ω' are space-like separated. |If (S1),(S2),(S4) and (S5) hold then the vacuum Ω is the only vector satisfying (4.3), i.e. the vacuum is unique|.

A more precise formulation and a proof of this basic theorem will be given elsewhere, |15|.

Some of the main tools in the proof of Theorem 4.1 not already used in |13,14| are: A result concerning the selfadjoint extensions of symmetric semigroups |16| that serves to construct the representation of the Poincaré group, and the observation that the Schwinger functionals determine a state $< \Omega, \cdot\ \Omega >$ which satisfies the KMS condition with respect to the Lorentz boosts, |17|. A somewhat novel approach to the results of |17| and to proving locality are consequences of that observation. See |15|.

Next we discuss a few physical properties coded directly into the Schwinger functionals. The first is a consequence of extended Osterwalder-Schrader positivity $|10^1|$ in Yang-Mills theory. In that theory d = 1, and $\Omega^{(d)}$ is the space of loops in $\mathbb{E}^\nu$ diffeomorphic to circles. Let ω_{LxT} be a (smoothed version of a) rectangular loop with sides of length L and T. Assume that ($S2^{ext.}$), $|10^1,15|$, holds. Then $S_1(Y(\omega_{LxT}))$ is log convex. Therefore

$$V_Y(L) := \lim_{T\to\infty} - \frac{1}{T} \log S_1 (Y(\omega_{LxT})) \qquad (4.4)$$

exists, and moreover one concludes

Proposition 4.2

$$V_Y(L) \leqq \text{const. } L, \text{ as } L \to \infty. \qquad (4.5)$$

For lattice theories this inequality has been established in |18|. Physically, it says that the potential between a static (infinitely heavy) quark and a static anti-quark cannot rise faster than linearly.

It was suggested in |7,11| that $S_1(Y(\omega))$ contains information about the boundstate spectrum of very heavy quarks, and $S_2(Y(\omega),Y(\omega'))$ about the low-lying mass spectrum of pure Yang-Mills theory.

Next, we sketch the notion of "disorder fields" |19-22|. We assume that, in addition to the "random fields" $Y(\omega), \omega \in \Omega^{(d)}$, there are "fields" $B(\gamma)$, $\gamma \in \Omega^{(\nu-2-d)}$, $(\Omega^{(0)} := \mathbb{E}^2)$, with joint Schwinger functions

$$\{ S_{n,m}(Y_1(\omega_1),\ldots,Y_n(\omega_n),B(\gamma_1),\ldots,B(\gamma_n)) \}_{n,m=0}^{\infty}. \qquad (4.6)$$

These Schwinger functions are supposed to have properties analogous to (S1)-(S4), but in addition they are required to have certain specific discontinuities (which cannot arise in standard field theories of the Wightman type):

Choose $\omega,\gamma \subset \{(\vec{x},t): t = \text{const.}\}$ and let $\nu(\omega,\gamma)$ be the linking number of ω,γ. Let ω_ε be the translation of ω in the t-direction. Then

$$\lim_{\varepsilon\searrow 0} S_{\cdot,\cdot}(\ldots,Y(\omega_\varepsilon),B(\gamma),\ldots) = z_{Y,B}^{\nu(\omega,\gamma)} \lim_{\varepsilon\searrow 0} S_{\cdot,\cdot}(\ldots,Y(\omega_{-\varepsilon}),B(\gamma),\ldots) \tag{4.7}$$

In two-dimensional scalar field theories with soliton behavior and three-or four dimensional Yang-Mills theory one can prove that $|z_{Y,B}| = 1$, |19-21|. In fact, in Yang-Mills theory, z_{YB} is an element of the center of the gauge group G which depends on Y and B, |20,22|. An extension of the reconstruction theorem, Theorem 4.1, provides us with fields $y(\omega;Y)$ and $b(\gamma,B)$, which for $\omega,\gamma \subset \{(\vec{x},t): t = 0\}$ satisfy the following formal time 0 commutation relations

$$y(\omega;Y)b(\gamma;B) = z_{YB}^{\nu(\omega;\gamma)}\, b(\gamma;B)y(\omega;Y) \tag{4.8}$$

(The field b is said to be "dual" to y).

For $\nu = 2$, $d = 0$, $\nu = 3$, $d = 1$, and $\nu = 4$, $d = 2$, (i.e. $\nu-2-d=0$) such commutation relations have been discussed in |19| and representations with $d = 0$ have been constructed for two-dimensional scalar field theories with soliton behavior, (the sine-Gordon and the $\lambda\phi^4$ models). For $\nu = 3$, $d = 1$, and $\nu = 4$, $d = 1$ they have been proposed and interpreted in |20|; see also 't Hooft's contribution to these proceedings. For $\nu = 2,3,4$, $d = 0$, certain "quasi-free" representations have been constructed in a series of remarkable papers by Jimbo, Miwa and Sato |21|. Their work shows how powerful relations like (4.7) may be and has resulted in the calculation of the correlation functions of the two-dimensional Ising model. For lattice theories representations of (4.8) have been constructed for $\nu = 2,3,4$, with $d = 0,1,1$, respectively, |22|.

In |19| properties of the representations of (4.8) when $\nu = 2$, $d = 0$ or $\nu = 3$, $d = 1$ (or $\nu = 4$, $d = 2$) have been related to the structure of super-selection sectors of the corresponding quantum field theories: If $\text{const.}_\omega y(\omega;Y)$ converges on $\mathcal{H}$ to a non-zero element of the center of the observable algebra, as $|\omega| \to \infty$, and $z_{YB} \neq 1$, then $b(x;B)$ intertwines disjoint superselection sectors of that algebra. (The two-dimensional case has been studied from first principles, whereas in higher dimensions one needs suitable technical assumptions). In |20| 't Hooft has suggested connections between properties of the representations of (4.8) in a gauge theory and quark-resp. monopole confinement. He argues that (4.8) rules out the possiblity that quarks and monopoles are both confined. This has been elaborated and tested in models in |22,23,10²|. See the contributions of 't Hooft and Mack to these proceedings.

§5. Existence of Quantized Gauge Fields

In this paragraph we quote some results concerning the existence of models satisfying the axiomatic scheme of §4. At present, the only models that fit into that scheme are models of quantized interacting gauge fields-and matter fields-on a lattice of arbitrary dimension (see end of §2 and Example (I), §3) and in a continuum space-time ($\mathbb{E}^\nu$) of dimension $\nu = 2$, and presumably $\nu = 3$. Of course, the free electromagnetic field in two, three or four dimensions satisfies (S1)-(S5). In the continuum only abelian gauge fields have been constructed so far. If the gauge group is abelian there are, in addition, lattice theories describing abelian gauge fields which are connections on bundles whose base space is e.g. the space of oriented loops in the lattice: To each plaquette p one assigns an element $e^{ia_{\partial p}} \in G, G \overset{e.g.}{=} U(1), \mathbb{Z}_n$, with a priori distribution the Haar measure on G. The action is given by $A_\Lambda = -\sum_{c\subset\Lambda} \cos\left(\sum_{p\in\partial c} a_{\partial p}\right)$. These theories have Schwinger functionals of random "holonomy operators" associated with closed lattice surfaces satisfying (S1)-(S3), (S4'), (in the limit $\Lambda = \mathbb{Z}^\nu$). The models of the type described in Example (III), §3, (without the constraint (3.16)) are not known to fit into the scheme of §4. (This may be related to the difficulties which are met in string theories |24|). For detailed studies of lattice theories see e.g. |9,10|.

Let $d\mu^{(\varepsilon)}(g)$ denote a limit of the measures $d\mu_\Lambda^{(\varepsilon)}(g)$, (Example (I), §3, (3.7)-(3.9)), as $\Lambda \uparrow \mathbb{Z}^\nu$.

For small β, the limit is unique and the Schwinger functionals have exponential cluster properties; detailed properties such as confinement can be investigated by means of high temperature expansions, |9|. Uniqueness of the $\Lambda \nearrow \mathbb{Z}^\nu$ limit can also be proven in a class of abelian models, for all β. A few physical properties of lattice gauge theories are sketched in the next paragraph.

The list of models of quantized, interacting gauge fields in the continuum satisfying (S1)-(S4) is still short:

1) The abelian Higgs model (scalar QED) in two space-time dimensions, |25|.

2) Spinor QED in two space-time dimensions with massless or massive fermions and massless or massive photons, |26|. (For a different approach see |27|).

|3) For spinor QED in three dimensions, a proof of stability of the theory is announced |28|.

4) For some super-renormalizable gauge theories (including non-abelian ones), T. Balaban has announced a proof of stability |29| based on a rigorous form of renormalization group - "block spin" - transformations for lattice theories, extending previous work due to Gallavotti, et al.|30| for the $\lambda\phi^4$ theory in three dimensions |31| |.

§6. Random Geometrical Methods in Lattice Gauge Theories.

In this paragraph we briefly discuss four examples in lattice gauge theory the analysis of which is based on estimating the joint distribution of random variables labelled by geometrical objects such as closed flux tubes or (interacting) oriented random paths with fixed endpoints. We sketch some typical steps in that analysis thereby providing examples for the uses of random-geometrical arguments in the study of lattice gauge theories.

Example 1.

We discuss the behavior of the expectation of the Wilson loop in a three-dimensional $\mathbb{Z}_2$ lattice theory, (i.e. G = $\mathbb{Z}_2$ is the gauge group). This model can be thought of as a Kindergarden theory of vortices in a type II superconductor. The Wilson loop - dual to the "vortex field" - is the non-integrable phase factor of the superconducting medium. The action of the model is

$$A = -\sum_p \sigma_{\partial p},\ \sigma_{\partial p} = \prod_{xy\subset\partial p} \sigma_{xy},\ \sigma_{xy} = \pm 1. \tag{6.1}$$

The infinite volume expectation in that model at inverse coupling β - see Example (I), §3, (3.8), (3.9) - is denoted $\langle - \rangle_\beta$. (It can be constructed by means of correlation inequalities, for all β |32|).

Let c be an arbitrary 3-cell (unit cube) in $\mathbb{Z}^3$. Then $\prod_{p\subset\partial c} \sigma_{\partial p} = 1$, since $\sigma_{xy}^2 = 1$, for all xy. We now introduce the random phase factors $\sigma_{\partial p}$ as a priori independent variables, inserting the constraint $\prod_{c\subset\mathbb{Z}^3} \delta_{\sigma_{\partial c},1}$, with $\sigma_{\partial c} = \prod_{p\subset\partial c} \sigma_{\partial p}$ (6.2)

We set $$\phi_{\partial p} = \begin{cases} 0 \text{ if } \sigma_{\partial p} = 1 \\ 1 \text{ if } \sigma_{\partial p} = -1 . \end{cases} \tag{6.3}$$

Let ω be a rectangle with sides of length L and T parallel to two coordinate axes. Let $\bar{\omega}$ be the planar surface bounded by ω, i.e. $\partial\bar{\omega} = \omega$. Since $\sigma_{xy}^2 = 1$, for all xy,

$$\sigma_\omega \equiv \prod_{xy \subset \omega} \sigma_{xy} = \prod_{p \subset \bar{\omega}} \sigma_{\partial p} . \tag{6.4}$$

This is the "non-integrable phase factor" (Wilson loop) observable of the medium.

<u>Theorem 6.1</u>.

For sufficiently small β

$$< \sigma_\omega >_\beta \leqq e^{-O(|\bar{\omega}|)}, \quad |\bar{\omega}| = \text{area of } \bar{\omega} .$$

For sufficiently large β,

$$< \sigma_\omega >_\beta \geqq e^{-O(|\omega|)}, \quad |\omega| = \text{perimeter of } \omega.$$

The result for small β follows from a standard high temperature expansion |9|. The large β result has first been proven in |33|; see also |34|.

We outline a simple proof.

Let $c_{1/0}(\sigma_{\partial p})$ be the characteristic function of $\{\sigma_{\partial p} = 1\}/\{\sigma_{\partial p} = -1\}$. Then

$$<\sigma_\omega>_\beta = \sum_\phi (-1)^{|\phi|} < \prod_{p \subset \bar{\omega}} c_{\phi_p} >_\beta, \tag{6.5}$$

with $\phi_p = 1$ or 0, $c_{\phi_p} \equiv c_{\phi_p}(\sigma_{\partial p})$, and $|\phi| = \sum_{p \subset \bar{\omega}} \phi_p$. The constraint (6.2) implies <u>flux conservation</u>, i.e. the total flux (= # of p's with $\phi_p = 1$) through each closed surface is 0, mod. 2. Thus all flux tubes, τ, are closed.

See Fig. 1:

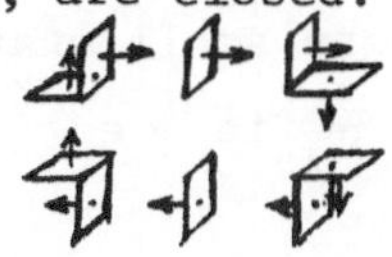

Given ω, each flux tube τ (closed loop in the dual lattice) can be assigned a linking number, $\nu(\omega,\tau)$, (with ω), defined mod. 2. Let

$$0 < \Pr_\omega(n) = \text{prob.}\ (\{\exists n \text{ flux tubes, } \tau_1,\ldots,\tau_n, \text{ with } \nu(\omega,\tau_i) = 1, \forall i\}).$$

By (6.5) $$<\sigma_\omega>_\beta = \Pr_\omega(0) - \Pr_\omega(1) + \Pr_\omega(2) - \Pr_\omega(3) + \ldots \qquad (6.6)$$

Now to each configuration $\phi = \{\phi_p\}$ contributing to $\Pr_\omega(2n+1)$ there is one contributing to $\Pr_\omega(2n)$ with one flux tube τ, $\nu(\omega,\tau) = 1$, less, (i.e. $\phi_p = 1 \rightarrow \phi_p = 0$, $\forall p \varepsilon \tau$). The statistical weight of one flux tube, τ, is

$$\propto e^{-\beta|\tau|}, \text{where } |\tau| = \#\ p\text{'s } \varepsilon\ \tau \text{ (with } \phi_p = 1). \qquad (6.7)$$

Thus $\Pr_\omega(2n) - \Pr_\omega(2n+1) \geqq \alpha \Pr_\omega(2n)$, with $\alpha \geqq 1 - e^{-\text{const.}\beta} > 0$. (6.8)

This yields with (6.6)

$$<\sigma_\omega>_\beta \geqq \alpha \Pr_\omega(0).$$

Let $\Pr'_p = \text{cond. prob.}\ (\{\exists\ \tau: p\ \varepsilon\ \tau, \phi_p = 1, \nu(\omega,\tau) = 1\})$,

given $\phi_{p'}$, for some $p' \neq p$. A simple argument shows that

$$\Pr_\omega(0) \geqq \prod_{p \varepsilon \bar{\omega}} (1 - \Pr'_p), \qquad (6.9)$$

and by (6.7) and standard arguments for counting closed flux tubes through p of a given length one finds

$$\Pr'_p \leqq e^{-\text{const.}\ \beta\ \text{dist.}(p,\omega)} \qquad (6.10)$$

<u>if β is large enough.</u> From (6.6)-(6.10) we obtain by a simple calculation

$$<\sigma_\omega>_\beta \geqq \text{const.}\ e^{-\text{const.}'\ |\omega|}, \qquad (6.11)$$

for large β which proves our contention.

Thus if flux tubes have a very small statistical weight, the non-integrable phase factor σ_ω is $\propto e^{-O(|\omega|)}$, in the average.

This situation is analogous to one met in the Ising model: If contours have a very small statistical weight then $\sigma_o \sigma_x \propto$ const., uniformly in x, in the average. Theorem 6.1 has been extended to the four dimensional U(1) model in |35|, (the proof being very different).

More realistic models of superconductors in three (and four) dimensions are discussed e. g. in $|10^2|$,and refs..

Example 2.

We consider pure Yang-Mills lattice theories with gauge group G = U(n) or SU(n), n = 2,3,.... See Example (I), §3, (3.5)-(3.9), and we set ε=1 and choose in (3.7) χ to be the character of the fundamental representation. Moreover,

$$Y(\omega) = \chi(g_\omega). \tag{6.12}$$

We study the behavior of $S_1(Y(\omega)) = < Y(\omega) >_\beta$ in β. Let $\omega = \omega_{L\times T}$ be a rectangle in the $(1,\nu)$ plane with sides of length L and T, and let $V_Y(L)$ be the function ($q\bar{q}$ potential) defined in (4.4). It is easy to show that, for ν=2, $V_Y(L) \geqq$ const. L, for all β; (i.e. permanent confinement by a linear potential). For ν=3, G=U(n), n=1,2,3,...,

$V_Y(L) \geqq$ const. log (L+1); see |36|.

There are arguments in support of

$V_Y(L) \geqq$ const. L, for G = U(n), SU(n),

n=2,3..., ν=3. An interesting case is G=SU(2), ν=3 or 4. In |7| the following somewhat remarkable identity has been proven: Let Σ be a family of oriented paths $\{\gamma_u^\Sigma : 1 \leqq u \leqq T\}$ starting at the site (0,...,0,u), ending at (L,0,...,0,u) and lying in the plane $\pi_u = \{x : x^\nu = u\}$. Let $(\gamma_u^\Sigma)^{-1}$ be the path obtained by reversing the orientation of γ_u^Σ. Then

$$S_1(Y(\omega)) \equiv < Y(\omega) >_\beta$$

$$= \sum_{\Sigma:\partial\Sigma=\omega} (2(\nu-1))^{-|\gamma_u^\Sigma|-1} < \prod_{u=1}^{T} F(g^h(u-1), g^h(u)|\gamma_u^\Sigma) . \tag{6.13}$$

$$\cdot \prod_{u=0}^{T} \chi(g_{\gamma_u^\Sigma} \circ (\gamma_{u+1}^\Sigma)^{-1}) >_\beta$$

where γ_o^Σ is the bottom face and γ_{T+1}^Σ the top face of ω, $g^h(u) = \{ g_{xy} : xy \in \pi_u \}$, and $F(g^h(u-1), g^h(u) | \gamma_u)$ is a gauge invariant function of the "horizontal" gauge fields $g^h(u-1), g^h(u)$ depending on γ_u. The r.s. of (6.13) can be viewed as a sum over joint correlations of interacting random paths (forming a "random surface" Σ). In mean

$$F(g^h(u-1), g^h(u) | \gamma_u) \sim e^{-\alpha|\gamma_u|}, \tag{6.14}$$

with $\alpha > -\ln \left[(2\nu-3)(\nu-1)(4\nu-4)^{-1}\beta\right]$.

$$\text{Thus if} \quad \beta < \frac{4}{3} \ (\nu=3), \text{ resp. } \beta < \frac{4}{5} \ (\nu=4) \tag{6.15}$$

$$< Y(\omega)_{L\times T} >_\beta \leqq e^{-0(|\bar{\omega}|)}, \text{ (area decay)}, \tag{6.16}$$

by (6.13) and (6.14). Condition (6.15) is not nearly necessary for area decay, because (6.14) is only a rough estimate and because the factor $\prod_{u=0}^{T} \chi(g_{\gamma_u^\Sigma \circ (\gamma_u^\Sigma)^{-1}})$ on the r.s. of (6.13) provides for strong additional damping of $< Y(\omega_{L\times T}) >_\beta$,

$$\simeq \prod_{u=0}^{T} \exp - 0\left(|\gamma_u^\Sigma \,\Delta\, \gamma_{u+1}^\Sigma|\right). \tag{6.17}$$

We expect that an improvement of the estimates in |7,37| taking into account that factor ought to permit to show that $< Y(\omega_{L\times T}) >_\beta \leqq e^{-0(|\bar{\omega}|)}$, for all β when $\nu=3$. (The situation for $\nu=4$ is technically less well understood).

Next, recall Example (II), §3, (3.13), (3.14). Choose $x=0$, $y=(L,0,\ldots,0)$, $t \propto T$ and approximate $\exp\left[-t(\mathcal{D}_1+V_\alpha)\right]$ by

$$\{ \exp\left[- t/T\, \mathcal{D}_1\right] \exp\left[- t/T\, V_\alpha\right] \}^T \tag{6.18}$$

If we write out (6.18) as a sum over products of matrix elements labelled by paths $\gamma_{xy} \in \Gamma_L(x,y)$ and compare with (6.13), (6.14) and (6.17) we see that for suitably small β and a proper choice of t and α

$$< Y(\omega_{L\times T}) >_\beta \lessapprox \exp\left[-t(\mathcal{D}_1+V_\alpha)\right] (\gamma_T^\Sigma, \gamma_o^\Sigma).$$

Connections between lattice gauge theories and the diffusion of strings or loops of this sort might have interesting consequences for the heuristic understanding of the string dynamics in Yang-Mills theory.

Example 3.

We consider the behavior of the disorder parameter in a three-(or four) dimensional SU(2) lattice gauge theory with distribution $d\tilde{\mu}(g)$ as proposed by Mack and Petkova |22|: $d\tilde{\mu}(g)$ is given by

$$d\tilde{\mu}(g) = \lim_{\Lambda \uparrow \mathbb{Z}^3} \tilde{Z}_\Lambda^{-1} \prod_{c \subset \Lambda} \Theta\left(\prod_{\partial p \subset c} \chi(g_{\partial p})\right) d\mu(g), \qquad (6.19)$$

with $d\mu(g) \equiv d\mu^{(1)}(g)$ as in (3.8), (3.9).

The expectation in $d\tilde{\mu}$ is denoted $< \text{---} >_{\tilde{\beta}}$. One is interested in the behavior of the expectation of the disorder parameter, $< B(\tau_{ox}) >_{\tilde{\beta}}$, with τ_{ox} as depicted in Fig. 2:

Let $\phi_p = \frac{1}{2}(1+\text{sgn}\, \chi\, (g_{\partial p}))$. The constraint $\prod_c \Theta\left(\prod_{\partial p \subset c} \chi(g_{\partial p})\right)$ enforces that $\sum_{\partial p \subset c} \phi_p = 0$, mod. 2. Thus ϕ_p may be interpreted as a $\mathbb{Z}_2$ flux through p, and only closed flux tubes are compatible with the constraint; as in Example 1. The statistical weight of a closed flux tube, τ, is bounded by $e^{-k(\beta)|\tau|}$, with $k(\beta) \nearrow \infty$ as $\beta \nearrow \infty$, as follows from a chessboard estimate |38|. Expanding $< - >_{\tilde{\beta}}$ in flux tube configurations it is a fairly simple matter of counting flux tubes of given lengths passing through τ_{ox} to prove that when $e^{-k(\beta)}$ is sufficiently small (i.e. β large)

$$< B(\tau_{ox}) >_{\tilde{\beta}} \leqq e^{-O(|x|)}, \qquad (6.20)$$

see |22,23|. (In outline we have followed here |23|). One can show, by comparison with the $\mathbb{Z}_2$ model,

$$< \prod_{p \subset \bar{\omega}} \sigma_p >_{\tilde{\beta}} \leqq e^{-O(|\bar{\omega}|)}, \quad \text{where} \quad \sigma_p = \text{sgn}\, \chi\, (g_{\partial p}),$$

for β small enough, see |22|, and by arguments very similar to those used in Example 1,

$$< \prod_{p\subset\bar{\omega}} \sigma_p >_{\tilde{\beta}} \geq e^{-O(|\omega|)}, \quad \partial\bar{\omega} = \omega,$$

for large β.

Example 4.

Let $G=U(N), N= 1,2,3,\ldots$. Let $Y_N(\omega) = \frac{1}{N} \chi_N(g_\omega)$ with χ_N the character of the fundamental representation of $U(N)$.

One is interested in an expansion of $S_n(Y_N(\omega_1),\ldots,Y_N(\omega_n)) = < \prod_{j=1}^{n} Y_N(\omega_j) >_\beta$ in powers of $\frac{1}{N}$.

To leading order in $\frac{1}{N}$, $< \prod_{j=1}^{n} Y_N(\omega_j) >_\beta$ factorizes, i.e. correlations are suppressed in the $N = \infty$ limit. The problem is to identify and compute the $N = \infty$ limit of $< Y_N(\omega) >_\beta$ and to then determine systematic corrections to S_n, in particular to $< Y_N(\omega) >_\beta$, in the form of power series in $\frac{1}{N}$. A somewhat heuristic calculation |39| yields

$$< Y_N(\omega) >_\beta = \sum_{\Sigma:\partial\Sigma=\omega} w(\beta,N,\Sigma), \tag{6.21}$$

where $\{ \Sigma: \partial\Sigma = \omega \}$ are all surfaces built of oriented plaquettes (2-cells in $\mathbb{Z}^\nu$) bounded by the loop ω, and $w(\beta,N,\Sigma)$ are the weights of these surfaces. Once can argue |39,40| that, to leading order in $\frac{1}{N}$, only simply connected, normal surfaces, Σ, with $\partial\Sigma = \omega$ contribute to the r.s. of (6.21). The weights of these surfaces are $\propto \exp\left[-d_\beta |\Sigma|\right]$, with $|\Sigma|$ the total area of Σ. Moreover surfaces of higher genus (with handles) are suppressed by powers of $1/N^2$, |39,40|.

In spite of these preliminary findings a systematic expansion in $\frac{1}{N}$ is missing. To do that one must first find geometrical characterizations of all surfaces contributing to a given order in $1/N$, determine their weights and sum up their contributions. This appears to raise very subtle problems in the combinatorial geometry of lattice surfaces and combinatorics. (An alternate approach based on the techniques sketched in Example 2 has been suggested in |7|).

For a more detailed analysis and refs. see E. Witten's contribution, and for results concerning the $\frac{1}{\nu}$-expansion Parisi's contribution to these proceedings.

Related problems arise in the statistical mechanics of discrete polymers, of crystal growth, etc. A great deal of knowledge in combinatorial geometry required for the solution of such problems seems to be missing, at least among physicists.

§7. Conclusions and Acknowledgments.

Here are some important open problems which are presumably central to the further development of quantized Yang-Mills theory.

(1) Proof of ultraviolet stability of quantized, non-abelian Yang-Mills theories and connections to renormalization group arguments. Use of "block spin" transformations. (Important progress in this direction in the super-renormalizable case has been announced by Bałaban |29|). See also |25|-|27|, |30|, |31|.

(2) Construction of algorithms permitting rigorous error estimates for the calculation of large scale (low energy) phenomena such as quark confinement, absence of coloured physical states (colour screening), Regge behavior of resonance spectrum, quark bound states, in QCD. (Along these lines one would like e.g. to test the validity of "instanton physics," set up calculable $1/_N$- and $\frac{1}{\nu}$-expansions and prove their asymptotic nature, extend the methods sketched in §6, Example 2, to the continuum limit, find rigorous connections to dual resonance models|see the contribution by J.-L. Gervais and A. Neveu|, etc.).

(3) Investigation of conservation laws and complete integrability (at the classical and quantum mechanical level) of pure, non-abelian Yang-Mills theory. (Existence of Bäcklund transformations, conserved currents?)

(4) Application and extension to theories with non-trivial S-matrix of the methods of Jimbo, Miwa and Sato to Yang-Mills theory. (Their methods are based on using Schwinger-Dyson equations for the Schwinger functionals discussed in §4 and the discontinuity properties (4.7), in conjunction with expressing the fields y in terms of the disorder fields b; see |21|).

In conclusion I wish to thank my collaborators, D. Brydges, B. Durhuus, E. Seiler and T. Spencer for all they have taught me and the joy of collaboration. They should have written these notes. Special thanks are due to H. Epstein, G. Mack and E. Seiler for numerous, very valuable discussions and encouragement. I also thank the organizers of the Cargèse School for inviting me to participate and lecture and for financial support.

REFERENCES

1. N. Steenrod, "The Topology of Fibre Bundles," Princeton University Press, Princeton, NJ, 1951.

2. J. Fröhlich, "On the Construction of Quantized Gauge Fields," Proceedings of the Kaiserslautern conference on gauge theory, 1979, W. Rühl (ed.); to appear.

3. For G = (S)O(n), (S)U(n) the result was found by the author (unpublished). B. Durhuus has extended it to general, compact Lie groups, (paper to appear).

4. I.M. Singer, Commun. math. Phys. 60, 7, (1978), and these proceedings.

5. D.G. Kendall and E.F. Harding (eds.), "Stochastic Geometry," Wiley, New York, 1977.

6. A. Connes, in "Mathematical Problems in Theoretical Physics," G. Dell'Antonio, S. Doplicher and G. Jona-Lasinio (eds.), Lecture Notes in Physics 80, Springer-Verlag, Berlin—Heidelberg-New York, 1978.

7. B. Durhuus and J. Fröhlich, "A Connection Between ν-Dimensional Yang-Mills Theory and $(\nu-1)$-Dimensional, Non-Linear σ-Models," Commun. math. Phys., to appear.

8. J. Fröhlich, "Random Geometry and Yang-Mills Theory," to appear in the Proceedings of the "Colloquium on Random Fields," Esztergom (Hungary) 1979; Lecture given at the "Strasbourg Rencontres," 1979.

9. K. Wilson, Phys. Rev. D10, 2445, (1974). R. Balian, J.M. Drouffe and C. Itzykson, Phys. Rev. D10, 3376, (1974), D11, 2098, (1975), D11, 2104, (1975).
K. Osterwalder and E. Seiler, Ann. Phys. (N.Y.) 110, 440, (1978).

10. D. Brydges, J. Fröhlich and E. Seiler, Ann. Phys. (N.Y.) 121, 227 (1979); Nucl. Phys. B152, 521, (1979).

11. M. Reed and B. Simon, "Methods of Modern Mathematical Physics," Vol. IV, Academic Press, New York, 1978; (page 209).

12.[1] See e.g. "Dual Theory," M. Jacob (ed.), North Holland, Amsterdam 1974; R. Giles and C.B. Thorn, Phys. Rev. D16, 366, (1977).

12.[2] Y. Nambu, Phys. Letters 80B, 372, (1979); J.-L. Gervais and A. Neveu, Phys. Letters 80B, 255, (1979); F. Gliozzi, T. Regge and M.A. Virasoro, Phys. Letters 81B, 178, (1979). The lattice approach presented in §3, (II),(III) and other examples in "random geometry" are also studied in unpublished work of the author briefly summarized in |8|.

13. K. Osterwalder and R. Schrader, Commun. math. Phys. 31, 83, (1973), 42, 281, (1975).

14. V. Glaser, Commun. math. Phys. 37, 257, (1974).

15. Details concerning this theorem and its proof will appear in work with H. Epstein, K. Osterwalder and E. Seiler.

16. J. Fröhlich, unpublished,(1974), and preprint to appear. The relevant results can also be inferred from: V. Glaser, in "Problems of Theoretical Physics," p. 69, "Nauka," Moscow, 1969.

17. J. Bisognano and E. Wichmann, J. Math. Phys. 16, 985, (1975).

18. E. Seiler, Phys. Rev. D18, 482, (1978).

19. J. Fröhlich, in "Les Méthodes Mathématiques de la Théorie Quantique des Champs," F. Guerra, D.W. Robinson, R. Stora, (eds.), Editions du C.N.R.S., Paris, 1976.
Commun. math. Phys. 47, 269, (1976) and 66, 223, (1979). (Results for two-and three-dimensional models go back to spring 1975).

20. G. 't Hooft, Nucl. Phys. B138, 1, (1978).

21. M. Jimbo, T. Miwa and M. Sato,"Holonomic Quantum Fields I, II, III,..." Publ. RIMS 14, 223, (1977), 15, 201, (1979), and Preprints, Kyoto 1977-1979.

22. G. Mack and V. Petkova,"Comparison of Lattice Gauge Theories with Gauge Group $\mathbb{Z}^2$ and SU(2),"Preprint 1978; and "$\mathbb{Z}^2$ Monopoles in the Standard SU(2) Lattice Gauge Theory Model,"Preprint 1979.

23. L. Yaffe,"Confinement in SU(N) Lattice Gauge Theories,"Preprint 1979.

24. See ref. 12, in particular 12^1 and refs. given there; also refs. 39,40 and the contribution of J.-L. Gervais and A. Neveu to these proceedings.

25. D. Brydges, J. Fröhlich and E. Seiler,"On the Construction of Quantised Gauge Fields, II,"to appear in Commun. math. Phys., III, Preprint 1979.

26. J. Fröhlich and E. Seiler, Helv. Phys. Acta 49, 889, (1976), J. Fröhlich, in "Renormalization Theory," G. Velo and A.S. Wightman, (eds.), Reidel, Dordrecht-Boston, 1976.

27. J. Challifour and D. Weingarten, Preprint 1979.

28. J. Magnen and R. Sénéor, IAMP Proceedings, 1979, to appear in Springer Lecture Notes in Physics, K. Osterwalder, (ed.)

29. T. Bałaban, see ref. 28.

30. G. Benfatto, M. Cassandro, G. Gallavotti, et al, Commun. math. Phys. 59, 143, (1978), and I.H.É.S. Preprint 1978.

31. J. Glimm and A. Jaffe, Fortschr. der Physik 21, 327, (1973).

32. G. F. DeAngelis, D. de Falco, F. Guerra and R. Marra, Acta Physica Austr., Suppl. XIX, 205, (1978), and refs. given there.

33. G. Gallavotti, F. Guerra and S. Miracle-Solé, in "Mathematical Problems...," see ref. 6.

34. R. Marra and S. Miracle-Solé, "On the Statistical Mechanics of the Gauge Invariant Ising Model," to appear in Commun. math. Phys.

35. A. Guth, "Existence Proof of a Non-Confining Phase in Four-dimensional U(1) Lattice Gauge Theory," Preprint 1979.

36. J. Fröhlich, Physics Letters 83B, 195, (1979).

37. J. Fröhlich and T. Spencer, unpublished. See also ref. 7 and K. Symanzik, in "Local Quantum Theory," R. Jost, (ed.), Academic Press, New York-London, 1969; D. Brydges and P. Federbush, Commun. math. Phys. 62, 79, (1978).

38. J. Fröhlich and E.H. Lieb, Commun. math. Phys. 60, 233, (1978).

39. I. Bars and F. Greene, "Complete Integration of U(N) Lattice Gauge Theory in a Large N Limit," Preprint 1979.

40. D. Förster, "Yang-Mills Theory — A String Theory in Disguise," Preprint 1979.

STRING STATES IN Q.C.D.

Jean-Loup Gervais and André Neveu
Laboratoire de Physique Théorique de l'Ecole Normale Supérieure
24, rue Lhomond - 75231 PARIS CEDEX 05 - FRANCE

INTRODUCTION

It has been conjectured for some time that there exists a correspondence between Yang-Mills fields and a dual-resonance model similar to the two-dimensional correspondence between sine-Gordon and Thirring models [1-2]. In a recent series of papers [3-6], we have investigated such a possibility more concretely.

Guided by Mandelstam's formulation of the sine-Gordon-Thirring correspondence [1], we consider the Yang-Mills theory to be analogous to sine-Gordon, with the dual model playing the role of the Thirring model. The dual string would be the analogue of the Thirring fermion, or sine-Gordon soliton. A difference arises from the fact that the soliton exists semi-classically, which is not the case of a string in pure Yang-Mills theory for example. But, consider that the slope α' of the dual Regge trajectories (assuming that they exist) should be given by a renormalization-group invariant formula of the type

$$\alpha' = C\mu^{-2} \exp\left(\frac{1}{ag^2}\right) g^{-b/a^2} \left(1 + O(g^2)\right). \qquad (I.1)$$

where μ is the subtraction point, g the coupling constant, c a pure number, and a and b the first terms in the expansion of the Callan-Symanzik β -function :

$$\beta(g) = -ag^3 + bg^5 + O(g^7) \tag{I.2}$$

Now, for $g \to 0$, and $a > 0$ (as is the case for pure Yang-Mills theories), $\alpha' \to \infty$ exponentially, and this would account for the absence of the string in the semi-classical approximation, which would give $\alpha' \sim g^2$.

Following these ideas, one should look for a string creation operator (analogue of the Thirring field), expressed in terms of the Yang-Mills field, and then look for its Heisenberg equations of motion (analogue of the Thirring field equations). These equations of motion should have some resemblance to those of the second quantized dual string [7]. It has been proposed by us [3] and independently by Nambu [8], to consider the Wilson loop operator for this purpose.

Consider a curve $x^\mu(\cdot)$ parametrized by σ ($0 \leq \sigma \leq 2\pi$), in Minkowskian space-time and define

$$T[x^\mu(\cdot)] = \mathrm{tr}\, U \equiv \mathrm{tr}\, P \exp\left(\int A_\mu dx^\mu\right) \tag{I.3}$$

where A_μ is the usual Lie algebra valued Yang-Mills field. The path ordered integral of (I.3) runs along the curve $x^\mu(\sigma)$. In a pure Yang-Mills theory, $x^\mu(\sigma)$ must be a closed curve if T is to be gauge invariant ; if the theory contains quarks, the path may be open and end on the quark world lines, and their creation or annihilation operators will enter (I.3) in an obvious fashion to restore gauge invariance.

In the dual model, a similar object, i.e. a functional over curves is introduced[7] together with its functional derivatives with respect to the path. We are thus led to consider variations of (I.3) with respect to $x^{\mu}(\eta)$. A simple calculation gives

$$\frac{\delta}{\delta x^{\mu}(\sigma)} T = \mathrm{Tr}\, P\left[\frac{dx^{\nu}}{d\sigma} G_{\mu\nu}(x(\sigma))\, U\right] \tag{I.4}$$

$$\frac{\delta^2}{\delta x^{\mu}(\sigma)\,\delta x^{\mu}(\sigma')} = \mathrm{Tr}\, P\left[\delta(\sigma-\sigma')\, D^{\rho} G_{\rho\nu}(x(\sigma))\, \frac{dx^{\nu}}{d\sigma}\, U\right]$$

$$+ \mathrm{Tr}\, P\left[G_{\nu\rho}(x(\sigma))\frac{dx^{\nu}}{d\sigma}\, G_{\tau\rho}(x(\sigma'))\frac{dx^{\tau}}{d\sigma'}\, U\right] \tag{I.5}$$

If in (I.5) we assume, in a first approximation, that the free Yang-Mills equations of motion are satisfied, the first term on the right-hand side vanishes ; it is also the singular term for $\sigma = \sigma'$. In the classical theory, it is then tempting to replace the non-linear field equations of motion by the simpler requirement that T be locally harmonic[4]. Speculations on whether a classical Yang-Mills field obeying the equations of motion can be retrieved from any locally harmonic T are presented in ref. 4 . At any rate, this would only be possible in the $N \rightarrow \infty$ limit, since for finite N, the number of degrees of freedom of A_{μ} is finite, while that of T is infinite.

In the quantum case, the discussion is made much more delicate by the existence of short-distance singularities. One should first worry about whether T itself can be defined with a finite number of counterterms. We shall return to this problem in section II. Also, for $\sigma \rightarrow \sigma'$ the second term of the right hand side of

(I.5) can be very singular. Nambu[8] observed that the simple replacement of $G_{\nu\rho} G_{\tau\rho}$ in this term by its expectation value, assumed to be a non-zero constant, leads to an equation identical with that of the quantized dual string[7], with finite slope. The trouble with such a replacement is that any Euclidean calculation would give the wrong sign for $\langle 0| G_{\nu\rho} G_{\tau\rho} |0\rangle$. As Nambu noticed, it would be more reasonable to take the expectation value inside the string, where the Yang-Mills field is electric, rather than magnetic in a Euclidean calculation, which changes the sign of $G_{\mu\nu}{}^2$. Whether this can be turned into a consistent calculational scheme (beyond Nambu's original remarks) is an open question, and we shall not pursue this approach in the rest of these lectures.

In section III, we shall propose another line of attack of the last term of eq. (I.5), based on the usual light-cone analysis of the dual resonance model[9]. In light cone variables, we shall reduce eq. (I.5) to the dynamics of a chain of atoms, or polymer, and show that the existence of a phonon spectrum, with a finite velocity of sound, is equivalent to the existence of linear Regge trajectories with finite slope, as already proposed by Thorn[10] in a simplified model. We shall introduce a short-range approximation by which only nearby points on the string interact, via a local δ-function potential in the transverse coordinates, instead of the harmonic potential of the conventional dual-resonance model. We shall show how this δ -function potential gives naturally rise to dimensional transmutation and to the appropriate phase structure of the theory around 4 dimensions. We shall also show how to relate the coefficient of the δ potential to deep-inelastic parameters, and hence, in principle, how to compute the slope of Regge trajectories in terms of the scaling violations in deep-inelastic scattering, and vice-versa.

Section IV is the conclusion in which we discuss extensions and further investigations of these ideas.

II. RENORMALIZATION OF LOOP OPERATOR

Before attacking the more difficult problem of the dynamics of the loop, it is necessary to examine whether its definition, eq. (I.3), makes sense in the quantum theory, or whether it should be improved by the introduction of counterterms. To our surprise, this well-defined and interesting question has not yet been examined seriously[11]. It is beyond the scope of these lectures, and beyond our present knowledge, to go into the details of this problem. We shall only indicate a line of attack, together with a few one-loop results.

First, it is convenient to disentangle the color indices in

a systematic way. This is done by reducing the loop to a one dimensional fermion living on the curve $x(\sigma)$ and coupled to the Yang-Mills field. For example, it can be proved[6] that

$$T = \frac{\delta^2}{\delta\lambda_\alpha(0)\delta\bar{\lambda}_\alpha(0)} \ln\left[\int \mathcal{D}\zeta\mathcal{D}\bar{\zeta} \exp\int d\sigma\left(\bar{\zeta}\partial_\sigma\zeta - ig\,\bar{\zeta}\frac{dx_\mu}{d\sigma}A^\mu\zeta + i\bar{\lambda}\zeta + \bar{\zeta}\lambda\right)\right]_{\lambda=\bar{\lambda}=0} \quad \text{(II.1)}$$

where $\zeta(\eta)$ is an anticommuting field[12] on the curve, in the fundamental representation, λ and $\bar{\lambda}$ the corresponding sources. For such a field, antiperiodic boundary conditions must be chosen[13]

$$\zeta(\sigma+2\pi) = -\zeta(\sigma) \quad \text{(II.2)}$$

The logarithm in (II.1) removes the sort of vacuum graphs coming from closed circular contractions of the color indices.

We are looking for an expression for the Green functions of Wilson loop operators. Basically, it is obtained by inserting (II.1) into the usual functional integral over A_μ and the Faddeev-Popov fields C and $\bar{C}$. We use the standard covariant gauge, where the action reads :

$$S = \mathrm{tr}\int d^4x\left[-\frac{1}{4}G_{\mu\nu}^2 - \frac{1}{2\alpha}(\partial_\mu A_\mu)^2 + \bar{C}\partial_\mu D_\mu C\right] \quad \text{(II.3)}$$

A crucial point is the gauge invariance of (II.1) if we transform ζ and $\bar{\zeta}$ according to

$$\delta\zeta = ig\Lambda\zeta$$
$$\delta\bar{\zeta} = -ig\bar{\zeta}\Lambda \quad \text{(II.4)}$$

where Λ is a Hermitian generator of the gauge group. Also, the action, including the loop, is invariant under the Becchi-Rouet-Stora transformation

$$\delta A_\mu = \varepsilon D_\mu C\omega$$
$$\delta\bar{C} = -\varepsilon/\alpha\ \partial_\mu A_\mu\omega$$
$$\delta C = -i\frac{\varepsilon}{2}g[C,C]\omega$$
$$\delta\zeta = ig\varepsilon C\omega\zeta$$
$$\delta\bar{\zeta} = -ig\bar{\zeta}C\varepsilon\omega.$$

(II.5)

where ω is an anticommuting c-number. The problem is then reduced to that of a Yang-Mills field coupled to a one-dimensional fermion, and the renormalization will proceed to a large extent as in Yang-Mills field theory with charged quark fields. In ref. 6, we have followed the elegant method summarized by B.W. Lee[14], and we have used dimensional regularization. We only quote the results of our investigations in lowest order for regular curves (for example curves with a finite number of Fourier modes) :

In four space-time dimension, besides the usual counterterms, the only extra divergence is of the form

$$\frac{g^2}{d-4}\int d\sigma\, \bar{z}\,\partial_\sigma z \tag{II.6}$$

This can be reabsorbed in a multiplicative renormalization of the fields z and $\bar{z}$. In three space-time dimensions, other counterterms are needed to cancel divergences of the form

$$\frac{g^2}{d-3}\int d\sigma\left[\left(\frac{dx_\mu}{d\sigma}\right)^2\right]^{1/2}\bar{z}z \tag{II.7}$$

and

$$\frac{g^2}{d-3}\int d\sigma\left[\left(\frac{dx_\mu}{d\sigma}\right)^2\right]^{1/2}\left(\bar{z}\,I^a z\right)^2$$

Although the general form of the counterterms is restricted by gauge and reparametrization invariance, a complete analysis is made non trivial by the dimensionlessness of z, which allows arbitrary powers as in the two-dimensional non-linear σ-model[15]; using the same methods, however, one can show that invariant multiplets are multiplicatively renormalized.

Besides the above-mentioned divergences, others can appear if the curve is irregular enough. For example a discontinuity in the first derivative (angle in the curve) gives rise to an extra divergence at $d = 4$, which depends on the angle. From past experience with dual string stationary states, one expects that they will be concentrated around very irregular curves. In the dimensional regularization scheme, fractal curves, with non integer Hansdorff dimension, exhibit divergences in non integer dimension[6]. Thus, the divergences and renormalization of the Wilson loop depend on its dynamics, which greatly complicates the discussion.

III. LOOP DYNAMICS IN LIGHT-CONE VARIABLES

The canonical quantization of the dual string was first achieved in light-cone variables[9], where all redundant degrees of freedom can be eliminated explicitly. Up to now, it is also the only known treatment of the second quantized interacting dual string[7]. In light-cone variables, we shall see that we obtain, instead of equations (I.4-5), a Schrödinger equation which is much easier to interpret.

The independent degrees of freedom of the dual string consist of $d-2$ transverse coordinates $x^i(\eta)$, together with the total momentum in the + direction. In the one-string formalism, the free quantum string is then described by a wave functional $\varphi_D(x^+, p^+, x^i(\cdot))$ which satisfies the Schrödinger equation

$$\left\{2ip_+\frac{\partial}{\partial x^+} - \frac{1}{2}\int d\sigma\left[\frac{-\delta^2}{\delta x^{i2}(\sigma)} + \frac{1}{\alpha'^2}\left(\frac{dx^i}{d\sigma}\right)^2\right] + \frac{\alpha_0}{\alpha'} + E_0\right\}\varphi_D = 0 \tag{III.1}$$

where α' is the slope, α_0 the intercept and E_0 the zero-point energy of the harmonic oscillators q^i_n and $\bar{q}^i_n$ which are the Fourier components of the transverse coordinates

$$x^i(\sigma) = q^i_0 + \sum_{n=1}^{\infty}\left[q^i_n \cos n\sigma + \bar{q}^i_n \sin n\sigma\right] \tag{III.2}$$

There are various ways of regularizing the functional Laplacian of (III.1). One is to keep only a finite number of Fourier modes. This is what is usually done for dual models, because Fourier modes diagonalize the potential term $\int d\sigma(dx/d\sigma)^2$. For a more transparent generalization to the Yang-Mills case, however, we find it more convenient to discretize the curve by replacing it by a polygon with M sides, and M vertices $x^i_1, \ldots\ldots x^i_M$. Defining $\varepsilon = 2\pi/M$ we then have[3]

$$\int_0^{2\pi} d\sigma \frac{\delta^2}{\delta x^i(\sigma)^2} = \lim_{\varepsilon\to 0} \frac{1}{\varepsilon}\sum_{m=1}^{M}\frac{\partial^2}{\partial x^{i2}_m}. \tag{III.3}$$

and, for M finite, (III.1) takes the form

$$\left\{2ip^+\frac{\partial}{\partial x^+} + \frac{1}{2\varepsilon}\left[\sum_{m=1}^{M}\frac{\partial^2}{\partial x^{i2}_m} - \frac{1}{\alpha'^2}\sum_m (x^i_{m+1}-x^i_m)^2\right] + \frac{\alpha_0}{\alpha'} + E_0\right\}\varphi_D = 0 \tag{III.4}$$

We see that we are dealing with a closed ring of M identical particles bound together by harmonic nearest neighbour forces, except for the factor $1/2\varepsilon$, in front of the Hamiltonian. This factor means that all finite energy excitations (for $\varepsilon\to 0$) of the chain disappear ultimately from the spectrum of physical dual string states. The only excitations of the chain that survive are those of energy $\sim \varepsilon$, i.e. the soft phonon modes. Hence, a harmonic potential between nearest neighbours is not necessary to prevent the string from breaking. Any two-body potential with at least one bound state with finite binding energy will do, as shown by Thorn[10], using a very simple variational argument. He also exhibited an approximate method, due to Goldstone, to compute the velocity of sound along the chain, which is directly related to the slope of Regge trajectories. Whether this slope is finite or not depends on whether the velocity of sound along the chain is finite or not. At this point, help can be obtained from standard facts of polymer physics.

A polymer[16] is a chain of identical atoms (or molecules) with a potential energy U of the form

$$U = U_c + U_I \tag{III.5}$$

The chain potential U_c is between nearest neighbours on the chain, and provides the binding force. The interaction potential U_I is between all pairs of atoms, and contains a short-range repulsive core, which prevents any two atoms from overlapping (excluded volume effect), together with a possible long range part. In the absence of a long-range part in U_I, the mean square size $\langle R^2 \rangle$ of the polymer behaves like $M^{2\nu}$ for the number of atoms M going to infinity[17]. For a polymer in four or more space-dimensions, or if $U_I = 0$, it can be shown that $\nu = 1/2$. The effect of the dimension of space comes via the excluded volume effect : in 4 and more dimensions, there is enough "space" that the polymer essentially never crosses itself. In $d < 4$ dimensions, ν can be computed by renormalization group arguments, in powers of $4-d$, and in general $\nu > 1/2$ for $d < 4$. In the case of interest to us, $d = 2$.

Another quantity of interest in polymer physics is the relaxation time τ of the polymer. For a harmonic U_c, $\tau \sim M^2$. This behaviour is actually expected for any U_c, as long as $U_I = 0$. It simply expresses the fact that a soft mode is independent of the details of the short-range binding potential. For $U_I \neq 0$, one finds however $\tau \sim M^{2\nu+1}$, with the same ν as defined above. Now, at least in the harmonic case, this relaxation time is obtained by solving the dynamics of the polymer by an equation of the form

$$-\frac{\partial U}{\partial x_s} = \kappa \frac{\partial x_s}{\partial t} \tag{III.6}$$

with the constant κ proportional to the viscosity of the liquid in which the polymer is in solution, and x_s the position of the

s^{th} atom. However, for the problem of interest here, the classical equations of the string are second order in time, rather than first order ; it is as though the polymer was oscillating freely in the vacuum rather than being in a highly viscous liquid. Because of that, the period T of the lowest excitation mode is expected to be obtained by solving a second order equation of the type

$$-\frac{\partial U}{\partial x_s} = \mu \frac{\partial^2 x_s}{\partial t^2} \tag{III.7}$$

For the harmonic case, $T \sim M$, and we expect this to be true for any U_c, as long as it binds, of course, at $U_I = 0$. For U_I 0, we would also expect $T \sim M^{\nu'+1/2}$, with ν' possibly different from ν. This proportionality of T and M means that the polymer has a phonon excitation spectrum, with a finite velocity of sound. But it is well-known in the dual model (harmonic case) that the leading Regge trajectory is obtained by building up the excited states of the lowest excitation mode. These have finite mass for $M \to \infty$ because of the extra factor $1/\varepsilon = M/2\pi$ in going from the polymer to the string (eq. III.4).

In Yang-Mills theory, the problem is then to find an object with the same number of degrees of freedom as φ_D and examine its equation of motion. This is done by considering eq. (I.3-5) more closely. We show in ref. 6 that the $D_\rho G^{\rho\nu}$ term vanishes, in spite of the facts that the quantum equations of motion have to include the gauge fixing and Faddeev-Popov ghost terms, and that the loop is a source which can interact with itself. To deal with the G^2 term, we introduce a short range approximation, which is to replace it by a short range potential $V(x(\sigma) - x(\sigma'))$. This short range potential could be obtained perturbatively by first replacing $G_{\mu\rho}(x)\, G_{\nu\rho}(x')$ by its expectation value, for example, or by using an appropriate operator product expansion.

The point of this short range approximation is to reduce the perturbation expansion of the right hand side of (I.5), so that for large M, i.e. in the continuum limit of the curve, the relevant part of all diagrams is entirely in the far ultraviolet region, and can be estimated from asymptotic freedom. Up to logarithms, the dominant terms can be computed from the lowest order graph, where the two operators $G_{\mu\rho}$ and $G_{\nu\rho}$ are replaced by their expectation value. Taking $\sigma \to \sigma'$, we get a linear equation of the form

$$\left[\frac{-\delta^2}{\delta x^\mu(\sigma)^2} + V_\sigma\,(x(\cdot))\right] T = 0 \tag{III.8}$$

For regular curves, we can determine the behaviour of V for

$M \to \infty$. It is given up to logarithms by the scaling properties of the lowest order diagram, and is of the form

$$V_\sigma \simeq \frac{M^4}{(\ln M)^\sigma} \tag{III.9}$$

Thus, the functional equation (III.8) does not possess a well-defined continuum limit in the usual functional sense, and we shall study it for finite M first, taking the limit $M \to \infty$ at the end. We discretize the curve by introducing a set of M points on the curve as was done in ref. 3 and at the beginning of this section. Also, to use light-cone coordinates, we define

$$\varphi_{YM}(x^+, p^+, x^i(\cdot)) \equiv \int \mathcal{D}x^- \, e^{i\frac{p^+}{2\pi}\int d\sigma\, x^-(\sigma)} \, T(x^+, x^-(\cdot), x^i(\cdot))$$

$$x^+(\sigma) = \text{cte} = x^+ \tag{III.10}$$

As indicated, φ_{YM} is defined from T by restriction to curves at fixed x^+ and Fourier transformation over x^- , with a constant density p^+ , which fixes the parametrization of the curve $x(\sigma)$. From (III.8), we now deduce

$$2ip_+ \frac{\partial \varphi_{YM}}{\partial x^+} = 2\pi \int d\sigma \left[-\frac{\delta^2}{\delta x^i(\sigma)^2} + V_\sigma \right] \varphi_{YM} \tag{III.11}$$

By Lorentz and translation invariances, V_σ does not depend on x^- if x^+ is a constant. The above equation, in discretized form, is a M body Schrödinger equation and the limit $M \to \infty$ can be discussed in the language of statistical mechanics : we have to decide whether this M body system has a phase where string states are completely stable objects. As mentioned above, this will be the case if the potential V_σ , in discretized form, can be reduced to a nearest neighbour interaction with at least one bound state.

By construction, V_σ depends only on a small piece of string around σ , of order M^{-1} . Taking it as a nearest neighbour interaction, we write

$$\int d\sigma V_\sigma \simeq \frac{2\pi}{M} \sum_s \bar{V}(x^i_{s+1} - x^i_s) \tag{III.12}$$

Ordinary dimensional analysis shows that $\bar{V}$ must be of the form.

$$\bar{V}(x^i) = \frac{1}{r^2} h(\bar{\mu} r) \qquad r^2 = x^{i2}$$

$$\bar{\mu} = \mu \exp - \int^g \frac{dg'}{\beta(g')} \tag{III.13}$$

μ is the subtraction point of the renormalization procedure, and β denotes the usual function of the Callan-Symanzik equation. $\bar{\mu}$ is renormalization group invariant.

No information can be gained in perturbation theory about the large distance behaviour of $\bar{V}$. Indeed, at finite distances, it would be necessary to determine β to all orders. In the spirit of the short range approximation, we shall simply take it to be a contact interaction, so that

$$\bar{V} = \left(\frac{M}{2\pi}\right)^2 V = \left(\frac{M}{2\pi}\right)^2 \left[- \frac{2\pi \delta(x^i)}{|\ln(\bar{\mu} r)|} \right] \tag{III.14}$$

and the discretized form of (III.11) is finally

$$2ip^+ \frac{\partial}{\partial x^+} \varphi_{Yn} = MH \varphi_{Yn}$$

$$H = \sum_{s=1}^{M} \left(-\frac{1}{2} \frac{\partial^2}{\partial x_s^{i\,2}} \right) + \sum_{s=1}^{M-1} V(x_{s+1} - x_s) \tag{III.15}$$

There are several justifications to the form (III.14) : indeed a zero-range potential is the only one that can be unambiguously determined by perturbation theory in low order thanks to asymtotic freedom. We shall see later that it is enough to give the loop a string spectrum. As usual, a logarithm must be included in the potential, and its exact power, together with the factor -2π in (III.14) will be explained below ; an explicit determination of this power and factor of the logarithm would provide a crucial test of the present scheme. Another test would come from the short distance behaviour of φ for $x_{s+1} \to x_s$, which could be determined from asymptotic freedom ; since the naive dimension is zero, only a power of $\ln \bar{\mu} r$ is allowed, and the only power compatible with a zero range potential is $\ln \bar{\mu} r$. See ref. 6 for details; we are presently investigating these points.

We shall now discuss the behaviour of the theory above and

below four dimensions, in the spirit of our short-range approximation and see how dimensional transmutation arises.

If the lattice arguments for confinement in a strong coupling phase and asymptotic freedom are to be compatible, it seems that $SU(N)$ gauge theories in $d \leq 4$ dimensions can have one phase only. In $4+\varepsilon$ dimensions, $\varepsilon > 0$, there is a non trivial fixed point for $g^2 = O(\varepsilon)$, and one expects both a strong coupling phase and a weak coupling, perturbation theory, unconfined phase.

In $4+\varepsilon$ dimensions, the β function becomes, in lowest order in ε and g^2

$$\beta(g) = \frac{\varepsilon}{2} g - a g^3 \tag{III.16}$$

and, in lowest order, eq. (I.1) for α' is replaced by

$$\alpha' = C \mu^{-2} \left(1 - \frac{\varepsilon}{2ag^2}\right)^{-2/\varepsilon} \tag{III.17}$$

For $\varepsilon > 0$, this formula requires $g^2 \gg \varepsilon/2a$, i.e. that one be in the strong coupling phase ($a > 0$ in an asymptotically free theory). For $\varepsilon \leq 0$, this formula is meaningful for all real values of the coupling constant, and, at fixed g^2, is a smooth function of ε at $\varepsilon = 0$. The number C may depend in a smooth way on ε.

Our choice of light-cone variables has turned the Yang-Mills string into a non-relativistic system in $2+\varepsilon$ dimensions. In less than two dimensions, a non-relativistic δ- function potential always has a bound state and hence, according to ref. 10, the string has a finite slope. This can be seen simply from the two-body radial Schrödinger equation

$$\psi'' + \frac{1+\varepsilon}{r} \psi' - m^2 \psi = 0 \tag{III.18}$$

where $-m^2$ is the energy of the bound state. At $\varepsilon = -1$ (δ potential on a line), the solution of (III.18) is $\exp(-mr)$. For general ε, the solution is

$$\psi(r) = \left(\frac{1}{2} m r\right)^{-\frac{1}{2}\varepsilon} K_{\frac{1}{2}\varepsilon}(mr) \tag{III.19}$$

where K is a modified Bessel function[18]. For small r, we have

$$\psi \simeq \frac{\pi}{2\sin\pi\varepsilon/2}\left[\frac{1}{\Gamma(1+\varepsilon/2)}-\left(\frac{mr}{2}\right)^{-\varepsilon}\frac{1}{\Gamma(1-\varepsilon/2)}\right]+\cdots \tag{III.20}$$

For $\varepsilon<0$, ψ is finite at r = 0, and we can write

$$-\Delta\psi - \xi g^2\mu^{-\varepsilon}\delta(x^i)\,\psi = -m^2\psi. \tag{III.21}$$

The relation between m , g , and μ is found by integration over a small sphere centered at the origin. The result is

$$m = 2\mu\left[\frac{\xi g^2\,\Gamma(-\varepsilon/2)}{2\pi^{1+\varepsilon/2}}\right]^{1/\varepsilon} \tag{III.22}$$

The right-hand side of this equation is singular for $\varepsilon\to 0$; but for ε = 0, we see on (III.20) that

$$\psi \simeq |\ln\bar{\mu}r| + O(r) \tag{III.23}$$

Hence, a modification is needed in eq. (III.21) in order to give a meaning to a δ -function potential in two dimensions. This has already been studied in the literature, in particular in three dimensions, as a model for nucleon-deuteron interactions. See the lecture at the 1967 Cargèse Summer School by G. Flamand for a review of the three-dimensional case, and a list of references. At an elementary level, we see that eq. (III.21) at ε = 0 can be reinterpreted formally as

$$-\Delta\psi - \frac{2\pi}{|\ln\bar{\mu}r|}\delta(x^i)\psi = -m^2\psi \tag{III.24}$$

with $\bar{\mu}=\frac{1}{2}me^{\gamma}$ (γ being Euler's constant). This equation simply means that the bare coupling constant $g^2\mu^{-2}$ actually vanishes logarithmically at small distances in a well-defined way. This explains the rather ad-hoc factors and power of logarithm that we chose for the potential V in eq. (III.14). This vanishing of the bare coupling constant is what happens in an asymptotically free theory, and, in the present case, is the source of dimensional transmutation.

If we impose invariance of (III.21) under the renormalization group, we are led to replace $g^2\mu^{-\varepsilon}\delta(x^i)$ by

$$g^2\mu^{-\varepsilon}\Big/\left(1-2ag^2/\varepsilon+2ag^2(\mu r)^{-\varepsilon}/\varepsilon\right)\delta(x^i) \tag{III.25}$$

Provided that $\zeta = 2\pi a$, one then recovers for m in eq. (III.22) an expression which is continuous at $\varepsilon = 0$, and essentially identical to (III.17). In our case of the Wilson loop, whether ζ assumes this exact value is a delicate question, which, however, we do not need to answer to obtain a first numerical estimate for α', as we shall explain. We also see that (III.25) allows the continuation (provided $\zeta = 2\pi a$) of the δ-function potential to $\varepsilon > 0$: in this case, the effective coupling constant in front of the δ-function goes to zero like a power at small distances, and, in the strong coupling phase, the formula for the bound state energy is just the analytic continuation of the expression at $\varepsilon < 0$. In the weak coupling phase, there is no bound state.

The above discussion on the Schrödinger equation with δ-function potential in 2 space dimensions is nothing but a small corner of the study of the Lagrangian field theory.

$$\mathcal{L} = -i\,\mathrm{Tr}\,\varphi^*\partial_t\varphi - \frac{1}{2}\mathrm{Tr}\,\partial_i\varphi^*\partial_i\varphi + \frac{\lambda}{4}\mathrm{Tr}(\varphi^*\varphi)^2 \tag{III.26}$$

in two space dimensions. This is a generalization of the non-linear Schrödinger equation by the fact that φ is allowed to be an N x N matrix. Just as for the non-linear Schrödinger equation, (III.26) is quantum mechanically equivalent to a non-relativistic system of M particles with contact interactions. The presence of matrix fields φ gives quantum numbers to these particles, so that there are various types of interactions. In one-space dimension, (III.26) has been shown[19] to be completely integrable classically, even in the matrix case. When the matrix becomes very large, one can recover a "large N" limit, and reduce the interaction between the non-relativistic particles to be of the nearest neighbour type, which is exactly the problem considered in this section. Also, note that in two space dimensions, (III.26) is just renormalizable, and, for $\lambda > 0$, asymptotically free.

The above discussions show that the renormalization group combined with a nearest neighbour short-range interaction on the loop, provides a dynamical confinement mechanism which, as a function of the coupling constant, subtraction point and space-time dimension has very appealing analytic properties. This is of course in sharp contrast with topological or semi-classical approaches, which generally put a very strong emphasis on a particular space-time dimension. We shall now see that the explicit form (I.3) of the string wave functional enables us to go further, and actually relates in a rather precise numerical fashion the slope of the leading trajectory to the mass scale characteristic of scaling violations.

In the deep inelastic region, the effective coupling constant vanishes logarithmically as

$$g^2(k) \sim \left(\ln k/\Lambda\right)^{-1} \tag{III.27}$$

where Λ can in principle be measured experimentally in scaling violation experiments. From Fourier transform theory[20], the effective coupling constant in position space vanishes logarithmically at small distances as

$$g^2(x) \sim \left|\ln \frac{\Lambda x}{2}\right|^{-1} \tag{III.28}$$

On the other hand, renormalization group theory tells us that

$$\bar{\mu} = \frac{\Lambda}{2} \tag{III.29}$$

Hence the two-body bound state energy

$$m^2 = \Lambda^2 e^{-2\gamma} \tag{III.30}$$

In a random-phase approximation due to Goldstone, Thorn[19] has computed the slope α' as

$$\alpha' = \left(2\pi m^2 \sqrt{3}\right)^{-1} \tag{III.31}$$

Finally we obtain

$$\Lambda^2 = \frac{e^{2\gamma}}{2\pi\sqrt{3}} \frac{1}{\alpha'} \tag{III.32}$$

Numerically, $\alpha' = 0.9\ \text{GeV}^{-2}$ gives $\Lambda \cong 0.55$ GeV. Experimental values for Λ lie between 0.4 and 0.7 GeV[21].

VI. CONCLUSION

Extension of the considerations of this paper to open strings with quarks at the ends, and to string splitting by quark pair creation poses no immediate problem[5]. A more difficult question is raised by the choice of constant p^+ made in section III. This choice forbids a finite accumulation of p^+ momentum on a point of the string. This causes problems, for example for low-lying states, in particular with heavy quarks, which clearly carry a finite amount of p^+, and in string splitting, as already noted in ref. 5. This will ultimately force us to allow for non-constant p^+ along the loop, and to study p^+ transfer between points on the loop. However, for long hadrons, i.e. far along the leading trajectory, the smooth string picture should be valid, and a constant p^+ should be a good approximation. As noted by Thorn[10], a constant p^+ also leads to the tachyon problem already present in the dual model. Whether or not this problem disappears when p^+ is allowed to vary in arbitrary ways is a completely open question.

Another question concerns the group structure of the non-Abelian theory. It is only in the limit of $N \to \infty$ for the gauge group SU_N that one expects a free string spectrum. Hence, it should be possible, in a more careful derivation of eq. (III.11) to see that this linear equation can be obtained only for large N. Another place where N would play a role is in the coefficient of the δ potential in eq. (III.14). Examination of these questions is under way, and will provide crucial tests for the ideas developed in these lectures.

REFERENCES

1 S. Mandelstam, Phys. Rep. 23C (1976) 307.
2 G. 't Hooft, EPS International Conference (Palermo 1975) ; ed. A. Zichichi.
3 J.L. Gervais and A. Neveu, Phys. Lett. 80B (1979) 255.
4 J.L. Gervais and A. Neveu, Nucl. Phys. B153 (1979) 445.
5 J.L. Gervais, M.T. Jaekel and A. Neveu, Nucl. Phys. B155 (1979) 75.
6 J.L. Gervais and A. Neveu, Nucl. Phys. in press.
7 M. Kaku and K. Kikkawa, Phys. Rev. D10 (1974) 1110, 1823 ; E. Cremmer and J.L. Gervais, Nucl. Phys. B 76 (1974) 209 ; B 90 (1975) 410.
8 Y. Nambu, Phys. Lett. 80B (1979) 372.
9 P. Goddard, J. Goldstone, C. Rebbi and C.B. Thorn, Nucl. Phys. B56 (1973) 109.
10 C.B. Thorn, Phys. Rev. D19 (1979) 639.
11 except by Symanzik (private communication).
12 S. Samuel, Nucl. Phys. B149 (1979) 517 ; J. Ishida and A. Hosoya, Prog. Theor. Phys. 62 (1979) 544.
13 R.F. Dashen, B. Hasslacher and A. Neveu, Phys. Rev. D12 (1975) 2443.
14 B.W. Lee, 1975 lecture at Les Houches Summer School.
15 E. Brezin, J.C. Le Guillou and J. Zinn-Justin, Phys. Rev. D14 (1976) 2615.
16 See e.g. J. des Cloizeaux, J. de Physique 37 (1976) C1-255.
17 P. Flory, Principles of Polymer Chemistry, Cornell U. Press.
18 See e.g. G. N. Watson "A treatise on the theory of Bessel functions" Cambridge University Press (1966).
19 D.V. and G.V. Choodnovsky, private communication.

20 L.M. Guelfand and G.E. Chilov, les distributions, tome I, Dunod, Paris.
21 L. Baulieu and C. Kounnas, Nucl. Phys. B155 (1979) 429.

WHY DO WE NEED LOCAL GAUGE INVARIANCE IN THEORIES WITH VECTOR PARTICLES? AN INTRODUCTION

G. 't Hooft

Institute for Theoretical Physics
Princetonplein 5, P.O.Box 80006
3508 TA Utrecht, The Netherlands

During the last decennium it has become more and more clear that gauge theories play a decisive role in elementary particle physics. Indeed, the most general possible theory for a finite class of particle types, with not too strong interactions, and valid to very high energies, *must* be a gauge theory with some scalar and spinor fields.

Renormalizable field theories with only scalar and spinor fields with Yukawa type interactions were known for a long time. How to introduce vector fields has for a long time been an outstanding problem, which was finally solved with the advent of the gauge theories. In this lecture I will give field theory in a nutshell, starting by "defining" functional integrals, applied to scalar field theories. Then I shall discuss the so-called cutting-relations. These crucial relations in this form were derived by M. Veltman in his pioneering work in a time that few believed in field theory and nobody in gauge theories. Finally we will see how unitarity and renormalizability can only be reconciled in a gauge theory.

I1. THE PRECURSORS OF FUNCTIONAL INTEGRALS

Consider integrals of the type:

$$Z(\vec{J},\vec{\lambda}) = \int \dots \int d\phi_1 \dots d\phi_n \exp\left[-\tfrac{1}{2}\,\phi_i M_{ij}\,\phi_j - \lambda_1 \sum_i \phi_i^3 - \lambda_2 \sum_i \phi_i^4 \dots + J_i\,\phi_i\right] . \qquad \text{(I1)}$$

Here $M_{ij} = M_{ji}$, and the coefficients M_{ij} and λ_i may either be real or imaginary with a small positive real part. Let us expand with respect to $\lambda_{1,2}$ and J_i:

$$Z(\vec{J},\vec{\lambda}) = \int \dots \int \prod_i d\phi_i \; e^{-\frac{1}{2}(\phi, M\phi)} \Big[1 - \lambda_1 \sum \phi_i^3 \dots$$
$$\dots + \tfrac{1}{2} \sum J_i J_j \phi_i \phi_j \dots - \frac{1}{3!} \sum J_i J_j J_k \phi_i \phi_j \phi_k \lambda_1 \sum_\ell \phi_\ell^3 \text{ etc.} \Big]. \qquad (I2)$$

The integrals can now be performed, for instance by diagonalizing M. But more conveniently, one can write

$$\prod_i \int d\phi_i \; \phi_i \phi_j \phi_k \dots e^{-\frac{1}{2}(\phi, M\phi)} =$$
$$= \frac{\partial}{\partial J_i} \frac{\partial}{\partial J_j} \frac{\partial}{\partial J_k} \dots \prod_i \int d\phi_i \; e^{-\frac{1}{2}(\phi, M\phi) + (J,\phi)} \Big|_{\vec{J}=o} =$$
$$= \frac{\partial}{\partial J_i} \frac{\partial}{\partial J_j} \frac{\partial}{\partial J_k} e^{+\frac{1}{2}(J, M^{-1}J)} Z(o,o) \; . \qquad (I3)$$

And it is easy to expand

$$e^{\frac{1}{2}(J, M^{-1}J)} = 1 + \tfrac{1}{2} J_i (M^{-1})_{ij} J_j + \frac{1}{2!4} J_i (M^{-1})_{ij} J_j J_k (M^{-1})_{k\ell} J_\ell \dots \, . \qquad (I4)$$

The resulting expressions are conveniently expressed in terms of diagrams. The right hand side of (I4) is to be written as

1 + 

\+ 

\+ ...

The factors $\frac{1}{2}$,... can often be incorporated in the diagrams. For instance one could write the third term of eq. (I4) by differentiating with respect to four different J terms:

$$\frac{\partial}{\partial J_1} \frac{\partial}{\partial J_2} \frac{\partial}{\partial J_3} \frac{\partial}{\partial J_4} (\dots) =$$

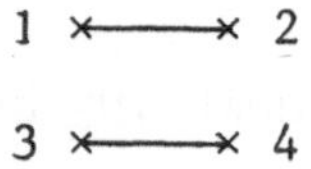

\+ 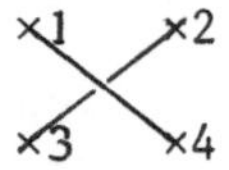

\+ 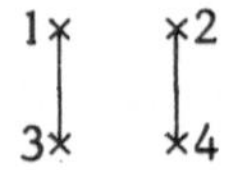

=

$$= (M^{-1})_{12} (M^{-1})_{34} + (M^{-1})_{14} (M^{-1})_{23} + (M^{-1})_{13} (M^{-1})_{24} \; . \qquad (I5)$$

It is now easy to compute the terms in (I2). For instance the last explicitly written term:

$- \frac{\lambda_1}{3!} \sum_\ell$ 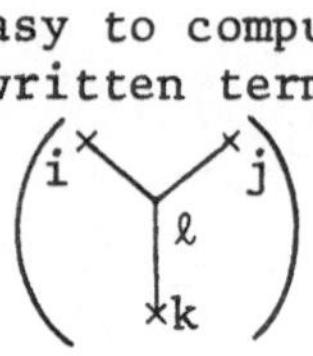

$= - \frac{1}{3!} J_i J_j J_k \sum_\ell (M^{-1})_{i\ell} (M^{-1})_{j\ell} (M^{-1})_{k\ell} \; .$

Still further, one gets diagrams such as

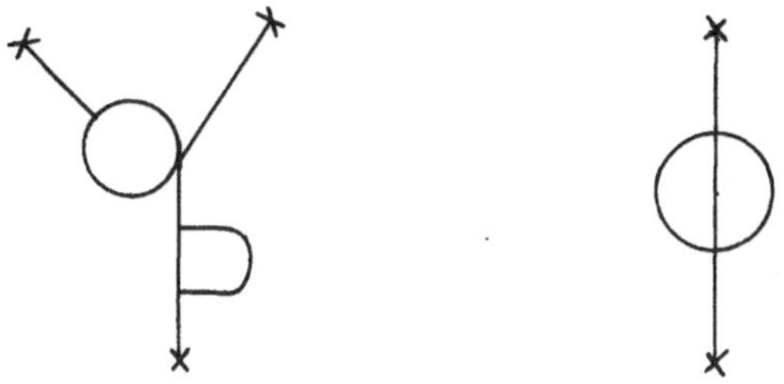

If one writes

$$Z(\vec{J},\vec{\lambda}) = Z(o,o)e^{W(\vec{J},\vec{\lambda})} , \tag{I6}$$

then W contains only *connected* diagrams. Further procedures are possible to obtain only *irreducible* diagrams, but we will not dwell any further on that. So far, exact and obvious mathematics.

I2. SCALAR FIELD THEORIES

In a scalar field theory the variables in the integrals to be considered are not a finite but an infinite set, namely the values of the field ϕ at each space-time point x. So we formally replace in the previous section

$$\phi_i \quad \text{by} \quad \phi(x) ,$$

where x is a vector in $\mathbb{R}^n$. To be explicit we consider $\mathbb{R}^n$ to be Minkowsky space. Usually we have

$$M = i\left(-\partial_x^2 + m^2\right) . \tag{I7}$$

The integral becomes an "integral" over functions $\phi(x)$:

$$Z\{J,\vec{\lambda}\} = \int D\phi \exp i\left[\int d^n x\left(-\tfrac{1}{2}\phi(x)(m^2-\partial^2)\phi(x) - \lambda_1\phi^3(x) - \lambda_2\phi^4(x) + \right.\right.$$
$$\left.\left. + J(x)\phi(x)\right)\right] , \tag{I8}$$

a functional integral which, particularly for $n > 3$, is devoid of any rigorous meaning. If $n = 4$ and only λ_1 and λ_2 are different from zero then our expressions can be suitably altered in such a way that they make sense. But it is not easy. We restrict ourselves always to perturbation expansions, which implies that we only consider the Feynman graphs, not the entire integral.

Let us consider $Z(J,\lambda)$ as a functional of the function J(x). Let us take $J(x) = \sum_\ell J_\ell(x)$ where $J_\ell(x)$ are functions with compact support D_ℓ and the domains D_ℓ move away from each other to infinity. Pictorially, we have the diagrams

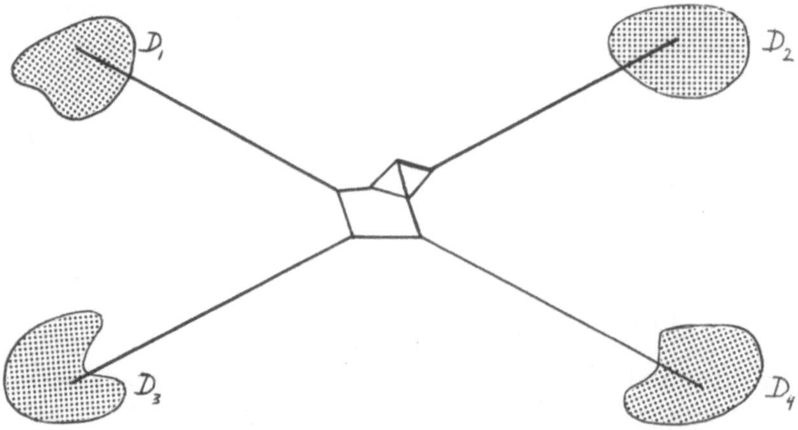

(I9)

Let us, for simplicity, ignore blobs that may occur in the external lines:

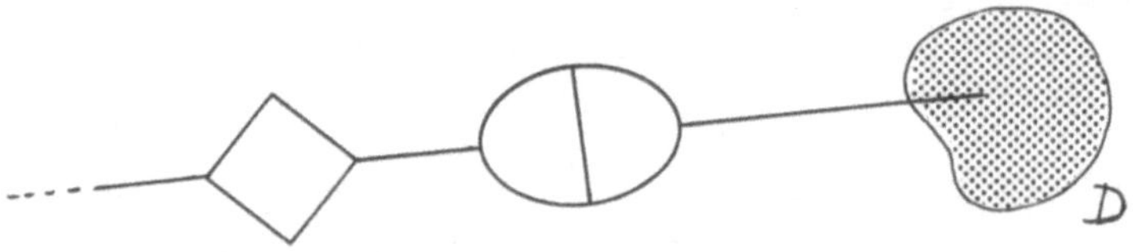

(I10)

(these can easily be treated later).
It is convenient to diagonalize the matrices M, which brings us to the space of Fourier transforms of $\phi(x)$, namely $\{\phi(k)\}$. In this space we have

$$M^{-1} = \frac{-i}{k^2+m^2} .$$

Now in Minkowsky space $k^2 + m^2$ can be zero. We improve the definition of our functional integral by replacing $k^2 + m^2$ by $k^2 + m^2 - i\varepsilon$, so that

$$e^{-\phi(-\partial^2+m^2)\phi}$$

is replaced by

$$e^{-i\phi(-\partial^2+m^2)\phi - \varepsilon\phi^2} , \tag{I11}$$

which converges if $-\partial^2+m^2$ happens to vanish.
Of course, ε is positive infinitesimal. We now have in x-space

$$M^{-1}(x_1,x_2) = -i(2\pi)^{-n} \int d^n k \frac{e^{ik(x_1-x_2)}}{k^2+m^2-i\varepsilon} . \tag{I12}$$

One may now ask what will be the asymptotic form of M^{-1} as $|x_1 - x_2| \to \infty$. It is easy to see that only that part of the integral (I11) survives where $k^2+m^2 \simeq o$. Physically that implies that the particles go to their mass shells. The sources J_ℓ at the domains D_ℓ can physically be interpreted as particle production or detection machines, far away from the interaction region. The amplitude approaches the scattering matrix S. We now short-circuit complicated lengthy definitions of S, defining it simply directly from the Feynman diagrams.

1) Consider all diagrams of the type I9 and replace in the external lines

$$\frac{1}{k^2+m^2-i\varepsilon} \quad \text{by} \quad \delta(k^2+m^2) , \tag{I13}$$

that is, we only consider external lines on mass shell, in the Fourier picture.

2) Look at the sign of $k_o = \pm\sqrt{m^2+\vec{k}^2}$. If $k_o > o$ we interpret this as an outgoing particle, if $k_o < o$ it is an ingoing particle.

3) The diagram is now a contribution to the matrix element

$$_{\text{outgoing}}\langle k_1,k_2,\ldots |S| k_3,k_4\ldots\rangle_{\text{ingoing}} . \tag{I14}$$

The normalization is to be fixed later.

Formally, this all can be derived from good old quantum mechanics by writing the evolution kernel

$$U(t_1,t_2) = e^{-i(t_1-t_2)H}$$

as a path integral, for the case of a set of anharmonic oscillators at all points x. This is why one expects that the S matrix defined this way will come out to be unitary. However, divergences, for instance in the diagram

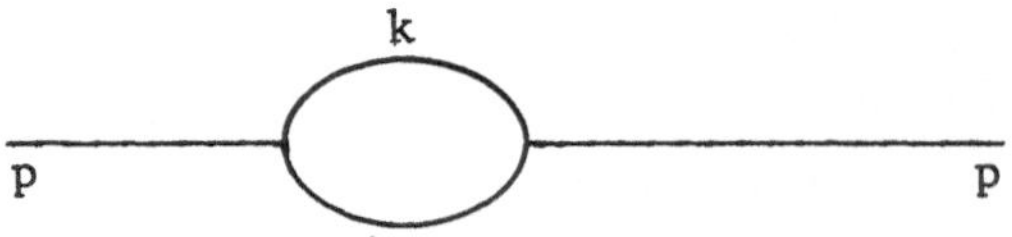

$$\int \frac{d^4k}{(k^2+m^2-i\varepsilon)((k+p)^2+m^2-i\varepsilon)} \tag{I15}$$

make redefinitions necessary. We must require that such redefinitions do not affect unitarity. Furthermore, we will want to write down functional integrals for the S matrix elements for particles with spin, in which case the canonical method is less direct. In short, we wish to understand explicitly how unitarity follows directly from the diagrams.

I3. CUTTING RELATIONS (SIMPLIFIED)

Consider the "propagator" in Fourier space

$$\frac{1}{(2\pi)^4 i}\,\frac{1}{k^2+m^2-i\varepsilon} \equiv \Delta(k) \ . \tag{I16a}$$

In x-space:

$$\Delta(x_i,x_j) = \int d^4k\, e^{ik(x_i-x_j)}\,\Delta(k) \equiv \Delta_{ij} \ . \tag{I16b}$$

Here, i and j are labels for two vertices in a diagram. x_i and x_j are the corresponding x variables. We write.

$$\Delta_{ij} = \theta(x_o)\Delta^+_{ij} + \theta(-x_o)\Delta^-_{ij} \ , \tag{I17}$$

$$\Delta^\pm_{ij} = \frac{1}{(2\pi)^3}\int d_4k\, e^{ikx}\,\theta(\pm k_o)\,\delta(k^2+m^2) \ , \tag{I18}$$

where $x = x_i - x_j$.
We have

$$\Delta^\pm_{ij} = (\Delta^\mp_{ij})^* = \Delta^\mp_{ji} \ . \tag{I19}$$

Therefore,

$$\Delta^*_{ij} = \theta(x_o)\Delta^-_{ij} + \theta(-x_o)\Delta^+_{ij} \ . \tag{I20}$$

The above identities are easily derived from

$$\theta(x) = \frac{1}{2\pi i}\int_{-\infty}^{\infty} d\tau\, \frac{e^{i\tau x}}{\tau - i\varepsilon} \ . \tag{I21}$$

Consider now a diagram in the x-representation

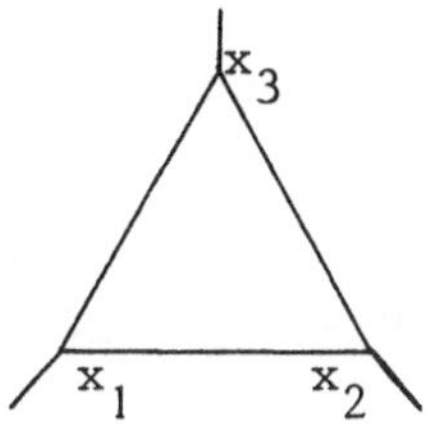

leaving out the δ functions for the external lines. Call it $F(x_1,x_2,x_3,\ldots)$ (here we sketched the diagram for $F(x_1,x_2,x_3) = (-i\lambda_1)(-i\lambda_2)(-i\lambda_3)\ \Delta_{12}\ \Delta_{23}\ \Delta_{31}$). This diagram may contribute to the S matrix after Fourier transformation. Now we define a more general class of diagrams

$$F(x_1,\ldots,\ \underline{x}_i,\ldots,\ \underline{x}_j,\ldots\ x_n)\ ,$$

with some of the arguments x underlined. In the diagram we denote this by drawing a circle around those vertices:

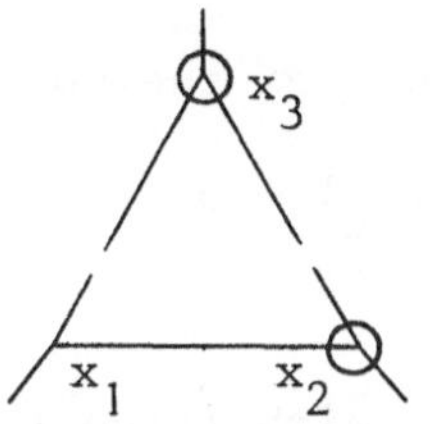

This function F is defined by the following replacements: replace Δ_{ij} by

1) Δ_{ij} if neither x_i nor x_j are underlined,
2) Δ^{+}_{ij} if x_i but not x_j is underlined,
3) Δ^{-}_{ij} if x_j but not x_i is underlined, (I22)
4) Δ^{*}_{ij} if both x_i and x_j are underlined,
5) Replace one factor i by $-i$ for every underlined x (remember they carry a factor $-i\lambda$; replace by $+i\lambda$, not $i\lambda^*$).

In the above example we have

$$F(x_1,\underline{x}_2,\underline{x}_3) = (-i\lambda_1)(i\lambda_2)(i\lambda_3)\ \Delta^{-}_{12}\ \Delta^{*}_{23}\ \Delta^{+}_{31}\ .$$

We now have <u>theorem 1</u> ("largest time equation"). Let x_1 be the x-coordinate with largest time component, $x_{1o} > x_{io}$ for all i. Then

$$F(x_1,\ldots) + F(\underline{x}_1,\ldots) = o \quad , \tag{I23}$$

where all other coordinates are underlined or not in the same way for both terms.

Proof: if x_j is not underlined then the first term contains Δ_{1j}, the second Δ^+_{1j}. But because $\theta(x_{1o}-x_{jo}) = 1$ we have (see eq. I17):

$\Delta_{1j} = \Delta^+_{1j}$, or 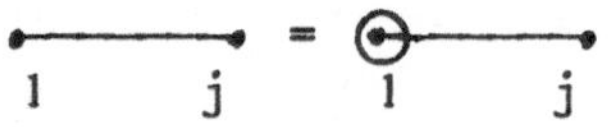

.

Similarly, if x_j is underlined we have Δ^-_{1j} in the first and Δ^*_{ij} in the second term. Because of eq. (I20),

$\Delta^-_{1j} = \Delta^*_{1j}$, or  .

Because of replacement # 5 we have one sign flip, from which follows (I23).

<u>Theorem 2</u>. For all values of the coordinates x_i we have

$$\sum_{\substack{\text{all possible}\\ \text{underlinings}}} F(x_1,\ldots, x_n) = o \tag{I24}$$

Proof: There is always one x with largest time component. The diagrams therefore combine in pairs of terms that cancel each other.

Next, let us deform all terms in (I24) in such a way that the underlined vertices occur at the right, the non-underlined at the left. We obtain:

$\sum$ 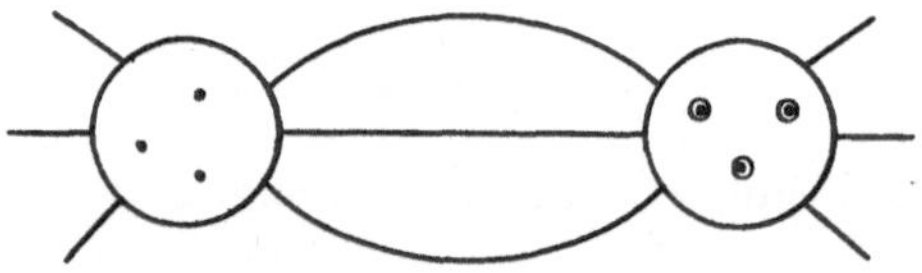 $= o$.

Add the trivial diagram (without any vertices) and sum formally over all diagrams. We see

$\sum$ 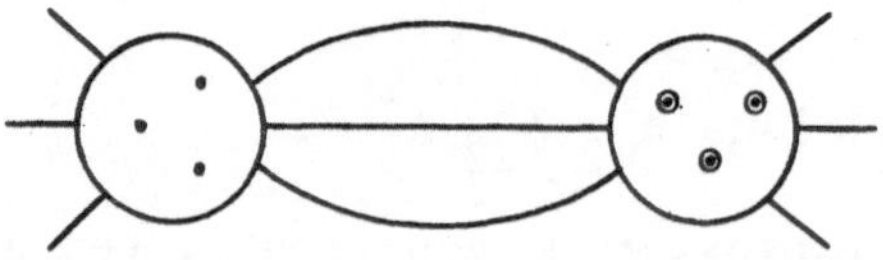 $= \sum$  .(I25)

How do we interprete this expression? Let the diagrams with all vertices encircled (all coordinates underlined) generate a matrix

$\tilde{S}$. The lines connecting S with $\tilde{S}$ are on mass shell with positive energy component k , because they all correspond to a term Δ^{+}_{ij}, see (I18). Our eq. (I25) says

$$\int dk_i \langle A|S|k_1 k_2 k_3, \ldots \rangle \prod_i \delta(k_i^2+m_i^2)\, \theta(k_{io}) \langle k_1 k_2 k_3, \ldots |\tilde{S}|B\rangle = \langle A|B\rangle , \tag{I26}$$

which is unitarity if $\tilde{S} = S^{\dagger}$. (I27)

One easily checks that indeed the diagrams for $\tilde{S}$ are the same as the ones for $S^{\dagger}$ provided that the coupling constants λ in eq. (I8) are real. Otherwise the replacement # 5 in (I22) would not imply complex conjungation.

What generalizations are allowed such that unitarity is not lost? Suppose our propagator was a matrix of the form

$$\frac{\rho_{ij}}{k^2+m^2-i\varepsilon} , \tag{I28}$$

then the intermediate lines (in I25) would carry an extra matrix ρ:

$$S \rho S^{\dagger} = I . \tag{I29}$$

We reobtain unitarity by redefining the norm of the particle states with a factor $\sqrt{\rho}$, but this only works if

<u>ρ is positive definite</u>. (I30)

If ρ has zero eigenvalues besides positive ones then one can write

$$\rho = \sum_i |\psi_i\rangle \rho_i \langle\psi_i| , \tag{I31}$$

where the number of states is less than the dimension of ρ. That describes a system with fewer types of particles than fields ϕ and is therefore acceptable.

I4. REGULATORS

One would like to change the theory further in such a way that integrals such as (I15) are made convergent (regularized). An example is the Pauli-Villars regulator. We replace all propagators $1/k^2+m^2-i\varepsilon$ by

$$\frac{1}{(k^2+m^2-i\varepsilon)(1+\frac{k^2}{\Lambda^2}-i\varepsilon)} = \frac{\Lambda^2}{\Lambda^2-m^2}\left(\frac{1}{k^2+m^2-i\varepsilon} - \frac{1}{k^2+\Lambda^2-i\varepsilon}\right) \tag{I32}$$

We immediately see that the minus sign destroys unitarity. A fictitious particle with mass Λ may be produced, but in view of the sign of its contribution the production "probability" for any odd number of these objects is negative. However we also observe that our identities can still be used. An immediate Λ line in (I25) is now associated with a factor

$$\delta(k^2+\Lambda^2)\ \theta(k_o)$$

and since $k_o = \pm\sqrt{\vec{k}^2+\Lambda^2}$ this only contributes if $k_o > \Lambda$. All other energies k_o in the intermediate lines are positive as well. Therefore, we find only violation of unitarity in channels where the total energy in the initial and final states $\langle A|$ and $|B\rangle$ exceeds Λ.

We can now take the limit $\Lambda \to \infty$ carefully, inserting explicitly Λ dependent extra terms ("counter terms") in the Lagrangian. A theory is called "renormalizable" if such a limit can be taken in such a way that the S matrix according to our definition remains finite. For scalar field theories this requires the absence of any couplings other than cubic or quartic in the fields ϕ.

I5. PARTICLES WITH SPIN ONE

We skip the case of spin $\frac{1}{2}$ particles which yields the well-known Dirac equation and Fermi-Dirac statistics. We wish to construct the theory for particles with spin one. Let us try to construct the propagator, first when the particle is at rest: $k = (k_o,o,o,o)$. (Relativistic covariance will give us the moving case.) The states $|\psi_i\rangle$ in (I31) must transform as vectors under the "little group" rotations (these are the SO(3) rotations that leave the form-vector k_μ invariant). Suppose that we choose for the matrix $\rho_{\mu\nu}$ in (I28)

$$\rho_{\mu\nu} = g_{\mu\nu} = \mathrm{diag}\ (-1,1,1,1)\ . \tag{I33}$$

Then unitarity would clearly be violated. The minus sign in (I33) would correspond to an unwanted particle with negative metric. This is why the Lagrangian (= term in the exponent of the functional integral)

$$\mathcal{L} = -\tfrac{1}{2}(\partial_\mu A_\nu)^2 - \frac{m^2}{2} A_\mu^2\ , \tag{I34}$$

for a massive vector particle is unacceptable. We want for the particle at rest

$$\rho_{\mu\nu}(k) = \mathrm{diag}\ (o,1,1,1)\ . \tag{I35}$$

For general k_μ that could be

$$\rho_{\mu\nu}(k) = g_{\mu\nu} - \frac{k_\mu k_\nu}{k^2 \pm i\varepsilon} . \tag{I36}$$

But, whatever sign we choose for $\pm i\varepsilon$, the extra part introduces an extra pole at $k^2 = o$ with wrong sign (the factors k_μ were not yet considered in my simplified cutting relations, but when they are carefully taken into account one verifies that the sign corresponds to particles with wrong metric).

The only correct Lorentz invariant propagator is

$$\frac{\delta_{\mu\nu} + k_\mu k_\nu / m^2}{k^2 + m^2 - i\varepsilon} . \tag{I37}$$

One verifies (I35) on mass shell. No extra spurious poles occur corresponding to unwanted particle states. The propagator has only one major disadvantage: it is badly divergent as $k \to \infty$. This could also imply complications in our unitarity relations as $x_i - x_j \to o$ because the θ functions have not yet been properly defined there. Suitable addition of regulators could cure such problems. However because of the $k_\mu k_\nu$ terms the limits $\Lambda \to \infty$ for these regulators are much harder to take. Indeed, in a general theory for [illegible] particles with such propagators the desease of high-momentum infinities would spread beyond control. Note that the Lagrangian giving this propagator is

$$\mathcal{L} = -\tfrac{1}{4}(\partial_\mu A_\nu - \partial_\nu A_\mu)^2 - \tfrac{1}{2}m^2 A_\mu^2$$
$$\mathcal{L} = -\tfrac{1}{4}F_{\mu\nu}^2 - \tfrac{1}{2}m^2 A_\mu^2 . \tag{I38}$$

I6. GAUGE TRANSFORMATIONS

We now wish to cure the infinity problems without destroying the good (= unitary) features of the Lagrangian (I38). Consider the functional integral

$$Z = \int DA_\mu \left(\det \frac{\partial C(A)}{\partial \Omega}\right) \exp i \int d^n x \left[\mathcal{L}^{inv}(A) - \tfrac{1}{2}(C(A))^2\right] . \tag{I39}$$

Here $\mathcal{L}^{inv}$ is a function of $A_\mu(x)$ and $\partial_\mu A_\nu(x)$ which is invariant under local gauge transformations Ω of $A_\mu(x)$:

$$A'_\mu(x) = \Omega(x)\left[A_\mu(x) + \frac{1}{g}\partial_\mu\right]\Omega^{-1}(x) . \tag{I40}$$

Here $A_\mu(x)$ is written as an antihermitian matrix. $C(A)$ is a gauge-fixing term. That is, the restriction $C(A) = o$ can always be satisfied for any field configuration A after performing gauge rotations of the type (I40). Finally,

$$\det\left(\frac{\partial C(A)}{\partial\Omega}\right)$$

is a formal functional determinant of an operator defined by subjecting C to an infinitesimal gauge rotation Ω. In fact, this determinant simply fixes a particular measure (through still superficially) for the functional integral. Let us give an example. The infinitesimal gauge rotations are:

$$A^a_\mu(x) \to A^a(x) - \partial_\mu \Lambda^a(x) + g f_{abc}\, \Lambda^b(x)\, A^c_\mu(x) \ ; \tag{I41}$$

$$C^a(x) = \partial_\mu A^a_\mu(x) \ , \tag{I42}$$

$$\mathcal{L}^{inv}(A) = -\tfrac{1}{4}\, G^a_{\mu\nu}(x)\, G^a_{\mu\nu}(x) \ , \tag{I43}$$

$$G^a_{\mu\nu}(x) = \partial_\mu A^a_\nu - \partial_\nu A^a_\mu + g f_{abc}\, A^b_\mu A^c_\nu \ , \tag{I44}$$

$$\frac{\partial C^a(x)}{\partial \Omega^b(x)} = \delta_{ab}\,\partial^2 + g f_{abc}\, \partial_\mu (A^c_\mu) \ . \tag{I45}$$

Here f_{abc} are structure constants of some compact group. The determinant is something new. But, we can give it a more familear appearance, by writing

$$\det{}^{-N} \frac{\partial C}{\partial\Omega} = \int D\phi D\phi^* \, e^{-\phi^*_i \frac{\partial C}{\partial\Omega} \phi_i} \ , \tag{I46}$$

where ϕ_i have N components. In our case:

$$-\phi^*_i \frac{\partial C}{\partial\Omega} \phi_i = \int dx \left[-\partial_\mu \phi^{a*}_i \partial_\mu \phi^a_i - g f_{abc}\, \partial_\mu \phi^{a*}_i\, A^b_\mu \phi^c_i\right] \tag{I47}$$

(where we performed a partial integration).
The Feynman rules for

$$\det \frac{\partial C}{\partial\Omega}$$

are now easily read off from this "Lagrangian" in the usual way. For the factor N, associated to each closed ϕ-loop, we must substitute -1.

<u>Theorem</u>: the function Z defined in (I39) is independent of the choice of the gauge fixing function C(A).
The proof of this theorem can be given in various ways, either by combinatorics with diagrams or by first proving the assertion for finite dimensional integrals. One of the basic ingredients of the original proofs of renormalizability of the gauge theories was to show that this theorem remains valid even if the expression (I39) for Z has been made finite by special regularization techniques (dimensional regularization). I will not go into this but simply

formulate how in principle this central theorem can be used to obtain a renormalizable theory.

I7. A GAUGE THEORY

Let us consider as a gauge group SU(2). Let the fields be A_μ^a and a complex doublet ξ_i.

$$\mathcal{L}^{inv}(A_\mu^a,\xi_i,\xi_i^*) = -\tfrac{1}{4}G_{\mu\nu}^a G_{\mu\nu}^a - D_\mu\xi^* D_\mu\xi$$

$$-\tfrac{1}{2}\lambda(\xi^*\xi-F^2)^2 \ . \qquad \text{(I48)}$$

Here F is a c-number, and

$$D_\mu\xi = \partial_\mu\xi - \tfrac{1}{2}ig\,A_\nu^a\,\tau^a\xi \ . \qquad \text{(I49)}$$

$\tau^{1,2,3}$ are the 2x2 Pauli matrices.
First choice of gauge (the so-called <u>unitary</u> gauge):

$$C^a(A,\xi) = \alpha\begin{pmatrix}\text{Re }\xi_2\\ \text{Im }\xi_1\\ \text{Im }\xi_2\end{pmatrix} \ ; \qquad \alpha\to\infty \qquad \text{(I50)}$$

It is easily seen that if $\alpha\to\infty$ this simply corresponds to freezing out the variables

$$\text{Re }\xi_2 \ , \ \text{Im }\xi_1 \ , \ \text{Im }\xi_2 \to o \ .$$

Let us write

$$\xi = \begin{pmatrix}F\\ o\end{pmatrix} + \frac{1}{\sqrt{2}}\begin{pmatrix}Z\\ o\end{pmatrix} \ . \qquad \text{(I51)}$$

where Z is now a single real scalar field. The functional determinant

$$\det\frac{\partial C}{\partial\Omega}$$

gets no space-time derivatives. This implies that the effects of this determinant, through highly divergent, vanish completely when we regularize and add counter terms (not explained any further here).
Substituting (I51) we obtain

$$\mathcal{L} = -\tfrac{1}{4}G_{\mu\nu}^a G_{\mu\nu}^a - \tfrac{1}{4}g^2F^2A_\mu^2 - \tfrac{1}{2}(\partial_\mu Z)^2 - \lambda F^2Z^2$$

$$+ \text{ local interaction terms.} \qquad \text{(I52)}$$

Observe that the vector field A_μ^a occurs precisely in the combination that gives a unitary propagator (see I38). The mass is $gF/\sqrt{2}$. The scalar field Z has mass $F\sqrt{2\lambda}$. We have a unitary theory with three- spin-one and one spin-sero particle types.

What makes this theory so special? Let us consider a second choice of gauge (the so-called renormalizable gauge):

$$C^a(A,\xi) = \partial_\mu A^a_\mu - gF\begin{pmatrix} I_m\ \xi_2 \\ -Re\ \xi_2 \\ Im\ \xi_1 \end{pmatrix} , \qquad (I53)$$

a special combination chosen only for convenience. We get

$$\mathcal{L}^{inv} = -\tfrac{1}{2}(\partial_\mu A^a_\nu)^2 - \tfrac{1}{4}g^2F^2A^2_\mu -$$

$$-\tfrac{1}{2}(\partial_\mu \psi^a)^2 - \tfrac{1}{4}g^2F^2(\psi^a)^2 - \tfrac{1}{2}\partial_\mu Z^2 - \lambda F^2 Z^2$$

$$+ \text{ total derivatives } + \text{ local interaction terms} \qquad (I45)$$

where

$$\xi_i = \begin{pmatrix} F \\ o \end{pmatrix} + \frac{1}{\sqrt{2}}\begin{bmatrix} i\psi^3 + Z \\ -\psi^2 + i\psi^1 \end{bmatrix} .$$

The total derivatives are irrelevant.
The vector field Lagrangian is now rather like (I34) and gives nicely convergent propagators:

$$\frac{g_{\mu\nu}}{k^2 + \tfrac{1}{2}g^2F^2 - i\varepsilon} . \qquad (I55)$$

Only due to gauge invariance the two theories are equivalent. The renormalizable gauge still shows infinities but they are not worse than in the scalar and spinor field theories. A renormalization procedure can remove them.

The g_{oo} part of the propagator (I55) describes a spurious particle or "ghost". Suppose we apply our unitarity criterion directly. We would notice

$$S\ K\ S^\dagger = I , \qquad (I56)$$

where K is an operator with eigenvalues ±1. K has a factor −1 for each of these spurious particles. Furthermore, the complex scalar field ϕ in (I47) now also contributes. It is produced in pairs. Because their multiplicity factor N is −1, they also come with wrong metric: a factor −1 for each ϕ, anti-ϕ pair. Finally, the "particles" described by the fields ϕ^a in (I54) have positive metric but should not be considered as real particles because they are absent in (I52). All these spurious states cancel in (I56). If we write Hilbert space as a product of a physical Hilbert space (containing only real particles) and a ghost space G (of all states with at least one unphysical particle), then

$$I = SKS^{\dagger} = \sum_{P} S|P\rangle\langle P|S^{\dagger} + \sum_{G} S|G\rangle(\pm)\langle G|S^{\dagger}$$

$$= \sum_{P} S|P\rangle\langle P|S^{\dagger} \ . \qquad \text{(I57)}$$

The ghost part can be shown to vanish when considered as an operator on the physical Hilbert space.

I showed you how such a proof can be set up by showing equivalence with a "unitary gauge", containing no ghost space G. In practice a more accurate procedure requires intermediate gauges, having only ghost particles with very high masses. Our cutting relations enable us to understand unitarity in all channels where the energy does not exceed the values of these masses.

I hope to have shown in this lecture why there is a strong theoretical argument (quite independent of the impressive experimental indications) in favor of a gauge theoretical structure of any field theory containing particles with spin one.

REFERENCES

More details on the subject of this lecture can be found in:

G. 't Hooft and M. Veltman, "DIAGRAMMAR", CERN report 73-9.

WHICH TOPOLOGICAL FEATURES OF A GAUGE THEORY CAN BE RESPONSIBLE FOR PERMANENT CONFINEMENT?

G. 't Hooft

Institute for Theoretical Physics
Utrecht, The Netherlands

III. INTRODUCTION

In the previous lecture a simple gauge model was considered with a scalar field doublet ξ. Perturbation expansion was considered not about the point $\xi = o$ but about the "vacuum value"

$$\xi = \begin{pmatrix} F \\ o \end{pmatrix} .$$

Such a theory is usually called a theory with "spontaneous symmetry breakdown" [1]. In contrast one might consider "unbroken gauge theories" where perturbation expansion is only performed about a symmetric "vacuum". These theories are characterized by the absence of a mass term for the gauge vector bosons in the Lagrangian. The physical consequences of that are quite serious. The propagators now have their poles at $k^2 = o$ and it will often happen that in the diagrams new divergences arise because such poles tend to coincide. These are fundamental infrared divergencies that imply a blow-up of the interactions at large distance scales. Often they make it nearly impossible to understand what the stable particle states are.

A particular example of such a system is "Quantum Chromodynamics", an unbroken gauge theory with gauge group SU(3), and in addition some fermions in the 3-representation of the group, called "quarks". We will investigate the possibility that these quarks are permanently confined inside bound structures that do not carry gauge quantum numbers. First of all this idea is not as absurd as it may seem. The converse would be equally difficult to understand. Gauge quantum numbers are a priori only defined up to local gauge transformations. The existence of *global* quantum numbers that would correspond to these local ones but would be detectable experimentally from a distance is not at all a prerequisite. We are nevertheless accustomed to attaching a global significance to local gauge transformation properties because we are familiar with the theories with spontaneous breakdown. The electron and its neutrino, for example, are usually said to form a gauge doublet, to be subjected to local gauge transformations. But actually these words are not properly used. Even the words "spontaneous breakdown" are formally not correct for local gauge theories (which is why I put them between quotation marks). The vacuum *never* breaks local gauge invariance because it itself is gauge invariant. All states in the physical Hilbert space are gauge-invariant. This may be confusing so let me illustrate what I mean by considering the familiar Weinberg-Salam-Ward model. The invariant Lagrangian is

$$\mathcal{L}^{inv} = -\tfrac{1}{4}G^a_{\mu\nu}G^a_{\mu\nu} - \tfrac{1}{4}F_{\mu\nu}F_{\mu\nu} - D_\mu\phi^* D_\mu\phi - V(|\phi|)$$

$$- \bar{\psi}_L\gamma D\psi_L - \bar{e}_R\gamma De_R - \kappa\bar{e}_R(\phi^*\psi_L) - \kappa(\bar{\psi}_L\phi)e_R \;. \qquad \text{(II1)}$$

Here ϕ is the scalar Higgs doublet. The gauge group is $SU(2)\times U(1)$, to which correspond A^a_μ $(G^a_{\mu\nu})$ and A^o_μ $(F_{\mu\nu})$. The subscripts L and R denote left and right handed components of a Dirac field, obtained by the projection operators $\frac{1}{2}(1\pm\gamma_5)$.

e_R is a singlet;

ψ_L is a doublet.

D_μ stands for covariant derivative.

The function $V(|\phi|)$ takes its minimum at $|\phi| = F$. Usually one takes

$$<\phi>_{vacuum} = \begin{pmatrix} F \\ o \end{pmatrix} \qquad \text{(II2)}$$

and perturbes around that value: $\phi = \begin{pmatrix} F+\tilde{\phi}_1 \\ \tilde{\phi}_2 \end{pmatrix}$.

One identifies the components of ψ with neutrino and electron:

$$\psi = \begin{pmatrix} \nu_L \\ e_L \end{pmatrix} \;. \qquad \text{(II3)}$$

However, this model is *not* fundamentally different from a model with "permanent confinement". One could interpret the same physical particles as being all gauge singlets, bound states of the fundamental fields with extremely strong confining forces, due to the gauge fields A_μ^a of the group SU(2). We have scalar quarks (the Higgs field ϕ) and fermionic quarks (the ψ_L field) both as fundamental doublets. Let us call them q. Then there are "mesons" ($q\bar{q}$) and "baryons" (qq). The neutrino is a "meson". Its field is the composite, SU(2)-invariant

$$\phi^* \psi_L = F\nu_L + \text{negligible higher order terms.}$$

The e_L field is a "baryon", created by the SU(2)-invariant

$$\varepsilon_{ij}\phi_i\psi_j = Fe_L + \ldots \qquad \text{(II4)}$$

the e_R field remains an SU(2) singlet.
Also bound states with angular momentum occur: The neutral intermediate vector boson is the "meson"

$$\phi^* D_\mu \phi = \frac{i}{2} gF^2 A_\mu^{(3)} + \text{total derivative} + \text{higher orders}, \qquad \text{(II5)}$$

if we split off the total derivative term (which corresponds to a spin-zero Higgs particle).
The $W_\mu^\pm$ are obtained from the "baryons" $\varepsilon_{ij}\phi_i D_\mu \phi_j$, and the Higgs particle can also be ontained from $\phi^*\phi$.
Apparently some mesonic and baryonic bound states survive perturbation expansion, most do not (only those containing a Higgs "quark" may survive).

Is there no fundamental difference then between a theory with spontaneous breakdown and a theory with confinement? Sometimes there is. In the above example the Higgs field was a faithful representation of SU(2). This is why the above procedure worked. But suppose that all scalar fields present were invariant under the center Z(N) of the gauge group SU(N), but some fermion fields were not. Then there are clearly two possibilites. The gauge symmetry is "broken" if physical objects exist that transform non-trivially under Z_N, such as the fundamental fermions. We call this the Higgs phase. If on the other hand all physical objects are invariant under Z_N, such as the mesons and the baryons, then we have permanent confinement.

Quantum Chromodynamics is such a theory where these distinct possibilities exist. It is unlikely that one will ever prove from first principles that permanent confinement takes place, simply because one can always imagine the Higgs mode to occur. If no fundamental scalar fields exist then one could introduce composite fields such as

$$\phi_{ab} = G_{\mu\nu}^a G_{\mu\nu}^b$$

or

$$\phi_i^j = \bar{\psi}_i \psi^j$$

and postulate nonvanishing vacuum expectation values for them:

$$\langle\phi_{ab}\rangle = F_1\, d_{ab8} + F_2\, d_{ab3} \qquad \text{or}$$

$$\phi_i^j = F_1\, \lambda_{8i}{}^j + F_2\, \lambda_{3i}{}^j \ .$$

In that case there would be no confinement. Whether or not $F_{1,2}$ are equal to zero will depend on details of the dynamics. Therefore, dynamics must be an ingredient of the confinement mechanism, not only topological arguments. What we will attempt in this lecture is to show that topological arguments imply for this theory the existence of phase regions, separated by sharp phase transition boundaries (usually of first order). One region corresponds to what is usually called "spontaneous breakdown", and will be referred to as Higgs phase. Another corresponds to absolute quark confinement. Still another phase exists which allows for long range Coulomb-like forces to occur. (Coulomb phase.)

II2. VORTICES IN THE PERIODIC BOX

We concentrate on long-range topological phenomena. One topological feature is the instanton, corresponding to a gauge field configuration with non-trivial Pontryagin or Second Chern Class number. This however has no direct implication for confinement. What is needed for confinement is something with the space-time structure of a string, i.e. a two dimensional manifold in 4 dim. space-time. Instantons are rather event-like, i.e. zero dimensional and can for instance give rise to new types of interactions that violate otherwise apparent symmetries. We will not consider these further here*. A topological structure which is extended in two dimensional sheets exists in gauge theories, as has been first observed by Nielsen, Olesen [2] and Zumino [3]. They are crucial. We will exhibit them by compactifying space-time. For the instanton it had been convenient to compactify space-time to a sphere $S^{(4)}$. For our purposes a hypertorus

$$S^{(1)} \times S^{(1)} \times S^{(1)} \times S^{(1)} \ ,$$

is more suitable [4]. One can also consider this to be a four dimensional cubic box with periodic boundary conditions. Inside, space-time is flat. The box may be arbitrarily large. To be ex-

*Surely, as is explained by C. Callan in his lectures, configurations such as the instanton gas will influence the *dynamics* and thereby give rise to a perhaps crucial force that causes the system to choose the confinement rather than the Higgs mode. The absoluteness of the confining force is however not explained that way.

plicit we put a pure SU(N) gauge theory in the box (no quarks yet). Now in the continuum theory the gauge fields themselves are representations of SU(N)/Z(N), where Z(N) is the center of the group SU(N):

$$Z(N) = \left\{ e^{2\pi i n/N} I \ ; \ n = o, 1, \ldots, N-1 \right\} . \tag{II6}$$

This is because any gauge transformation of the type (II6) leaves $A_\mu(x)$ invariant. A consequence of this is the existence of another class of topological quantum numbers in this box besides the familiar Pontryagin number. Consider the most general possible periodic boundary condition for $A_\mu(x)$ in the box. Take first a plane (x_1, x_2) in the 12 direction with fixed values of x_3 and x_4. One may have

$$\begin{aligned} A_\mu(a_1, x_2) &= \Omega_1(x_2)\, A_\mu(o, x_2) \ , \\ A_\mu(x_1, a_2) &= \Omega_2(x_1)\, A_\mu(x_1, o) \ . \end{aligned} \tag{II7}$$

Here, a_1, a_2 are the periods.
ΩA_μ stands short for $\Omega A_\mu \Omega^{-1} + \frac{1}{gi} \Omega \partial_\mu \Omega^{-1}$.
The periodicity conditions for $\Omega_{1,2}(x)$ follow by considering (II7) at the corners of the box:

$$\Omega_1(a_2)\, \Omega_2(o) = \Omega_2(a_1)\, \Omega_1(o) Z \ , \tag{II8}$$

where Z is some element of Z(N).
One may now perform continuous gauge transformations on $A_\mu(x)$.

$$A_\mu(x_1, x_2) \to \Omega(x_1, x_2)\, A_\mu(x_1, x_2) \ , \tag{II9}$$

where $\Omega(x_1, x_2)$ (non-periodic) can be arranged either such that $\Omega_2(x_1) = I$ or such that $\Omega_1(x_2) = I$, but not both, because Z in (II8) remains invariant under (II9) as one can easily verify. We call this element Z(1,2) because the 12 plane was chosen. By continuity Z(1,2) cannot depend on x_3 or x_4. For each $(\mu\nu)$ direction such a Z element exist, to be labeled by integers

$$n_{\mu\nu} = -n_{\nu\mu} \ , \tag{II10}$$

defined modulo N. Clearly this gives

$$N^{\frac{d(d-1)}{2}} = N^6 \tag{II11}$$

topological classes of gauge field configurations. Note that these classes disappear if a field in the fundamental representation of SU(N) is added to the system (these fields would make unacceptable jumps at the boundary). Indeed, to understand quark confinement it is necessary to understand pure gauge systems without quarks first. As we shall see, the new topological classes will imply the existence of new vacuum parameters besides the well-known instanton angle θ.

II3. ORDER AND DISORDER LOOP INTEGRALS

To elucidate the physical significance of the topological numbers $n_{\mu\nu}$ we first concentrate on gauge field theory in a three dimensional periodic box with time running from $-\infty$ to ∞. To be specific we will choose the temporal gauge,

$$A_4 = o \tag{II12}$$

(this is the gauge in which rotation towards Euclidean space is particularly elegant). Space has the topology $S(1)^3$. There is an infinite set of homotopy classes of closed oriented curves C in this space: C may wind any number of times in each of the three principal directions. For each curve C at each time t there is a quantum mechanical operator A(C,t) defined by

$$A(C,t) = \mathrm{Tr}\, P \exp \oint_C ig\vec{A}(\vec{x},t).d\vec{x} \ , \tag{II13}$$

called Wilson loop or order parameter.
Here P stands for path ordering of the factors $\vec{A}(\vec{x},t)$ when the exponents are expanded. The ordering is done with respect to the matrix indices. The $A(\vec{x},t)$ are also operators in Hilbert space, but for different $\vec{x}$, same t, all $A(\vec{x},t)$ commute with each other. By analogy with ordinary electromagnetism we say that A(C) *measures* magnetic flux *through* C, and in the same time *creates* an electric flux line *along* C. Since A(C) is gauge-invariant under purely periodic gauge transformations, our versions of magnetic and electric flux are gauge-invariant. Therefore they are not directly linked to the gauge *co*variant curl $G^a_{\mu\nu}(x)$.

There exists a dual analogon of A(C) which will be called B(C) or disorder loop operator [5]. C is again a closed oriented curve in $S(1)^3$. A simple definition of B(C) could be made by postulating its equal-time communication rules with A(C):

$$[A(C),\ A(C')] = o \ ;$$

$$[B(C),\ B(C')] = o \ ; \tag{II14}$$

$$A(C)\ B(C') = B(C')\ A(C) \exp 2\pi in/N \ ,$$

where n is the number of times C' winds around C in a certain direction. Note that n is only well defined if either C or C' is in the trivial homotopy class (that is, can be shrunk to a point by continuous deformations). Therefore, if C' is in a nontrivial class we must choose C to be in a trivial class. Since these commutation rules (II14) determine B(C) only up to factors that commute with A and B, we could make further requirements, for instance that B(C) be a unitary operator.

An explicit definition of B(C) can be given as follows. In the temporal gauge, $A_o = o$, one must distinguish a "large Hilbert space" $\mathcal{H}$ of all field configurations $\vec{A}(\vec{x})$ from a "physical Hilbert space $H \subset \mathcal{H}$. This H is defined to be the subspace of $\mathcal{H}$ of all gauge invariant states:

$$H = \left\{ |\psi>, \ <\vec{A}(\vec{x})|\psi> = <\Omega\vec{A}(\vec{x})|\psi> \right\} \qquad \text{(II15)}$$

where Ω is any infinitesimal gauge transformation in 3 dim. space. Often we will also write Ω for the corresponding rotation in $\mathcal{H}$:

$$H = \left\{ |\psi>, \ \Omega|\psi> = |\psi>, \ \Omega \text{ infinitesimal} \right\} . \qquad \text{(II16)}$$

Now consider a pseudo-gauge transformation $\Omega^{[C']}$ defined to be a genuine gauge transformation at all points $x \notin C'$, but singular on C'. For any closed path $x(\theta)$ with $o \leqslant \theta < 2\pi$ twisting n times around C' we require

$$\Omega^{[C']}(x(2\pi)) = \Omega^{[C']}(x(o)) \, e^{2\pi i n/N} . \qquad \text{(II17)}$$

This discontinuity is not felt by the fields $A(\vec{x},t)$ which are invariant under Z(N). They do feel the singularity at C' however. We define B(C') as

$$\Omega^{[C']}$$

but with the singularity at C' smoothened; this corresponds to some form of regularization, and implies that the operator differs from an ordinary gauge transformation. Therefore, even for $|\psi> \in H$ we have

$$B(C') \ |\psi> \neq |\psi> . \qquad \text{(II18)}$$

For any regular gauge transformation Ω we have an Ω' such that

$$\Omega\Omega^{[C']} = \Omega^{[C']} \, \Omega' . \qquad \text{(II19)}$$

Therefore, if $|\psi> \in H$ then $B(C') \ |\psi> \in H$, and B(C') is gauge-invariant. We say that B(C') *measures* electric flux *through* C' and *creates* a magnetic flux line *along* C'.

II4. NON-ABELIAN GAUGE-INVARIANT MAGNETIC FLUX IN THE BOX

We now want to find a conserved variety of Non-Abelian gauge-invariant magnetic flux in the 3-direction in the 3 dimensional periodic box. One might be tempted to look for some curve C enclosing the box in the 12 direction so that A(C) measures the flux through the box. That turns out not to work because such a flux is not guaranteed to be conserved. It is better to consider a curve C' in the 3-direction winding over the torus exactly once:

$$C' = \left\{ \vec{x}(s)\ , o \leqslant s < 1\ ;\ \vec{x}(1) = \vec{x}(o) + (o,o,a_3) \right\} . \tag{II20}$$

B(C') creates one magnetic flux line. But B(C') also changes the number n_{12} into $n_{12} + 1$. This is because

$$\Omega^{[C']}$$

makes a Z(N) jump according to (II17). If $\Omega_{1,2}(x)$ in (II7) are still defined to be continuous then Z in (II8) changes by one unit. Clearly, n_{12} measures the number of times an operator of the type B(C') has acted, i.e. the number of magnetic flux lines created. n_{12} is also conserved by continuity. We simply define

$$n_{ij} = \varepsilon_{ijk}\, m_k\ , \tag{II21}$$

with m_k the total magnetic flux in the k-direction. Note that $\vec{m}$ corresponds to the usual magnetic flux (apart from a numerical constant) in the Abelian case. Here, $\vec{m}$ is only defined as an integer modulo N.

II5. NON-ABELIAN GAUGE-INVARIANT ELECTRIC FLUX IN THE BOX

As in the magnetic case, there exists no simple curve C such that the total electric flux through C, measured by B(C), corresponds to a conserved total flux through the box. We consider a curve C winding once over the torus in the 3-direction and consider the electric flux creation operator A(C). But first we must study a new conserved quantum number.

Let $|\psi\rangle$ be a state in the before mentioned little Hilbert space H. Then, according to eq. (II16), $|\psi\rangle$ is invariant under *infinitesimal* gauge transformations Ω. But we also have some non-trivial homotopy classes of gauge transformations Ω. These are the pseudo-periodic ones:

$$\begin{aligned}
\Omega(a_1,x_2,x_3) &= \Omega(o,x_2,x_3)Z_1\ , \\
\Omega(x_1,a_2,x_3) &= \Omega(x_1,o,x_3)Z_2\ , \\
\Omega(x_1,x_2,a_3) &= \Omega(x_1,x_2,o)Z_3\ , \\
Z_{1,2,3} &\in \text{center } Z(N) \text{ of } SU(N)\ .
\end{aligned} \tag{II22}$$

Notice that not only do $A_\mu(x)$ transform smoothly under these Ω (they are invariant under Z(N) transformations), but their boundary conditions do not change. These Ω therefore commute with the magnetic flux $\vec{m}$. If two different Ω satisfy the same equation (II22) they act differently on states of the big Hilbert space $\mathcal{H}$, but since they only differ by regular gauge transformations* they act identically on states in H, defined in II16. Thus, $Z_{1,2,3}$, characterized by three integers k_1, k_2, k_3:

$$Z_t = e^{ik_t 2\pi/N} , \tag{II23}$$

define a set of N^3 different operators in H, under which the Hamiltonian H is invariant. Let us call these operators $\Omega[\vec{k}]$. Their eigenstates satisfy

$$\Omega[\vec{k}]\,|\psi\rangle = e^{i\omega(\vec{k})}\,|\psi\rangle , \tag{II24}$$

where $\omega(\vec{k})$ are strictly conserved numbers. Since $\Omega[\vec{k}]$ form a (finite) group $(Z(N))^3$ we have

$$\omega(\vec{k}_1) + \omega(\vec{k}_2) = \omega(\vec{k}_1 + \vec{k}_2 \bmod N) \bmod 2\pi , \tag{II25}$$

and

$$\omega(o) = o , \tag{II26}$$

because $|\psi\rangle \in H$.

Therefore,

$$\omega(\vec{k}) = \frac{2\pi}{N} \sum_i e_i k_i \pmod{2\pi} , \tag{II27}$$

where e_i are three fixed integers, defined modulo N. They are three conserved numbers, to be compared with the instanton angle θ.

Now let us turn back to A(C), defined in eq. (II13). If C is the curve considered in the beginning of this section, A(C) is not invariant under $\Omega[\vec{k}]$, because

---

*Here we left aside for sake of simplicity the fourth integer, the Pontryagin number of the gauge transformations. It gives interesting complications of which Witten's result for magnetic monopoles is an example[6]. To avoid these complications we must restrict ourselves to θ (the vacuum angle associated with instantons) $= o$, a subspace of H. Clearly, generalization to $\theta \neq o$ is possible.

$$A(C) \to \mathrm{Tr}\, \Omega(\vec{x}_1)\, P(\exp \int_C ig\vec{A}\,d\vec{x})\, \Omega^{-1}(\vec{x}_1+\vec{a}_3)$$

$$= e^{-2\pi i k_3/N} A(C) . \tag{II28}$$

Therefore,

$$A(C)\, \Omega[\vec{k}]\,|\psi\rangle = \Omega[\vec{k}]\, e^{-2\pi i k_3/N} A(C)\,|\psi\rangle . \tag{II29}$$

If $\quad \Omega[\vec{k}]\,|\psi\rangle = e^{i\omega(\vec{k})}|\psi\rangle ,$ (II30)

and $\quad A(C)\,|\psi\rangle = |\psi'\rangle ,$ (II31)

then $\Omega[\vec{k}]\,|\psi'\rangle = e^{i\omega(\vec{k})+2\pi i k_3/N}|\psi'\rangle .$ (II32)

Therefore $A(C)$ increases e_3 by one unit:

$$e_3\, A(C)|\psi\rangle = A(C)\, (e_3+1)|\psi\rangle . \tag{II33}$$

e_3 is a good indicator for electric flux in the 3-direction. It is strictly conserved. The physical interpretation of the three integers e_i (mod N) is electric flux. It is gauge-invariant and conserved. Notice that neither electric nor magnetic flux can be properly defined if fields in the fundamental reprosentation are present.

II6. THE FREE ENERGY OF A GIVEN FLUX CONFIGURATION AT LOW BUT FINITE TEMPERATURE

The free energy F of a system with given flux quantum $(\vec{m},\vec{e})$ at temperature $T = 1/k\beta$ is given by

$$e^{-\beta F} = \mathrm{Tr}_{H}\, P_e(\vec{e})\, P_m(\vec{m})\, e^{-\beta H} . \tag{II34}$$

Here H is the Hamiltonian, and H is the little Hilbert space. P are projection operators. $P_m(\vec{m})$ is simply defined to select a given set of $n_{ij} = \varepsilon_{ijk}\, m_k$, the three space-like indices of eq. (II10). How is $P_e(\vec{e})$ defined? We must select states $|\psi\rangle$ with

$$\Omega[\vec{k}]\,|\psi\rangle = e^{\frac{2\pi i}{N}(\vec{k}\vec{e})}|\psi\rangle , \tag{II35}$$

therefore

$$P_e(\vec{e}) = \frac{1}{N^3}\sum_{\vec{k}} e^{-\frac{2\pi i}{N}(\vec{k}\vec{e})}\,\Omega(\vec{k}) \ . \tag{II36}$$

Now $e^{-\beta H}$ is the evolution operator in imaginary time direction at interval β, expressed by a functional integral over a Euclidean box with sides (a_1,a_2,a_3,β):

$$<\vec{A}_{(1)}(\vec{x})\ e^{-\beta H}\ \vec{A}_{(2)}(\vec{x})> = \int \mathcal{D}A\ e^{S(A)} \Bigg|_{\vec{A}(\vec{x},o) = A_{(2)}(\vec{x})}^{\vec{A}(\vec{x},\beta) = A_{(1)}(\vec{x})} \tag{II37}$$

We may fix the gauge for $A_2(\vec{x})$ for instance by choosing

$$\begin{aligned} A_{(2)3}(\vec{x}) &= o, \\ A_{(2)2}(x,y,o) &= o \ , \\ A_{(2)1}(x,o,o) &= o \ . \end{aligned} \tag{II38}$$

We already had $A_4(\vec{x},t) = o$. Since only states in H are considered, we insert also a projection operator

$$\int_{\Omega\in I} \mathcal{D}\Omega \ ,$$

where I is the trivial homotopy class*.

"Trace" means that we integrate over all $A_{(1)} = A_{(2)}$, therefore we get periodic boundary conditions in the 4-direction. Insertions of $\int_{\Omega\in I} \mathcal{D}\Omega$ means that we have periodicity up to gauge transformations, in the completely unique gauge

$$A_4(\vec{x},\beta) = A_3(\vec{x},o) = A_2(x,y,o,o) = A_1(x,o,o,o) = o \ . \tag{II39}$$

Eq. (II36) tells us that we have to consider twisted boundary conditions in the 41, 42, 43 directions and Fourier transform:

$$e^{-\beta F(\vec{e},\vec{m},\vec{a},\beta)} = \frac{1}{N^3}\sum_{\vec{k}} e^{-\frac{2\pi i}{N}(\vec{k}\vec{e})}\ W\{\vec{k},\vec{m},a_\mu\} \ . \tag{II40}$$

Here $W\{\vec{k},\vec{m},a_\mu\}$ is the Euclidean functional integral with boundary conditions fixed by choosing

* See footnote page II21. Read: I is the class of purely periodic Ω.

$$n_{ij} = \varepsilon_{ijk}\, m_k ; \quad n_{i4} = k_i ; \quad a_4 = \beta \, .$$

Because of the gauge choice (II39) this functional integral must include integration over the Ω belonging to the given homotopy classes as they determine the boundary conditions such as (II7). The definition of W is completely Euclidean symmetric. In the next chapter I show how to make use of this symmetry with respect to rotation over 90° in Euclidean space.

II7. DUALITY

The Euclidean symmetry in eq. (II40) suggests to consider the following SO(4) rotation:

$$\begin{bmatrix} 0 & -1 & & \\ 1 & 0 & & \\ & & 0 & 1 \\ & & -1 & 0 \end{bmatrix} . \qquad \text{(II41)}$$

Let us introduce a notation for the first two components of a vector:

$$x_\mu = (\vec{x}, x_4) ;$$
$$\tilde{x} = (x_1, x_2) ,$$
$$\hat{x} = (x_2, x_1) . \qquad \text{(II42)}$$

We have, from eq. (II40):

$$\exp\left[-\beta F(\tilde{e}, e_3, \tilde{m}, m_3, \tilde{a}, a_3, \beta)\right] =$$

$$\frac{1}{N^2} \sum_{\tilde{k},\tilde{\ell}} \exp\left[\frac{2\pi i}{N}\left[-(\tilde{k}.\tilde{e}) + (\tilde{\ell}.\tilde{m})\right] - a_3 F(\tilde{\ell}, -e_3, \tilde{k}, -m_3, \hat{a}, \beta, a_3)\right] . \qquad \text{(II43)}$$

Notice that in this formula the transverse electric and magnetic fluxes are Fourier transformed and interchange positions. Notice also that, apart from a sign difference, there is a complete electric-magnetic symmetry in this expression, in spite of the fact that the definition of F in terms of W was not so symmetric. Eq. (II43) is an exact property of our system. No approximation was made. We refer to it as "duality".

II8. LONG-DISTANCE BEHAVIOR COMPATIBLE WITH DUALITY

Let us now assume that the theory has a mass gap. No

massless particles occur. Then asymptotic behavior at large distances will be approached exponentially. Then it is exluded that

$$F(\vec{e},\vec{m},\vec{a},\beta) \to o, \text{ exponentially as } \vec{a},\ \beta \to \infty ,$$

for all $\vec{e}$ and $\vec{m}$, which would clearly contradict (II43). This means that at least some of the flux configurations must get a large energy content as $\vec{a}$, $\beta \to \infty$. These flux lines apparently cannot spread out and because they were created along curves C it is practically inescapable that they get a total energy which will be proportional to their length:

$$E = \lim_{\beta\to o} F = Ca \quad . \tag{II44}$$

However, duality will never enable us to determine whether it are the electric or the magnetic flux lines that behave this way. From the requirement that W in II40 is always positive one can deduce the impossibility of a third option, namely that only exotic combinations of electric and magnetic fluxes behave as strings[4].

For further information we must make the physically quite plausible assumption of "factorizability":

$$F(\vec{e},\vec{m}) \Rightarrow F_e(\vec{e}) + F_m(\vec{m}) \quad \text{when } \vec{a},\ \beta \to \infty \ . \tag{II45}$$

Suppose that we have confinement in the electric domain:

$$F_e(o,o,1) \Rightarrow \rho a_3 \tag{II46}$$

where ρ is the fundamental string constant. Then we can derive from duality the behavior of $F_m(\vec{m})$.

First we improve (II46) by applying statistical mechanics to obtain F_e for large but finite β. One obtains:

$$e^{-\beta F_e(e_1,e_2,o,\vec{a},\beta) + C(\vec{a},\beta)}$$

$$= \sum_{n_1^\pm,n_2^\pm} \frac{1}{n_1^+!n_1^-!n_2^+!n_2^-!} \gamma_1^{\,n_1^++n_1^-} \gamma_2^{\,n_2^+-n_2^-} \delta_N(n_1^+-n_1^--e_1)\ \delta_N(n_2^+-n_2^--e_2) \tag{II47}$$

Here

$$\gamma_1 = \lambda\, a_2 a_3\, e^{-\beta\rho a_1} ,$$

$$\gamma_2 = \lambda\, a_1 a_3\, e^{\beta\rho a_2} ,$$

$$\delta_N(x) = \frac{1}{N}\sum_{k=o}^{N-1} e^{2\pi ikx/N} ,$$

$$\delta_N(x) = \begin{cases} 1 & \text{if } x = o \pmod N, \\ o & \text{otherwise.} \end{cases} \tag{II48}$$

The sum is over all nonnegative integer values of $n_i^{\pm}$ (the orientations $\pm$ are needed if $N \geqslant 3$). The γ's are Boltzmann factors associated with each string-like flux tube.

We now insert this, with (II45), into (II43) putting $e_3 = m_3 = o$. One obtains

$$e^{-\beta F_m(m_1,m_2,o,\vec{a},\beta)} = C' e^{2\sum_a \gamma'_a \cos(2m_a\pi/N)} \tag{II49}$$

where C' is again a constant and

$$\gamma'_1 = \lambda\, a_1 \beta\, e^{-\rho a_2 a_3} ,$$

$$\gamma'_2 = \lambda\, a_2 \beta\, e^{-\rho a_1 a_3} . \tag{II50}$$

At $\beta \to \infty$ we get

$$F_m(\tilde{m},o,\vec{a},\beta) \to E_m(\tilde{m},o,\vec{a}) = \sum_i E_i(m_i,\vec{a})$$

with

$$E_1(m_1,\vec{a}) = 2\lambda\left(1 - \cos\frac{2\pi m_1}{N}\right) a_1\, e^{-\rho a_2 a_3} . \tag{II51}$$

and similarly for E_2 and E_3.

One reads off from eq. (II51) that there will be no magnetic confinement, because if we let the box become wider the exponential factor

$$e^{-\rho a_2 a_3}$$

causes a rapid decrease of the energy of the magnetic flux. Notice the occurrence of the string constant ρ in there.

Of course we could equally well have started from the presumption that there were magnetic confinement. One then would conclude that there would be no electric confinement, because then the electric flux would have an energy given by (II51).

II9. THE COULOMB PHASE

To see what might happen in the absence of a mass gap one could study the (first) Georgi-Glashow model[7]. Here SU(2) is "broken spontaneously" into U(1) by an isospin one Higgs field. Ordinary perturbation expansion tells us what happens in the infrared limit. There are electrically charged particles: $W^{\pm}$ (the charged vector particles). They carry two fundamental electric flux units ("quarks" with isospin $\frac{1}{2}$ would have the fundamental flux unit $q_o = \pm\frac{1}{2}e$). There are also magnetically charged particles (monopoles,[8]). They also carry two fundamental magnetic flux units:

$$g = \frac{2\pi}{q_o} = \frac{4\pi}{e} \ . \tag{II52}$$

A given electric flux configuration of k flux units would have an energy

$$E = \frac{q_o^2 k^2 a_1}{2a_2 a_3} \ . \tag{II53}$$

A finite β however pair creation of $W^{\pm}$ takes place, so that we should take a statistical average over various values of the flux. Flux is only rigorously defined modulo $2q_o$. We have

$$e^{-\beta F_e(1,o,o)} = \frac{\sum_{k=-\infty}^{\infty} \exp\left[-\beta \frac{e^2 a_1}{2a_2 a_3} (k+\frac{1}{2})^2\right]}{\sum_{k=-\infty}^{\infty} \exp\left[-\beta \frac{e^2 a_1}{2a_2 a_3} k^2\right]} \ . \tag{II54}$$

Similarly, because of pair creation of magnetic monopoles

$$e^{-\beta F_m(1,o,o)} = \frac{\sum_{k=-\infty}^{\infty} \exp\left[-\beta \frac{8\pi^2 a_1}{g^2 a_2 a_3} (k+\frac{1}{2})^2\right]}{\sum_{k=-\infty}^{\infty} \exp\left[-\beta \frac{8\pi^2 a_1}{g^2 a_2 a_3} k^2\right]} \ . \tag{II55}$$

These expressions do satisfy duality, eq. (II43). This is easily verified when one observes that

$$\sum_{k=-\infty}^{\infty} e^{-\lambda k^2} = \sqrt{\frac{\pi}{\lambda}} \sum_{k=-\infty}^{\infty} e^{-\pi^2 k^2/\lambda} ,$$

and

$$\sum_{k=-\infty}^{\infty} (-1)^k e^{-\lambda k^2} = \sqrt{\frac{\pi}{\lambda}} \sum_{k=-\infty}^{\infty} e^{-\pi^2 (k+\frac{1}{2})^2/\lambda} .$$

Notice now that this model realizes the dual formula in a symmetric way, contrary to the case that there is a mass gap. This dually symmetric mode will be referred to as the "Coulomb phase" or "Georgi-Glashow phase".

Suppose that Quantum Chromodynamics would be enriched with two free parameters that would not destroy the basic topological features (for instance the mass of some heavy scalar fields in the adjoint representation). Then we would have a phase diagram as in the Figure below.

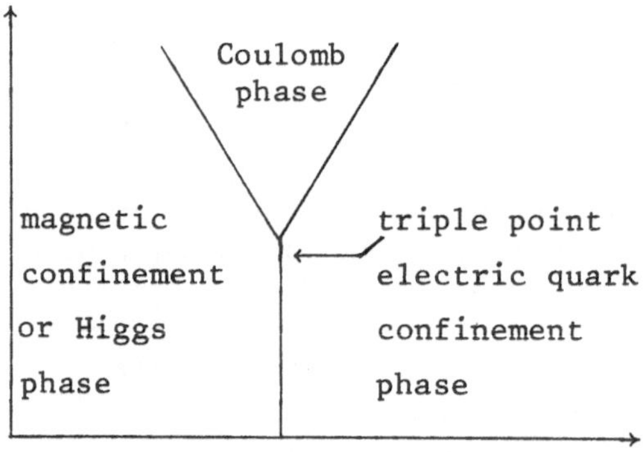

Numerical calculations[9)] suggest that the phase transition between the two confinement modes is a first order one. Real QCD is represented by one point in this diagram. Where will that point be? If it were in the Coulomb phase there would be long range, strongly interacting Abelian gluons contrary to experiment. In the Higgs mode quarks would have finite mass and escape easily. It could be still in the Higgs phase but very close to the border line with the confinement mode. If the phase transition were a second order one then that would imply long range correlation effects requiring light physical gluons. Again: they are not observed experimentally. If, which is more likely, the phase transition is a first order one then even close to the border line not even approximate confinement would take place: quarks would be produced copiously. There is only one possibility: we are in the confinement mode. Electric flux lines cannot spread out. Quark confinement is absolute.

REFERENCES

1. E.S. Abers and B.W. Lee, Physics Reports 9C, no 1.
   S. Coleman in: Laws of Hadronic Matter, Erice July 1973, ed. by A. Zichichi, Academic Press, New York and London.
2. H.B. Nielsen and P. Olesen, Nucl. Phys. B61 (1973) 45, ibidem B160 (1979) 380.
3. B. Zumino, in Renormalization and Invariance in Quantum Field Theory, ed. E.R. Caianiello, Plenum Press, New York, p. 367.
4. G. 't Hooft, Nucl. Physics B153 (1979) 141.
5. G. 't Hooft, Nucl. Physics B138 (1978) 1.
6. E. Witten, Harvard preprint (1979).
7. M. Georgi and S.L. Glashow, Phys. Rev. Lett. 28 (19732) 1494.
8. G. 't Hooft, Nucl. Physics B79 (1974) 276.
   A.M. Polyakov, JETP Lett. 20 (1974) 194.
9. M. Creutz, L. Jacobs and C. Rebbi, Phys. Rev. D20 (1979) 1915.

NATURALNESS, CHIRAL SYMMETRY, AND SPONTANEOUS CHIRAL SYMMETRY BREAKING

G. 't Hooft

Institute for Theoretical Fysics

Utrecht, The Netherlands

ABSTRACT

A properly called "naturalness" is imposed on gauge theories. It is an order-of-magnitude restriction that must hold at all energy scales μ. To construct models with complete naturalness for elementary particles one needs more types of confining gauge theories besides quantum chromodynamics. We propose a search program for models with improved naturalness and concentrate on the possibility that presently elementary fermions can be considered as composite. Chiral symmetry must then be responsible for the masslessness of these fermions. Thus we search for QCD-like models where chiral symmetry is not or only partly broken spontaneously. They are restricted by index relations that often cannot be satisfied by other than unphysical fractional indices. This difficulty made the author's own search unsuccessful so far. As a by-product we find yet another reason why in ordinary QCD chiral symmetry must be broken spontaneously.

III1. INTRODUCTION

The concept of causality requires that macroscopic phenomena follow from microscopic equations. Thus the properties of liquids and solids follow from the microscopic properties of molecules and atoms. One may either consider these microscopic properties to have been chosen at random by Nature, or attempt to deduce these from even more fundamental equations at still smaller length and time scales. In either case, it is unlikely that the microscopic equations contain various free parameters that are carefully adjusted by Nature to give cancelling effects such that the macroscopic systems have some special properties. This is a philosophy which we would like to apply to the unified gauge theories: the effective interactions at a large length scale, corresponding to a low energy scale μ_1, should follow from the properties at a much smaller length scale, or higher energy scale μ_2, without the requirement that various different parameters at the energy scale μ_2 match with an accuracy of the order of μ_1/μ_2. That would be unnatural. On the other hand, if at the energy scale μ_2 some parameters would be very small, say

$$\alpha(\mu_2) = \mathcal{O}(\mu_1/\mu_2) , \qquad \text{(III1)}$$

then this may still be natural, provided that this property would not be spoilt by any higher order effects. We now conjecture that the following dogma should be followed:
- at any energy scale μ, a physical parameter or set of physical parameters $\alpha_i(\mu)$ is allowed to be very small only if the replacement $\alpha_i(\mu) = 0$ would increase the symmetry of the system. -
In what follows this is what we mean by naturalness. It is clearly a weaker requirement than that of P. Dirac[1)] who insists on having no small numbers at all. It is what one expects if at any mass scale $\mu > \mu_0$ some ununderstood theory with strong interactions determines a spectrum of particles with various good or bad symmetry properties. If at $\mu = \mu_0$ certain parameters come out to be small, say 10^{-5}, then that cannot be an accident; it must be the consequence of a near symmetry.

For instance, at a mass scale

$$\mu = 50 \text{ GeV},$$

the electron mass m_e is 10^{-5}. This is a small parameter. It is acceptable because $m_e = 0$ would imply an additional chiral symmetry corresponding to separate conservation of left handed and right handed electron-like leptons. This guarantees that all renormalizations of m_e are proportional to m_e itself. In sects. III2 and III3 we compare naturalness for quantum electrodynamics and ϕ^4 theory.

Gauge coupling constants and other (sets of) interaction constants may be small because putting them equal to zero would turn the gauge bosons or other particles into free particles so that they are separately conserved.

If within a set of small parameters one is several orders of magnitude smaller than another then the smallest must satisfy our "dogma" separately. As we will see, naturalness will put the severest restriction on the occurrence of scalar particles in renormalizable theories. In fact we conjecture that this is the reason why light, weakly interacting scalar particles are not seen.

It is our aim to use naturalness as a new guideline to construct models of elementary particles (sect. III4). In practice naturalness will be lost beyond a certain mass scale μ_0, to be referred to as "Naturalness Breakdown Mass Scale" (NBMS). This simply means that unknown particles with masses beyond that scale are ignored in our model. The NBMS is only defined as an order of magnitude and can be obtained for each renormalizable field theory. For present "unified theories", including the existing grand unified schemes, it is only about 1000 GeV. In sect. 5 we attempt to construct realistic models with an NBMS some orders of magnitude higher.

One parameter in our world is unnatural, according to our definition, already at a very low mass scale ($\mu_0 \sim 10^{-2}$ eV). This is the cosmological constant. Putting it equal to zero does not seem to increase the symmetry. Apparently gravitational effects do not obey naturalness in our formulation. We have nothing to say about this fundamental problem, accept to suggest that *only* gravitational effects violate naturalness. Quantum gravity is not understood anyhow so we exclude it from our naturalness requirements.

On the other hand it is quite remarkable that all other elementary particle interactions have a high degree of naturalness. No unnatural parameters occur in that energy range where our popular field theories could be checked experimentally. We consider this as important evidence in favor of the general hypothesis of naturalness. Pursuing naturalness beyond 1000 GeV will require theories that are immensely complex compared with some of the grand unified schemes.

A remarkable attempt towards a natural theory was made by Dimopoulos and Susskind [2)]. These authors employ various kinds of confining gauge forces to obtain scalar bound states which may substitute the Higgs fields in the conventional schemes. In their model the observed fermions are still considered to be elementary.

Most likely a complete model of this kind has to be constructed step by step. One starts with the experimentally accessible aspects of the Glashow-Weinberg-Salam-Ward model. This model is natural if one restricts oneself to mass-energy scales below 1000 GeV. Beyond 1000 GeV one has to assume, as Dimopoulos and Susskind do, that the Higgs field is actually a fermion-antifermion composite field. Coupling this field to quarks and leptons in order to produce their mass, requires new scalar fields that cause naturalness to break down at 30 TeV or so. Dimopoulos and Susskind speculate further on how to remedy this. To supplement such ideas, we toyed with the idea that (some of) the presently "elementary" fermions may turn out to be bound states of an odd number of fermions when considered beyond 30 TeV. The binding mechanism would be similar to the one that keeps quarks inside the proton. However, the proton is not particularly light compared with the characteristic mass scale of quantum chromodynamics (QCD). Clearly our idea is only viable if something prevented our "baryons" from obtaining a mass (eventually a small mass may be due to some secondary perturbation).

The proton ows its mass to spontaneous breakdown of chiral symmetry, or so it seems according to a simple, fairly successful model of the mesonic and baryonic states in QCD: the Gell-Mann-Lévy sigma model[3]. Is it possible then that in some variant of QCD chiral symmetry is not spontaneously broken, or only partly, so that at least some chiral symmetry remains in the spectrum of fermionic bound states? In this article we will see that in general in SU(N) binding theories this is not allowed to happen, i.e. chiral symmetry must be broken spontaneously.

III2. NATURALNESS IN QUANTUM ELECTRODYNAMICS

Quantum Electrodynamics as a renormalizable model of electrons (and muons if desired) and photons is an example of a "natural" field theory. The parameters α, m_e (and m_μ) may be small independently. In particular m_e (and m_μ) are very small at large μ. The relevant symmetry here is chiral symmetry, for the electron and the muon separately. We need not be concerned about the Adler-Bell-Jackiw anomaly here because the photon field being Abelian cannot acquire non-trivial topological winding numbers[4].

There is a value of μ where Quantum Electrodynamics ceases to be useful, even as a model. The model is not asymptotically free, so there is an energy scale where all interactions become strong:

$$\mu_o \simeq m_e \exp(6\pi^2/e^2 N_f) \ , \tag{III2}$$

where N_f is the number of light fermions. If some world would be described by such a theory at low energies, then a replacement of the theory would be necessary at or below energies of order μ_o.

III3. ϕ^4-THEORY

A renormalizable scalar field theory is described by the Lagrangian

$$\mathcal{L} = -\tfrac{1}{2}(\partial_\mu\phi)^2 - \tfrac{1}{2}m^2\phi^2 - \frac{1}{4!}\lambda\phi^4 \ . \tag{III3}$$

the interactions become strong at

$$\mu \simeq m \exp(16\pi^2/3\lambda) \ , \tag{III4}$$

but is it still natural there?

There are two parameters, λ and m. Of these, λ may be small because $\lambda = o$ would correspond to a non-interacting theory with total number of ϕ particles conserved. But is small m allowed? If we put $m = o$ in the Lagrangian (III3) then the symmetry is not enhanced*). However we can take both m and λ to be small, because if $\lambda = m = o$ we have invariance under

$$\phi(x) \to \phi(x) + \Lambda \ . \tag{III5}$$

This would be an approximate symmetry of a new underlying theory at energies of order μ_o. Let the symmetry be broken by effects described by a dimensionless parameter ε. Both the mass term and the interaction term in the effective Lagrangian (III3) result from these symmetry breaking effects. Both are expected to be of order ε. Substituting the correct powers of μ_o to account for the dimensions of these parameters we have

$$\begin{aligned} \lambda &= \mathcal{O}(\varepsilon) \ , \\ m^2 &= \mathcal{O}(\varepsilon\mu_o^2) \ . \end{aligned} \tag{III6}$$

Therefore,

$$\mu_o = \mathcal{O}(m/\sqrt{\lambda}) \ . \tag{III7}$$

This value is much lower than eq. (III4). We now turn the argument around: if any "natural" underlying theory is to describe a scalar particle whose *effective* Lagrangian at low energies will be eq. (III3), then its energy scale cannot be given by (III4) but at best by (III7). We say that naturalness breaks down beyond $m/\sqrt{\lambda}$. It must be stressed that these are orders of magnitude. For instance one might prefer to consider λ/π^2 rather than λ to be the relevant parameter. μ_o then has to be multiplied by π. Furthermore, λ could be much smaller than ε because $\lambda = o$ separately also enhances the symmetry. Therefore, apart from factors π, eq. (III7) indicates a maximum value for μ_o.

Another way of looking at the problem of naturalness is by comparing field theory with statistical physics. The parameter m/μ would correspond to $(T-T_c)/T$ in a statistical ensemble. Why would the temperature T chosen by Nature to describe the elementary particles be so close to a critical temperature T_c? If $T_c \neq o$ then T may not be close to T_c just by accident.

III4. NATURALNESS IN THE WEINBERG-SALAM-GIM MODEL

The difficulties with the unnatural mass parameters only occur in theories with scalar fields. The only fundamental scalar

*) Conformal symmetry is violated at the quantum level.

field that occurs in the presently fashionable models is the Higgs field in the extended Weinberg-Salam model. The Higgs mass-squared, m_H^2, is up to a coefficient a fundamental parameter in the Lagrangian. It is small at energy scales $\mu \gg m_H$. Is there an approximate symmetry if $m_H \to o$? With some stretch of imagination we might consider a Goldstone-type symmetry:

$$\phi(x) \to \phi(x) + \text{const.} \tag{III8}$$

However we also had the local gauge transformations:

$$\phi(x) \to \Omega(x)\ \phi(x)\ . \tag{III9}$$

The transformations (III8) and (III9) only form a closed group if we also have invariance under

$$\phi(x) \to \phi(x) + C(x)\ . \tag{III10}$$

But then it becomes possible to transform ϕ away completely. The Higgs field would then become an unphysical field and that is not what we want. Alternatively, we could have that (III8) is an approximate symmetry only, and it is broken by all interactions that have to do with the symmetry (III9) which are the weak gauge field interactions. Their strength is $g^2/4\pi = \mathcal{O}(1/137)$. So at best we can have that the symmetry is broken by $\mathcal{O}(1/137)$ effects. Therefore

$$m_H^2/\mu^2 \gtrsim \mathcal{O}(1/137)\ .$$

Also the $\lambda\phi^4$ term in the Higgs field interactions breaks this symmetry. Therefore

$$m_H^2/\mu^2 \gtrsim \mathcal{O}(\lambda) \gtrsim \mathcal{O}(1/137)\ . \tag{III11}$$

Now

$$m_H^2 = \mathcal{O}(\lambda F_H^2)\ , \tag{III12}$$

where F_H is the vacuum expectation value of the Higgs field, known to be*)

$$F_H = (2G\sqrt{2})^{-1/2} = 174\ \text{GeV}\ . \tag{III13}$$

We now read off that

$$\mu \lesssim \mathcal{O}(F_H) = \mathcal{O}(174\ \text{GeV})\ . \tag{III14}$$

*) Some numerical values given during the lecture were incorrect. I here give corrected values.

This means that at energy scales much beyond F_H our model becomes more and more unnatural. Actually, factors of π have been omitted. In practice one factor of 5 or 10 is still not totally unacceptable. Notice that the actual value of m_H dropped out, except that

$$m_H = \mathcal{O}\left(\frac{\sqrt{\lambda}}{g} M_W\right) \gtrsim \mathcal{O}(M_W) \quad . \tag{III15}$$

Values for m_H of just a few GeV are unnatural.

III5. EXTENDING NATURALNESS

Equation (III14) tells us that at energy scales much beyond 174 GeV the standard model becomes unnatural. As long as the Higgs field H remains a fundamental scalar nothing much can be done about that. We therefore conclude, with Dimopoulos and Susskind[2)] that the "observed" Higgs field must be composite. A non-trivial strongly interacting field theory must be operative at 1000 GeV or so. An obvious and indeed likely possibility is that the Higgs field H can be written as

$$H = Z\bar{\psi}\psi \quad , \tag{III16}$$

where Z is a renormalization factor and ψ is a new quark-like object, a fermion with a new color-like interaction [2)]. We will refer to the object as meta-quark having meta-color. The theory will have all features of QCD so that we can copy the nomenclature of QCD with the prefix "meta-". The Higgs field is a meta-meson.

It is now tempting to assume that the meta-quarks transform the same way under weak SU(2) x U(1) as ordinary quarks. Take a doublet with left-handed components forming one gauge doublet and right handed components forming two gauge singlets. The meta-quarks are massless. Suppose that the meta-chiral symmetry is broken spontaneously just as in ordinary QCD. What would happen?

What happens is in ordinary QCD well described by the Gell-Mann-Lévy sigma model. The lightest mesons form a quartet of real fields, ϕ_{ij}, transforming as a

$$2^{\text{left}} \otimes 2^{\text{right}}$$

representation of

$$SU(2)^{\text{left}} \otimes SU(2)^{\text{right}} .$$

Since the weak interaction only deals with $SU(2)^{\text{left}}$ this quartet can also be considered as one complex doublet representation of weak SU(2). In ordinary QCD we have

$$\phi_{ij} = \sigma\delta_{ij} + i\tau^a_{ij}\pi^a \quad , \tag{III17}$$

and

$$<\sigma>_{vacuum} = \frac{1}{\sqrt{2}} f_\pi = 91 \text{ MeV} . \tag{III18}$$

The complex doublet is then

$$\phi_i = \frac{1}{\sqrt{2}} \begin{pmatrix} \sigma + i\pi^3 \\ \pi^2 + i\pi^1 \end{pmatrix} , \tag{III19}$$

and

$$<\phi_i>_{vacuum} = \begin{pmatrix} 1 \\ o \end{pmatrix} \times 64 \text{ MeV} . \tag{III20}$$

We conclude that if we transplant this theory to the TeV range then we get a scalar doublet field with a non-vanishing vacuum expectation value for free. All we have to do now is to match the numbers. If we scale all QCD masses by a scaling factor κ then we match

$$F_H = 174 \text{ GeV} = \kappa \; 64 \text{ MeV} ;$$

$$\kappa = 2700 . \tag{III21}$$

Now the mesonic sector of QCD is usually assumed to be reproduced in the 1/N expansion [5] where N is the number of colors (in QCD we have N = 3). The 4-meson coupling constant goes like 1/N. Then one would expect

$$f_\pi \propto \sqrt{N} . \tag{III22}$$

Therefore

$$\kappa = 2700 \sqrt{\frac{3}{N}} , \tag{III23}$$

if the metacolor group is SU(N).

Thus we obtain a model that reproduces the W-mass and predicts the Higgs mass. The Higgs is the meta-sigma particle. The ordinary sigma is a wide resonance at about 700 MeV[3], so that we predict

$$m_H = \kappa m_\sigma = 1900 \sqrt{\frac{3}{N}} \text{ GeV} , \tag{III24}$$

and it will be extremely difficult to detect among other strongly interacting objects.

III6. WHAT NEXT?

The model of the previous section is to our mind nearly inevitable, but there are problems. These have to do with the observed fermion masses. All leptons and quarks owe their masses to an interaction term of the form

$$g\bar{\psi}H\psi \ , \qquad \text{(III25)}$$

where g is a coupling constant, ψ is the lepton or quark and H is the Higgs field. With (III16) this becomes a four-fermion interaction, a fundamental interaction in the new theory. Because it is non-renormalizable further structure is needed. In ref. 2 the obvious choice is made: a new "meta-weak interaction" gauge theory enters with new super-heavy intermediate vector bosons. But since H is a scalar this boson must be in the crossed channel, a rather awkward situation. (See option a in Figure 1.) A simpler theory is that a new scalar particle is exchanged in the direct channel. (See option b in Figure 1.)

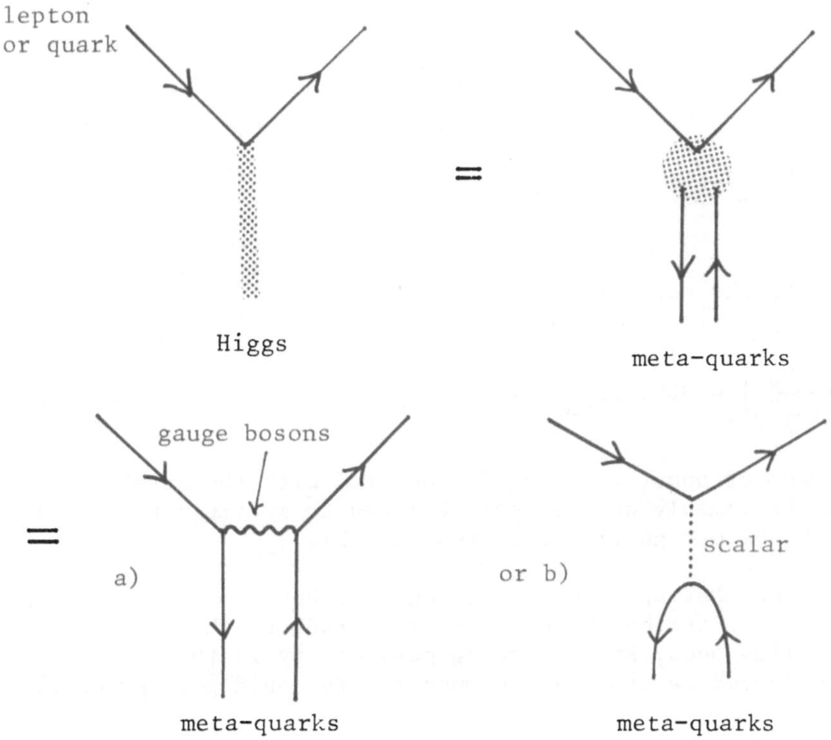

Figure 1.

Notice that in both cases new scalar fields are needed because in case a) something must cause the "spontaneous breakdown" of the new gauge symmetries. Therefore choice b) is simpler.
We removed a Higgs scalar and we get a scalar back. Does naturalness improve? The answer is yes. The coupling constant g in the interaction (III25) satisfies

$$g = g_1 g_2 / M_s^2 Z \ . \tag{III26}$$

Here g_1 and g_2 are the couplings at the new vertices, M_s is the new scalar's mass, and Z is from (III16) and is of order

$$Z \sim \frac{1}{\sqrt{\frac{N}{3}}\,(\kappa\, m_\rho)^2} = \frac{\sqrt{N/3}}{(1800\ \mathrm{GeV})^2} \ . \tag{III27}$$

Suppose that the heaviest lepton or quark is about 10 GeV. For that fermion the coupling constant g is

$$g = \frac{m_f}{F} \simeq 1/20 \ .$$

We get

$$g_1 g_2 \simeq \left(\frac{M_s}{1800\ \mathrm{GeV}}\right)^2 \sqrt{\frac{N}{3}} \cdot \frac{1}{20} \ .$$

Naturalness breaks down at

$$\mu = \mathcal{O}\left(\frac{M_s}{g_{1,2}}\right) = 8000 \ \sqrt[4]{\frac{3}{N}}\ \mathrm{GeV} \ ,$$

an improvement of about a factor 50 compared with the situation in sect. III4. Presumably we are again allowed to multiply by factors like 5 or 10, before getting into real trouble.

Before speculating on how to go on from here to improve naturalness still further we must assure ourselves that all other alleys are blind ones. An intriguing possibility is that the presently observed fermions are composite. We would get option c), Figure 2.

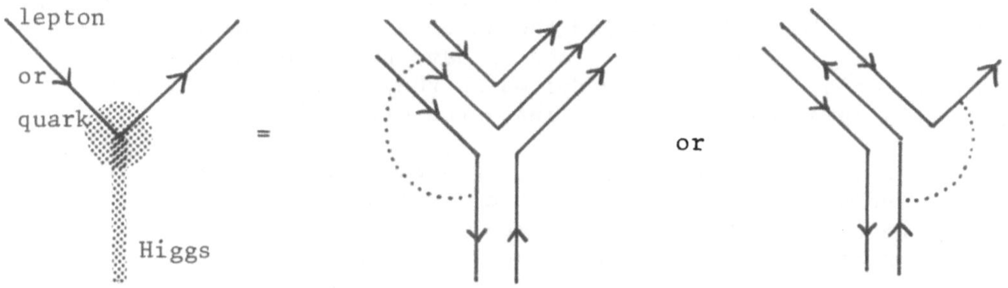

Fig. 2

The dotted line could be an ordinary weak interaction W or photon, that breaks an internal symmetry in the binding force for the new components. The new binding force could either act at the 1 TeV or at the 10-100 TeV range. It could either be an extension of metacolor or be a (color)" or paracolor force. Is such an idea viable?

Clearly, compared with the energy scale on which the binding forces take place, the composite fermions must be nearly massless. Again, this cannot be an accident. The chiral symmetry responsible for this must be present in the underlying theory. Apparently then, the underlying theory will possess a chiral symmetry which is not (or not completely) spontaneously broken, but reflected in the bound state spectrum in the Wigner mode: some massless chiral objects and parity doubled massive fermions. This possibility is most clearly described by the σ-model as a model for the lowest bound states occurring in ordinary quantum chromodynamics.

III7. THE σ MODEL

The fermion system in quantum chromodynamics shows an axial symmetry. To illuminate our problem let us consider the case of two flavors. The local color group is $SU(3)_c$. The subscript c here stands for color. The flavor symmetry group is $SU(2)_L \otimes SU(2)_R \otimes U(1)$ where the subscripts L and R stands for left and right and the group elements must be chosen to be space-time independent. We split the fermion fields ψ into left and right components:

$$\psi = \tfrac{1}{2}(1+\gamma_5)\psi_L + \tfrac{1}{2}(1-\gamma_5)\psi_R \ . \qquad \text{(III28)}$$

ψ_L transforms as a $3_c \otimes 2_L \otimes 1_R \otimes 2_{\mathcal{L}}$ (III29)

and ψ_R transforms as a $3_c \otimes 1_L \otimes 2_R \otimes \bar{2}_{\mathcal{L}}$ (III30)

where the indices refer to the various groups. $\mathcal{L}$ stands for the Lorentzgroup SO(3,1), locally equivalent to SL(2,c) which has two different complex doublet representations $2_{\mathcal{L}}$ and $\bar{2}_{\mathcal{L}}$ (corresponding to the transformation law for the neutrino and antineutrino, respectively). The fields ψ_L and ψ_R have the same charge under U(1), whereas axial U(1) group (under which they would have opposite charges) is absent because of instanton effects[4].

The effect of the color gauge fields is to bind these fermions into mesons and baryons all of which must be color singlets. It would be nice if one could describe these hadronic fields as representations of $SU(2)_L \otimes SU(2)_R \otimes U(1)$ and the Lorentz group, and then cast their mutual interactions in the form of an effective Lagrangian, invariant under the flavor symmetry group. In the case at hand this is possible and the resulting construction is a successful and one-time popular model for pions and nucleons: the σ model[3]. We have a nucleon doublet

$$N = \tfrac{1}{2}(1+\gamma_5)N_L + \tfrac{1}{2}(1-\gamma_5)N_R \ , \tag{III31}$$

where

$$N_L \text{ transforms as a } 1_c \otimes 2_L \otimes 1_R \otimes 2 \ , \tag{III32a}$$

and $$N_R \text{ transforms as a } 1_c \otimes 1_L \otimes 2_R \otimes \bar{2}_{\mathcal{L}} \ . \tag{III32b}$$

Further we have a quartet of real scalar fields $(\sigma,\vec{\pi})$ which transform as a $1_c \otimes 2_L \otimes 2_{\mathcal{L}} \otimes 1_{\mathcal{L}}$. The Lagrangian is

$$\mathcal{L} = -\bar{N}[\gamma\partial+g_o(\sigma+i\vec{\tau}.\vec{\pi}\gamma_5)]N - \tfrac{1}{2}(\partial\pi)^2 - \tfrac{1}{2}(\partial\sigma)^2 - V(\sigma^2+\vec{\pi}^2) \ . \tag{III33}$$

Here V must be a rotationally invariant function.

Usually V is chosen such that its absolute minimum is away from the origin. Let V be minimal at $\sigma = v$ and $\vec{\pi} = o$. Here v is just a c-number. To obtain the physical particle spectrum we write

$$\sigma = v + s \tag{III34}$$

and we find

$$\mathcal{L} = -\tfrac{1}{2}\bar{N}(\gamma\partial+ g_o v)N - \tfrac{1}{2}(\partial\vec{\pi})^2 - \tfrac{1}{2}(\partial s)^2 - 2v^2V''(v^2)s^2 + \text{interaction terms} \ . \tag{III35}$$

Clearly, in this case the nucleons acquire a mass term $m_s = g_o v$ and the s particle has a mass $m_s^2 = 4v^2V''(v^2)$, whereas the pion remains strictly massless. The entire mass of the pion must be due to effects that explicitly break $SU(2)_L \times SU(2)_R$, such as a small mass term $m_q\bar{\psi}\psi$ for the quarks (III28). We say that in this case the flavor group $SU(2)_L \otimes SU(2)_R$ is spontaneously broken into the isospin group SU(2).

Another possibility however, apparently not realised in ordinary quantum chromodynamics, would be that $SU(2)_L \otimes SU(2)_R$ is *not* spontaneously broken. We would read off from the Lagrangian (III33) that the nucleons N would form a massless doublet and that the four fields $(\sigma,\vec{\pi})$ could be heavy. The dynamics of other confining gauge theories could differ sufficiently from ordinary QCD so that, rather than a spontaneous symmetry breakdown, massless "baryons" develop. The principle question we will concentrate on is why do these massless baryons form the representation (III32), and how does this generalize to other systems. We would let future generations worry about the question where exactly the absolute minimum of the effective potential V will appear.

III8. INDICES

We now consider any color group G_c. The fundamental fermions in our system must be non-trivial representation of G_c and we assume "confinement" to occur: all physical particles are bound states that are singlets under G_c. Assume that the fermions are all massless (later mass terms can be considered as a perturbation). We will have automatically some global symmetry which we call the flavor group G_F. (We only consider exact flavor symmetries, not spoilt by instanton effects.) Assume that G_F is not spontaneously broken. Which and how many representations of G_F will occur in the massless fermion spectrum of the baryonic bound states? We must formulate the problem more precisely. The massless nucleons in (III33) being bound states, may have many massive excitations. However, massive Fermion fields cannot transform as a $2_\mathcal{L}$ under Lorentz transformations; they must go as a $2_\mathcal{L} \oplus \bar{2}_\mathcal{L}$. That is because a mass term being a Lorentz invariant product of two fields at one point only links $2_\mathcal{L}$ representations with $\bar{2}_\mathcal{L}$ representations. Consider a given representation r of G_F. Let p be the number of field multiplets transforming as $r \otimes 2_\mathcal{L}$ and q be the number of field multiplets $r \otimes \bar{2}_\mathcal{L}$. Mass terms that link the $2_\mathcal{L}$ with $\bar{2}_\mathcal{L}$ fields are completely invariant and in general to be expected in the effective Lagrangian. But the absolute value of

$$\ell = p - q \tag{III36}$$

is the minimal number of surviving massless chiral field multiplets. We will call ℓ the index corresponding to the representation r of G_F. By definition this index must be a (positive or negative) integer. In the sigma model it is postulated that

$$\text{index } (2_L \otimes 1_R) = 1 \tag{III37}$$

$$\text{index } (1_L \otimes 2_R) = -1$$

index (r) = o for all other representations r.

This tells us that if chiral symmetry is not broken spontaneously one massless nucleon doublet emerges. We wish to find out what massless fermionic bound states will come out in more general theories. Our problem is: how does (III37) generalize?

III9. ABSENCE OF MASSLESS BOUND STATES WITH SPIN 3/2 OR HIGHER

In the foregoing we only considered spin o and spin 1/2 bound states. Is it not possible that fundamentally massless bound states develop with higher spin? I believe to have strong arguments that this is indeed not possible. Let us consider the case of spin 3/2. Massive spin 3/2 fermions are described by a Lagrangian of the form

$$\mathcal{L} = \tfrac{1}{2}\bar{\psi}_\mu [\sigma_{\mu\nu}(\gamma\partial+m) + (\gamma\partial+m)\sigma_{\mu\nu}]\psi_\nu \ . \tag{III38}$$

Just like spin-one particles, this has a gauge-invariance if $m \to o$:

$$\psi_\mu \to \psi_\mu + \partial_\mu \eta(x) \ , \tag{III39}$$

where $\eta(x)$ is arbitrary. Indeed, massless spin 3/2 particles only occur in locally supersymmetric field theories. The field $\eta(x)$ is fundamentally unobservable.

Now in our model ψ_μ would be shorthand for some composite field: $\psi_\mu \to \psi\psi\psi$. However, then all components of this, including η, would be observables. If m = o we would be forced to add a gauge fixing term that would turn η into an unacceptable ghost particle*).

We believe, therefore, that unitarity and locality forbid the occurrence of massless bound states with spin 3/2. The case for higher spin will not be any better. And so we concentrate on a bound state spectrum of spin 1/2 particles only.

*) Note added: during the lectures it was suggested by one attendant to consider only gauge-invariant fields as $\Psi_{\mu\nu} = \partial_\mu\psi_\nu - \partial_\nu\psi_\mu$. However, such fields must satisfy constraints: $\partial_{[\alpha}\Psi_{\mu\nu]} = o$. Composite field will never automatically satisfy such constraints.

III10. SPECTATOR GAUGE FIELDS AND -FERMIONS

So far, our model consisted of a strong interaction color gauge theory with gauge group G_c, coupled to chiral fermions in various representations r of G_c but of course in such a way that the anomalies cancel. The fermions are all massless and form multiplets of a global symmetry group, called G_F. For QCD this would be the flavor group. In the metacolor theory G_F would include all other fermion symmetries besides metacolor.

In order to study the mathematical problem raised above we will add another gauge connection field that turns G_F into a local symmetry group. The associated coupling constants may all be arbitrarily small, so that the dynamics of the strong color gauge interactions is not much affected. In particular the massless bound state spectrum should not change. One may either think of this new gauge field as a completely quantized field or simply as an artificial background field with possibly non-trivial topology. We will study the behavior of our system in the presence of this "spectator gauge field". As stated, its gauge group is G_F.

Note however, that some flavor transformations could be associated with anomalies. There are two types of anomalies:

i) those associated with $G_c \times G_F$, only occurring where the color field has a winding number. Only U(1) invariant subgroups of G_F contribute here. They simply correspond to small explicit violations of the G_F symmetry. From now on we will take as G_F only the anomaly-free part. Thus, for QCD with N flavors, G_F is not $U(N) \times U(N)$ but

$$G_F = SU(N) \otimes SU(N) \otimes U(1) \ .$$

ii) those associated with G_F alone. They only occur if the spectator gauge field is quantized. To remedy these we simply add "spectator fermions" coupled to G_F alone. Again, since these interactions are weak they should not influence the bound state spectrum.

Here, the spectator gauge fields and fermions are introduced as mathematical tools only. It just happens to be that they really do occur in Nature, for instance the weak and electromagnetic $SU(2) \times U(1)$ gauge fields coupled to quarks in QCD. The leptons then play the role of spectator fermions.

III11. ANOMALY CANCELLATION FOR THE BOUND STATE SPECTRUM

Let us now resume the particle content of our theory. At small distances we have a gauge group $G_c \otimes G_F$ with chiral fermions in several representations of this group. Those fermions which are trivial under G_C are only coupled weakly and are called "spectator fermions". All anomalies cancel, by construction.

At low energies, much lower than the mass scale where color binding occurs, we see only the G_F gauge group with its gauge fields. Coupled to these gauge fields are the massless bound states, forming new representations r of G_F, with either left- or right handed chirality. The numbers of left minus right handed fermion fields in the representations r are given by the as yet unknown indices $\ell(r)$. And finally we have the spectator fermions which are unchanged.

We now expect these very light objects to be described by a new local field theory, that is, a theory local with respect to the large distance scale that we now use. The central theme of our reasoning is now that this new theory must again be anomaly free. We simply cannot allow the contradictions that would arise if this were not so. Nature must arrange its new particle spectrum in such a way that unitarity is obeyed, and because of the large distance scale used the effective interactions are either vanishingly small or renormalizable. The requirement of anomaly cancellation in the new particle spectrum gives us equations for the indices $\ell(r)$, as we will see.

The reason why these equations are sometimes difficult or impossible to solve is that the new representations r must be different from the old ones; if $G_C = SU(N)$ then r must also be faithful representations of $G_F/Z(N)$. For instance in QCD we only allow for octet or decuplet representations of $(SU(3))_{flavor}$, whereas the original quarks were triplets.

However, the anomaly cancellation requirement, restrictive as it may be, does not fix the values of $\ell(r)$ completely. We must look for additional limitations.

III12 APPELQUIST-CARAZZONE DECOUPLING AND N-INDEPENDENCE

A further limitation is found by the following argument. Suppose we add a mass term for one of the colored fermions.

$$\Delta\mathcal{L} = m\,\bar{\psi}_{1L}\,\psi_{1R} + \text{h.c.}$$

Clearly this links one of the left handed fermions with one of the right handed ones and thus reduces the flavor group G_F into $G_F' \subset G_F$. Now let us gradually vary m from o to infinity. A famous theorem [5] tells us that in the limit $m \to \infty$ all effects due to this massive quark disappear. All bound states containing this quark should also disappear which they can only do by becoming very heavy. And they can only become heavy if they form representations r' of G_F' with total index $\ell'(r') = o$. Each representation r of G_F forms an array of representations r' of G_F'. Therefore

$$\ell'(r') = \sum_{r \text{ with } r' \subset r} \ell(r) \; . \qquad \text{(III40)}$$

Apparently this expression must vanish.

Thus we found another requirement for the indices $\ell(r)$. The indices will be nearly but not quite uniquely determined now. Calculations show that this second requirement makes our indices $\ell(r)$ practically independent of the dimensions n_i of G_F. For instance, if G_c = SU(3) and if we have left- and righthanded quarks forming triplets and sextets then

$$G_F = SU(n_1)_L \otimes SU(n_2)_R \otimes SU(n_3)_L \otimes SU(n_4)_R \otimes U(1)^3 \qquad \text{(III41)}$$

where $n_{1,2}$ refer to the triplets and $n_{3,4}$ to the sextets. G_c is anomaly-free if

$$n_1 - n_2 + 7(n_3 - n_4) = o \; . \qquad \text{(III42)}$$

Here we have three independent numbers n_i.
If we write the representations r as Young tableaus then $\ell(r)$ could still depend explicitly on n_i.

However, suppose that someone would start as approximation of Bethe-Salpeter type to discover the zero mass bound state spectrum. He would study diagrams such as Fig. 3

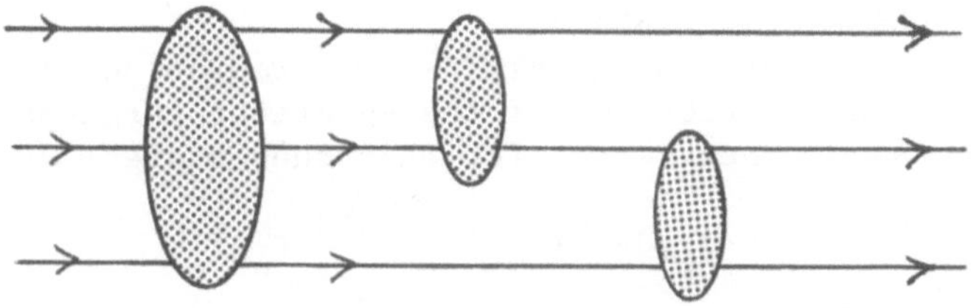

Fig. 3

The resulting indices $\ell(r)$ would follow from topological properties of the interactions represented by the blobs. It is unlikely that this topology would be seriously effected by details such as the contributions of diagrams containing additional closed fermion loops. However, that is the only way in which explicit n-dependence enters. It is therefore natural to assume $\ell(r)$ to be n-independent. This latter assumption fixes $\ell(r)$ completely. What is the result of these calculations?

III13. CALCULATIONS

Let G be any (reducible or irreducible) gauge group. Let chiral fermions in a representation r be coupled to the gauge fields by the covariant derivative

$$D_\mu = \partial_\mu + i\lambda^a(r) A_\mu^a , \tag{III43}$$

where A_μ^a are the gauge fields and $\lambda^a(r)$ a set of matrices depending on the representation r. Let the left-handed fermions be in the representations r_L and the right-handed ones in r_R. Then the anomalies cancel if

$$\sum_L \mathrm{Tr}\{\lambda^a(r_L), \lambda^b(r_L)\} \lambda^c(r_L) = \sum_R \mathrm{Tr}\{\lambda^a(r_R), {}^b(r_R)\} \lambda^c(r_R) . \tag{III44}$$

The object $d^{abc}(r) = \mathrm{Tr}\{\lambda^a(r), \lambda^b(r)\} \lambda^c(r)$ can be computed for any r. In table 1 we give some examples. The fundamental representation r_o is represented by a Young tableau: □ . Let it have n components. We take the case that $\mathrm{Tr}\, \lambda(r_o) = o$. Write

$$\mathrm{Tr}\, I(r_o) = n , \qquad \mathrm{Tr}\, I(r) = N(r) ,$$

$$\mathrm{Tr}\, \lambda(r) = o ,$$

$$\mathrm{Tr}\, \lambda^a(r) \lambda^b(r) = C(r)\, \mathrm{Tr}\, \lambda^a(r_o) \lambda^b(r_o) ,$$

$$d^{abc}(r) = K(r)\, d^{abc}(r_o) . \tag{III45}$$

We read off C and K from table 1.
Now III44 must hold both in the high energy region and in the low energy region. The contribution of the spectator fermions in both regions is the same. Thus we get for the bound states

$$\left(\sum_L - \sum_R\right) d^{abc}(r) = n_c \left(d^{abc}(r_{oL}) - d^{abc}(r_{oR})\right) \tag{III46}$$

where a,b,c are indices of G_F and r_o is the fundamental representation of G_F. We have the factor n_c written explicitly, being the number of color components.

Let us now consider the case $G_C = SU(3)$; $G_F = SU_L(n) \otimes SU_R(n) \otimes U(1)$. We have n "quarks" in the fundamental representations. The representations r of the bound states must be in $G_F/Z(3)$. They are assumed to be built from three quarks, but we are free to choose their chirality. The expected representations

Table 1

r	N(r)	C(r)	K(r)
	n	1	1
	n	1	-1
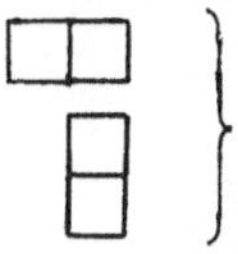	$\frac{n(n\pm1)}{2}$	$n\pm2$	$n\pm4$
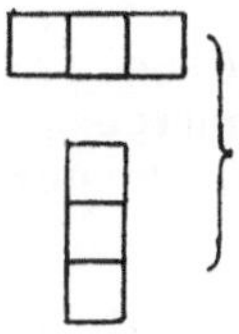	$\frac{n(n\pm1)(n\pm2)}{6}$	$\frac{(n\pm2)(n\pm3)}{2}$	$\frac{(n\pm3)(n\pm6)}{2}$
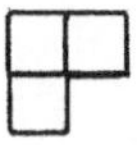	$\frac{n(n^2-1)}{3}$	n^2-3	n^2-9
$A \otimes B$	N(A)N(B)	C(A)N(B) + C(B)N(A)	K(A)N(B) + K(B)N(A)

are given in table 2, where also their indices are defined. Because of left-right symmetry these numbers change sign under interchange of left ↔ right.

Table 2

representation	index	representation	index
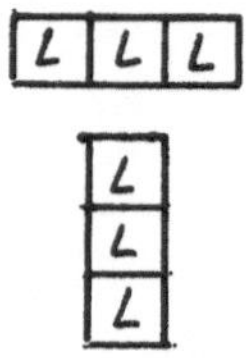	$\ell_{1\pm}$	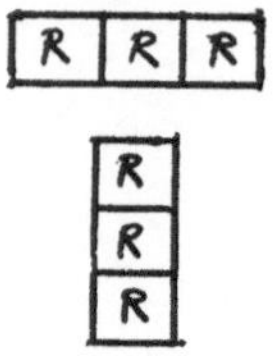	$-\ell_{1\pm}$
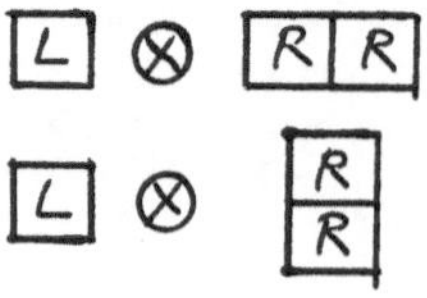	$\ell_{2\pm}$	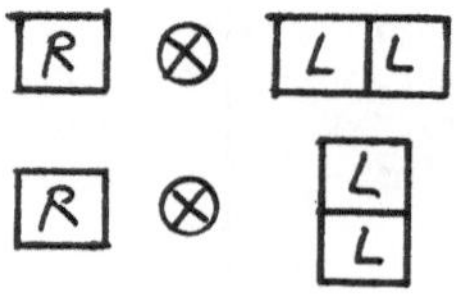	$-\ell_{2\pm}$
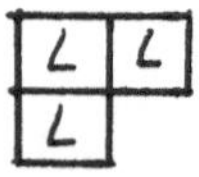	ℓ_3	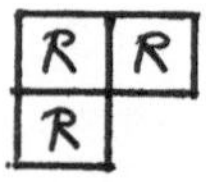	$-\ell_3$

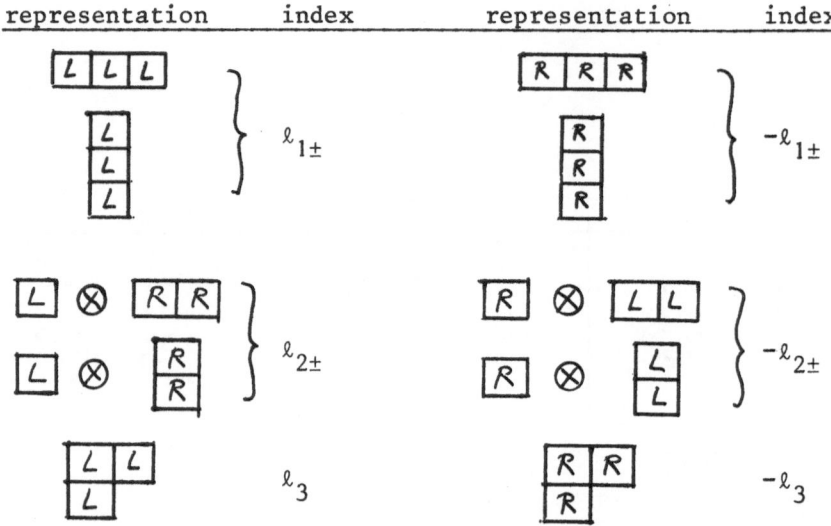

For the time being we assume no other representations. In eq. III46 we may either choose a, b and c all to be $SU(n)_L$ indices, or choose a and b to be $SU(n)_L$ indices and c the U(1) index. We get two independent equations:

$$\sum_\pm \tfrac{1}{2}(n\pm3)(n\pm6)\ell_{1\pm} - \sum_\pm \tfrac{1}{2}n(n\pm7)\ell_{2\pm} + (n^2-9)\ell_3 = 3, \text{ if } n > 2 ,$$

and

$$\sum_\pm \tfrac{1}{2}(n\pm2)(n\pm3)\ell_{1\pm} - \sum_\pm \tfrac{1}{2}n(n\pm3)\ell_{2\pm} + (n^2-3)\ell_3 = 1, \text{ if } n > 1 . \tag{III47}$$

The Appelquist-Carazzone decoupling requirement, eq. (III40), gives us in addition two other equations:

$$\ell_{1+} - \ell_{2+} + \ell_3 = o ,$$

$$\ell_{1-} - \ell_{2-} + \ell_3 = o , \text{ both if } n > 2 . \tag{III48}$$

For $n > 2$ the general solution is

$$\ell_{1+} = \ell_{1-} = \ell ,$$
$$\ell_{2+} = \ell_{2-} = 3\ell-\frac{1}{3} ,$$
$$\ell_{3} = 2\ell-\frac{1}{3} . \qquad \text{(III49)}$$

Here ℓ is still arbitrary. Clearly this result is unacceptable. We cannot allow any of the indices ℓ to be non-integer. Only for the case n = 2 (QCD with just two flavors) there is another solution. In that case ℓ_{2-} and ℓ_3 describe the same representation, and ℓ_{1-} an empty representation. We get

$$\ell_{2-} + \ell_3 = k = 1 - 10\ \ell_{1+} + 5\ \ell_{2+} . \qquad \text{(III50)}$$

According to the σ-model, $\ell_{1+} = \ell_{2+} = o$; $k = 1$. The σ-model is therefore a correct solution to our equations.

In the previous section we promised to determine the indices completely. This is done by imposing n-independence for the more general case including also other color representations such as sextets besides triplets. The resulting equations are not very illuminating, with rather ugly coefficients. One finds that in general no solution exists except when one assumes that all mixed representations have vanishing indices. With mixed representations we mean a product of two or more non-trivial representations of two or more non-Abelian invariant subgroups of G_F. If now we assume n-independence this must also hold if the number of sextets is zero. So ℓ_{2+} and ℓ_{2-} must vanish. We get

$$\ell_{1+} = \ell_{1-} = 1/9 ,$$
$$\ell_3 = -1/9 . \qquad \text{(III51)}$$

If all quarks were sextets, not triplets, we would get

$$\ell_{1+} = \ell_{1-} = 2/9 ,$$
$$\ell_3 = -2/9 . \qquad \text{(III52)}$$

In the case G_C = SU(5) the indices were also found. See table 3.

Table 3
indices for $G_c = SU(5)$

$\ell_{1+} = 1/25$

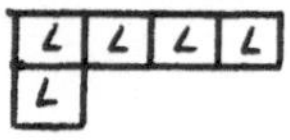
$\ell_{2+} = -1/25$

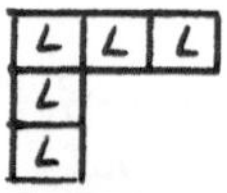
$\ell_3 = 1/25$

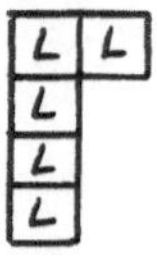
$\ell_{2-} = -1/25$

$\ell_{1-} = 1/25$

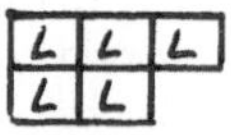
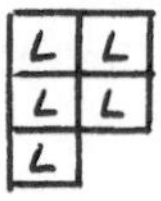
$\ell_{4\pm} = o$

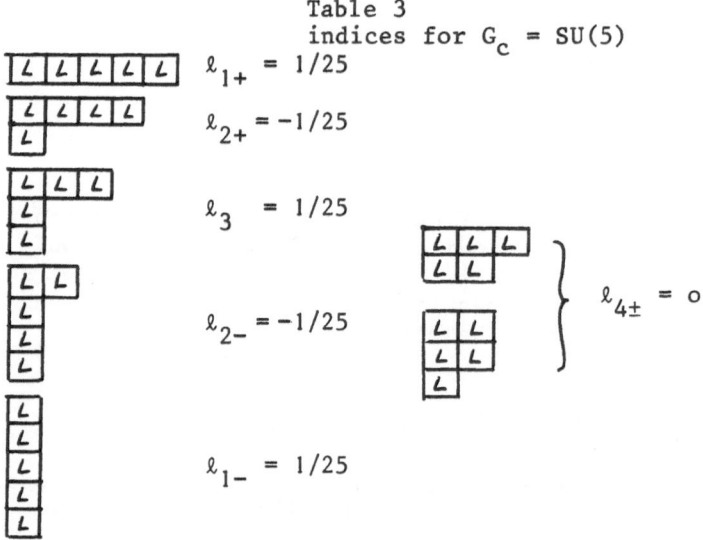

This clearly suggests a general tendency for SU(N) color groups to produce indices $\pm 1/N^2$ or o.

III14. CONCLUSIONS

Our result that the indices we searched for are fractional is clearly absurd. We nevertheless pursued this calculation in order to exhibit the general philosophy of this approach and to find out what a possible cure might be. Our starting point was that chiral symmetry is not broken spontaneously. Most likely this is untenable, as several authors have argued[6]. We find that explicit chiral symmetry in QCD leads to trouble in particular if the number of flavors is more than two. A daring conjecture is then that in QCD the strange quark, being rather light, is responsible for the spontaneous breakdown of chiral symmetry.
An interesting possibility is that in some generalized versions of QCD chiral symmetry is broken only partly, leaving a few massless chiral bound states. Indeed there are examples of models where our philosophy would then give integer indices, but since we must drop the requirement of n-dependence our result was not unique and it was always ugly. No such model seems to reproduce anything resembling the observed quark-lepton spectrum.

Finally there is the remote possibility that the paradoxes associated with higher spin massless bound states can be resolved. Perhaps the $\Delta(1236)$ plays a more subtle role in the σ-model than assumed so far (we took it to be a parity doublet).

We conclude that we are unable to construct a bound state theory for the presently fundamental fermions along the lines suggested above.

We thank R. van Damme for a calculation yielding the indices in the case $G_c = SU(5)$.

REFERENCES

1. P.A.M. Dirac, Nature 139 (1937) 323, Proc. Roy. Soc. A165 (1938) 199, and in: Current Trends in the Theory of Fields, (Tallahassee 1978) AIP Conf. Proc. No 48, Particles and Fields Subseries No 15, ed. by Lannuti and Williams, p. 169.
2. S. Dimopoulos and L. Susskind, Nucl. Phys. B155 (1979) 237.
3. M. Gell-Mann and M. Lévy, Nuovo Cim. 16 (1960) 705.
   B.W. Lee, Chiral Dynamics, Gordon and Breach, New York, London, Paris 1972.
4. G. 't Hooft, Phys. Rev. Lett. 37 (1976) 8; Phys. Rev. D14 (1976) 3432.
   S. Coleman, "The Uses of Instantons", Erice Lectures 1977.
   R. Jackiw and C. Rebbi, Phys. Rev. Lett. 37 (1976) 172.
   C. Callan, R. Dashen and D. Gross, Phys. Lett. 63B (1976) 334.
5. G. 't Hooft, Nucl. Phys. B72 (1974) 461.
6. T. Appelquist and J. Carazzone, Phys. Rev. D11 (1975) 2856.
7. A. Casher, Chiral Symmetry Breaking in Quark Confining Theories, Tel Aviv preprint TAUP 734/79 (1979).

INTRODUCTION TO LATTICE GAUGE THEORIES

C. Itzykson

Dph-T, CEN Saclay, BP n° 2 91190 Gif-sur-Yvette

1. INTRODUCTION

Since 1974 when K. Wilson[1] proposed to study the large coupling behaviour of a Yang-Mills theory using a lattice to provide an ultraviolet cut-off, the literature on the subject has developped very fast. It is not however possible to claim at this moment that realistic answers have been provided to the central problem of quark confinement in quantum chromodynamics, including particle spectrum, partial conservation of the axial current... As it sometimes occur, certain aspects of questions have led to unexpected developments. The most striking is a closer relation between quantum field theory and statistical mechanics, including the behavior of disordered systems, the concept of frustration, dual transformations to describe new types of excitations... Numerical studies have been reported or are in progress. Together with various rigorous results they enable us to get a preliminary view of the phase diagram, particularly of the existence of transitions.

Lattice field theory allows an investigation of a large range of coupling constants. One can locate critical points where hopefully some contact with continuous field theory can be achieved. The methods of the renormalization group come then to bear, and permit to study the behavior of the physical quantities as the lattice spacing shrinks to zero and is replaced by a measurable scale, correlation length or inverse physical mass. This program has been carried with great succes in the study of statistical models with global symmetrics and has led to remarkable agreement with experimental data.

The existence of a local invariance in the gauge models creates a specific difficulty. Indeed any transition will lead a system to a new phase (or set of phases) which cannot be characterized by the non vanishing of some local order parameter analogous to the magnetization in ferromagnetic systems(Elitzur theorem)[2]. Hence we have no a priori candidate for a Landau (or mean-field) theory to describe the gross features of the system. Said otherwise it is not possible to give a simple characterization of the order which might appear in such systems,even though a criterion such as the Wilson non-local loop average can be used to distinguish the various phases. This is analogous to the situation prevailing in amorphous media as examplified spin glass models. Here again the distinction between paramagnetic (high temperature, disordered) phase and the low temperature "ordered" or spin-glass phase is not easily described[3]. This suggests to use as substitute of mean field theory a study of models with large coordination number. Some attempts of this type will be briefly described.

Not only do we have a poor understanding of the distinction between high coupling and low coupling phases by even the nature of the transition according to space time dimension is not yet fully elucidated. Recent Monte Carlo simulations by a Brookhaven group[4] tend to confirm earlier suggestions[5]of a first order transition for high enough dimension and the value of zero coupling as the unique transition point for non abelian gauge theories in space-time dimension four (lower critical dimension) in agreement with the indications of continuous perturbation theory (asymptotic freedom).

These lectures are meant only as an introduction to more advanced topics to be treated by other lecturers. Hence we shall not enter in deep proofs nor very elaborate discussions but only prepare the way in presenting the vocabulary and some simple results.

Topics to be covered if time permits include

- Lattice quantization
- Local invariance
- Relation between Lagrangian and Hamiltonian approach
- Perturbation theory for large coupling
- Duality transformations for abelian systems
- Phase transitions
- Peierls' inquality and large coordination number models.

2. LATTICE QUANTIZATION

A quantum field theory is characterized by the set of vacuum expectation values of observables. The latter can be obtained using the (a priori ill-defined) methods of paths integrals. After a Wick rotation to Euclidean space-time

$$< O_1 \ldots O_N > = Z^{-1} \int \mathcal{D}\varphi \, e^{-S(\varphi)} \, O_1 \ldots O_N \tag{1}$$

with φ a generic name for the (classical) field, $O_1 \ldots O_N$ functionals of φ and Z^{-1} a (formal) normalizing factor such that

$$< 1 > = 1 \qquad Z = \int \mathcal{D}\varphi \, e^{-S(\varphi)} \tag{2}$$

we have used units such that $c = \hbar = 1$.

The action $S(\varphi)$ in the exponential is the integral over a Lagrangian density. In the case of a self-interacting scalar field it is the sum of a kinetic and potential term involving a coupling constant

$$S(\varphi) = \int d^d x \left\{ \frac{1}{2}(\partial\varphi)^2 + \frac{1}{g^2} V(g\varphi) \right\} \tag{3}$$

We assume that $V(0)$ together with $\frac{dV}{d\varphi}(0)$ vanish. To give a meaning to such infinite integrals several methods are used. The familiar small g perturbation expansion is based on the possibility of computing infinite dimensional Gaussian integrals (free field theory). The use of Wick's theorem leads to the Feynman diagrammatical rules, supplemented by additional prescriptions (in large enough dimensions) to remove the ultraviolet cut-off necessary to keep the short distance singularities under control. In spite of the successes of this approach in Quantum-Electrodynamics and recent advances to control the behavior of such series in the coupling constant for large orders, this method is generally of little use in particular when the particle (or excitation) spectrum of the model is strongly altered by the anharmonic terms in the interaction.

An alternative method (in fact very close in spirit to the original derivation of path integral by Feynman using time-slices) is to consider first the degrees of freedoms pertaining to a discrete space-time lattice. This is the analogous step that one performs in deriving Riemanian integrals as limits of (finite) discrete sums. By introducing first a volume cut-off (V) one deals therefore with a genuine multiple integral. We shall use for simplicity an hypercubical lattice with lattice spacing a, and ignore for the time being boundary conditions (by assuming for instance periodic ones).

The treatment of the kinetic term which couples nearby sites may involve some care. In the scalar case we may simply use

$$\frac{\partial}{\partial x^\mu} \varphi \rightarrow \frac{1}{a} [\varphi(x+n_\mu a) - \varphi(x)] \tag{4}$$

where n_μ is a unit vector along one of the lattice axis. Thus we find (φ_i^μ denotes the field at point x_i)

$$Z \to Z_{a,V} = \int \prod_i d\varphi_i \; e^{-\sum_i \left\{ a^{d-2} \sum_{\mu=1}^{d} (\varphi(x_i+n_\mu a) - \varphi(x_i))^2 + \frac{a^d}{g^2} V(g\varphi_i) \right\}} \tag{5}$$

which exhibits a natural correspondence with a classical statistical lattice model. This is even more striking if we rescale φ into φ/g. Then Z appears as a partition function, i.e. a sum over field configurations of Boltzman factors. The temperature is proportional to g^2 and the energy of a configuration to the Euclidean action $g^2 S(\varphi/g)$. Of course Euclîdean space-time should then be thought as physical space with $d \leqslant 3$ while particle physicists are interested in the case where $d = 4$. This justifies a double language where one mixes indiferently the concepts borrowed from the two domains. For instance we shall call "free energy" the limit $\frac{1}{V} \ell n\, Z_{a,V}$ which differs by a simple factor $-1/g^2$ from the similar quantity in thermodynamics. Provided $V(\varphi)$ makes sense (i.e. $V(\varphi) \to +\infty$ for $|\varphi| \to \infty$) the thermodynamic limit $V \to \infty$ exists, for appropriate boundary conditions and defines for g large enough a unique (vacuum) state. Precisely if some transition occurs, (in our simple case when $V(\varphi) = V(-\varphi)$ has two degenerate minima, and $d \geqslant 2$) then boundary conditions matter when g is small enough and define the various equivalent vacuum states (here two states corresponding to the spontaneous breaking of the symmetry $\varphi \to -\varphi$).

We should stress the fundamental interchange in the role of kinetic and potential term in going from the field theory action to the statistical model energy. The potential term now appears as a simple weight on each individual variable φ_i which defines its effective range. For instance $e^{-V(\varphi)}$ could be strongly peaked for $\varphi = \pm \varphi_0$ (i.e. $V(\varphi)$ could have two sharp minima at these points) in such a way that we recover an Ising model. The kinetic term on the other hand couples nearby degrees of freedom in such a way that from the statistical point of view it is the interaction responsible for the collective behavior of the system.

The introduction of a lattice naturally breaks Euclidean invariance, in particular we only have a discrete translational invariance. As a consequence Fourier analysis restricts the range of momentum within a Brillouin zone, $|k_\mu| \leqslant \frac{\pi}{a}$ for instance

$$x_\mu = a r_\mu \qquad r_\mu \text{ integer}$$

$$\varphi(x) = \int_{-\frac{\pi}{a}}^{+\frac{\pi}{a}} \prod_1^d \left(\frac{dk_\mu}{2\pi}\right) e^{i\sum_1^d k_\mu x_\mu} \tilde{\varphi}(k) \tag{6}$$

$$\varphi(x + an_\mu) \leftrightarrow e^{ika_\mu} \tilde{\varphi}(k) \ .$$

For a free scalar field the lattice action will then be equal to

$$\frac{1}{2} \int_{|k_\mu| < \frac{\pi}{a}} d^d k \quad \tilde{\varphi}^+(k) \left\{ \sum_{\mu=1}^{d} \frac{4}{a^2} \sin^2 ak_\mu + m^2 \right\} \tilde{\varphi}(k) \ . \tag{7}$$

Each k mode will contribute a term which will reduce in the long wave length limit to the familiar rotationally invariant form k^2+m^2 and admits a unique minimum at $k = 0$.

This is not the case if we reproduce naïvely the same steps for fermions. Since this difficulty has some bearing on chiral invariance we include here a brief discussion of this case.

First we remind ourselves the need of auticommuting (Grassmann) variables to define fermionic path integrals. With ψ and $\bar{\psi}$ the conjugate spinors we have a contribution

$$\bar{\psi}(k) \left[i \sum_\mu \gamma_\mu \frac{\sin ak_\mu}{a} + m \right] \psi(k) \ , \tag{8}$$

for each mode in the free fied case. The hermitian Dirac matrices γ_μ fulfill $[\gamma_\mu, \gamma_\nu]_+ = 2\delta_{\mu\nu}$. This assumes of course that we have naïvely replaced derivatives by the corresponding discrete difference operators. The corresponding propagators will have as dominator the c- number quantity $a^{-2} \sum_\mu \sin^2 ak_\mu + m^2$ wich reduces again to $k^2 + m^2$ for small k but admits a secondary minimum at the boundary of the Brillouin zone (Fig.1). These modes will carry infinite momentum in the continous limit but might nevertheless be excited for finite a no matter how small. Something is needed in order to avoid this unpleasant phenomenon. Wilson suggests to add a k-dependent

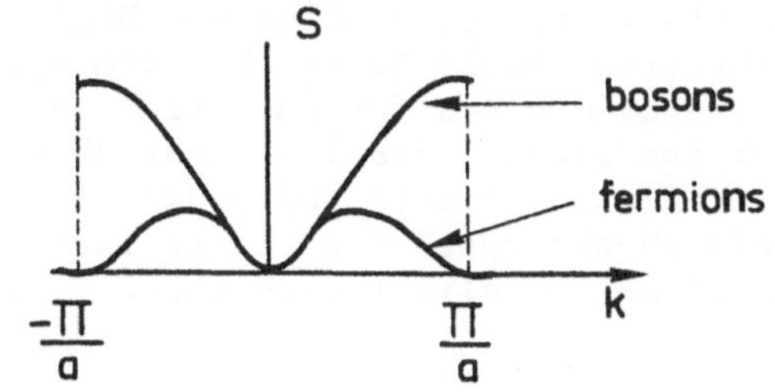

Figure 1.

Free Field Lattice Action for Bosons and Fermions.

term proportional to the unit matrix to lift the secondary minimum at the Brillouin zone boundary. Susskind and his collaborators distribute the components of the spinors on alternate sites to decrease the number of degrees of freedom breaking therefore chiral invariance in the zero mass case. The Stanford school favors the idea of introducing long range "interactions" by modifying the derivative operator in such a way that it remains k_μ in momentum space.

This discussion illustrates the unpleasant fact that a lattice theory admits a large arbitrariness which has to be fixed partly by symmetry arguments. One may however hope that the ideas of university classes will describe the limiting continuous theories in the neighborhood of transition points. In other words that several (most) parameters of the discrete theory, become irrelevant in this domain where critical theories are essentially characterized by geometric concepts like invariance groups and representations.

3. LOCAL INVARIANCE

We are interested in models involving a local invariance under a compact group G (the color symmetry of QCD). We are not concerned with the an approximate flavour symmetry which does not yet seem to play a fundamental role in the standard understanding of the mechanisms of strong interactions. As far as it is known, G should then be identified with the non abelian group SU(3). We shall however keep in mind the possibility of having an arbitrary group G in order to keep the discussion general, to include the possibility of extending these ideas to QED (U(1)) or even the Weinberg Salam unified model of weak interactions (U(1) × SU(2)) to make contact with other domains.

Wilson suggested to implement exactly the Yang Mills idea of local invariance in the context of lattice field theory. Assume that the dynamical "matter" variables at a site i, collectively denoted ϕ_i carry a (unitary) representation $g \to D(g)$ of the group. A typical quadratic invariant would be of the form $\bar{\phi}_i\phi_i$. Polynomials of such quantities are candidates for potential terms in the action and when summed over the sites i are invariant under group transformations which are allowed to vary independently from point to point. More precisely in the case of a finite lattice with V points the gauge group apears as the direct product of V copies of G(G, G,...,G).

However kinetic terms couple ϕ's at nearby points and are only invariant under global transformations unless one introduces a connection in the form of additional degrees of freedom $g_{ij} = g_{ji}^{-1}$ attached to the links of the lattice, and taking values in G(a compact manifold) in such a way that

$$\phi_i^+ \phi_j \rightarrow \phi_i^+ D(g_{ij}) \phi_j \tag{1}$$

Gauge invariance is now implemented if

$$\phi_i \rightarrow D(g_i) \phi_i \qquad g_{ij} \rightarrow g_{ij} g_{ij} g_j^{-1} \tag{2}$$

The group element g_{ij} is the lattice analog of the path dependant phase factor in the case of a continous Lie group

$$g_{ij} \sim T \exp i \int_{x_i}^{x_j} dx^\mu A_\mu(x) \tag{3}$$

Here $A_\mu(x)$ is the Yang Mills vector potential (with value in the Lie algebra of the group) and T denotes an ordering of the exponential along a path going from x_i to x_j (typically a straight path). The prescription implied by eq.(1) (minimal coupling) is analogous to the introduction of a covariant derivative in the continum case. If G is a Lie group,for slowly varying configurations of the fields, a difference of fields at nearby points will now be replaced by

$$\frac{\phi_i - \phi_j}{a} \rightarrow \frac{D(g_{ij}) \phi_i - \phi_i}{a} \sim \frac{\phi_i - \phi_j}{a} + i A_{ij} \phi_j$$

$$\sim \frac{1}{a}(x_i - x_j)^\mu D_\mu \phi(x_j), \tag{4}$$

with, $A_{ij} = A_\mu \dfrac{(x_i - x_j)^\mu}{a}$ and $D_\mu = \partial_\mu + i A_\mu$ the covariant derivative.

The dynamics of the gauge fields is then prescribed by adding to the action a (discrete) translation invariant term involving g_{ij} and invariant under local transformations(2). The simplest choice is obtained as follows. Let χ denote a real character of the group (i.e. a function $g \rightarrow \chi(g)$ with the property that $\chi(g_1 g_2 g_1^{-1}) = \chi(g_2)$ $\chi(g) = \chi(g^{-1}) = \chi(g)$). Then for any closed path C on the lattice going through the points $x_1, x_2, \ldots x_n$

$$\chi_C = \chi(g_{12}\, g_{23} \cdots g_{n1}) \tag{5}$$

is (i) invariant under (2) (ii) independent of the starting point on the path (iii) independent of the orientation of C. On a cubic lattice the smallest path circles around an elementary square (a plaquette) and therefore up to a coefficient the added piece to the action will be proportional to the sum over all plaquettes p

$$\Sigma\ \chi_p(g_p) \qquad g_p = g_{12}\, g_{23}\, g_{34}\, g_{41}\ . \tag{6}$$

Any continuous class function on the group (a fortiori square integrable) is a sum of elementary characters, i.e. traces in irreducible representations. One generally selects one of them, or if the representation is complex its real part. Eventhough it seems superficially that such a choice is irrelevant in the continous limit (see below) this might not be the case in particular if the group is not simple in the strict case i.e. admits a non trivial (abelian) center.

For a smooth configuration of gauge fields i.e. g_{ij} close to the identity, $g_{ij} = T \exp i\int_{x_i}^{x_j} A^\mu(x)\, dx_\mu$ one finds for a vanishing spacing

$$\Sigma_p\ \chi(g_p) \sim C_1 V - C_2\ a^{4-d} \int d^d x\ \mathrm{Tr}\ F^2 + \ldots \tag{7}$$

where C_1 and C_2 are finite constants depending on the character, the trace on the r.h.s. is taken in the Lie algebra and F is the usual fields strength

$$F_{\mu\nu} = \partial_\mu A_\nu - \partial_\nu A_\mu + i[A_\mu, A_\nu]\ . \tag{7'}$$

The omitted terms are of higher oder in the lattice spacing a. We therefore recognize the Yang Mills Lagrangian which arises very naturally from the discrete theory. A coupled theory involving (bosonic) matter fields ϕ will therefore characterized by a partition function

$$Z = \int \prod_i d\phi_i\ e^{-\sum_i V(\phi_i^+ \phi_i)} \prod_{(ij)} dg_{ij}\ \exp \beta_\ell \sum_\ell \phi_i^+ D(g_{ij})\phi_j + \beta_p \sum_p \chi(g_p) . \tag{8}$$

The argument of the exponential is, up to a constant and a sign, what is to be thought of as the lattice action, dg is the (normalized) invariant measure on the group and β_p is proportional to a^{d-4}/g^2 where g is the Yang Mills coupling constant. In (8) we did not consider the coupling to fermions which should of course be included in a realistic model for QCD.

The above discussion of the limit of zero spacing for a continuous group is of course much too naive. It clearly assumes that only smooth configurations close to the identity do contribute to the integral (8) which obviously can only occur as $\beta_p \to \infty$ or $g \to 0$ if $d = 4$ (the case of interest). Would there occur a transition at finite g the argument would have to be reconsidered. It ignores of course the possibility of short distance singularities and accompanying renormalization.

A distinctive feature of the lattice approach is that it uses gauge variables taking their values in the group i.e. on a compact manifold, instead of a linear tangent space, the Lie algebra, as in the continuous case. Consequently the necessity of gauge fixing is eliminated. Indeed the over abundant degrees of freedom corresponding to the transformations (2) are simply integrated over in the partition function and with our choice of normalization contribute even a factor unity to Z and not at all to expectation values of gauge invariant quantities. It is clear that all gauge fixing are artefacts even in the continuous theory only designed to enable us to perform calculations using known local procedures.

For simplicity we shall concentrate in the sequel on the pure gauge part of the action eventhough it is highly irrealistic, i.e. we shall set $\beta_\ell = 0$ in (8) and omit the index p on β_p.

An important criterion devised by Wilson to study the behavior of the theory is the loop average

$$W_C^{\chi'} = \langle \chi'(g_C) \rangle = Z^{-1} \int \prod_{(ij)} dg_{ij}\; e^{\beta_p \sum_p \chi_p(g_p)} \chi'(g_C), \qquad (9)$$

where C denotes any closed curve, g_C is the ordered product of g_{ij}'s along C and χ' a (real) character, possibly different from χ. Such a quantity would be related to the physical interaction energy of two static sources (corresponding to colour degrees of freedom pertaning to the representation which yields χ' as a character) in the case where C is a rectangle of sides T and L say. Then

$$\mathcal{E}(L) - \mathcal{E}_{\text{vacuum}} = -\lim_{T\to\infty} \frac{1}{T}(\ln W_C) \ . \qquad (10)$$

When $\beta \to 0$ and all plaquette variables become independant one expects $W_C \sim \exp -\rho TL$ and therefore $\mathcal{E}(L)$ to grow linearly with L. This is the confining property. On the other hand for large β(or $g \to 0$) an argument based on a continuous approximation yields $W_C \sim e^{-(aT+bL)}$ and therefore (L) is expected to go to a constant for large L. A thorough analysis has of course to be performed in order to give a firm basis to these statements which suggest the possibility of transitions in gauge systems. One should be warned that the relative properties of χ and χ' might play a role in this discussion.

An important remark made by Elitzar is the impossibility of spontaneous breaking of gauge symmetry even when the system undergoes a transition. To be simple we choose the symmetry group Z_2 (the multiplicative group with two elements ± 1). Let us compare the Ising model with its local counterpart the Z_2 gauge model. In dimension two or greater for the Ising model we can proceed in two equivalent ways. In the first we select a finite lattice impose that all spins be aligned at the boundary i.e. all equal to +1 or -1. Then we watch the behavior of $\langle\sigma_0\rangle$ (where 0 denotes a fixed site) as we increase the size of the lattice. If the temperature is above the critical one ($\beta < \beta_C$) then $\langle\sigma_0\rangle \to 0$. If on the contrary

$\beta > \beta_C$ then $<\sigma_0>$ goes to a finite value (the spontaneous magnetization) of the same sign as the one imposed on the boundary. Hence below $T_C(\beta_C^{-1})$ we have spontaneous symmetry breaking and equivalent phases (here two) characterized by a direction in group space (or rather in the representation space).

One can even be more drastic than selecting given invariance breaking boundary conditions. One can introduce a volume symmetry breaking term in the form of an external magnetic field H (i.e. add a term $H \sum_i \sigma_i$ in the action) and consider the limiting behaviour $\lim_{H\to 0} <\sigma_0>$ below T_C. Again according to the sign of H the two opposite values of the spontaneous magnetization will be reached.

Consider next the Z_2 gauge model with

$$Z = \frac{1}{2^{Nd}} \sum_{\sigma_{ij}=\pm 1} \exp \beta \sum_p \sigma_p \tag{11}$$

$$\sigma_p = \sigma_{12}\, \sigma_{23}\, \sigma_{34}\, \sigma_{41} \qquad \sigma_{ij} = \pm 1$$

N number of sites.

Could it happen that $<\sigma_{ij}>$ takes a non zero value below some transition point ? The answer is no. This seems almost obvious from the first point of view by selecting specific boundary conditions. Indeed the existence of a local invariance is unaffected by boundary conditions (we have only to require that g_i is the identity on the boundary) and therefore $<\sigma_{01}> = <\sigma_0\sigma_{01}\sigma_1>$ where σ_0, σ_1 are arbitrary group elements ± 1. The average is therefore always zero. But even if we add a volume symmetry breaking term in the form $H \sum_{(ij)} \sigma_{ij}$ the conclusion remains true in the limit $H \to 0$. Indeed we can perform a gauge transformation which is identically one except at site 0 therefore

$$<\sigma_{01}> = Z^{-1}(\beta,H) \frac{1}{2^{Nd}} \sum_{\sigma_{ij}=\pm 1} e^{\beta \sum_p \sigma_p + H \Sigma' \sigma_{ij}}$$

$$\times \frac{1}{2} \sum_{\sigma=\pm 1} e^{H\sigma(\sigma_{0,1}+\ldots+\sigma_{0,2d})} \sigma\, \sigma_{01} , \tag{12}$$

where Σ' is a sum overall links except those going through 0, while $\sigma_{01},\ldots\sigma_{02d}$ denote precisely those which go through zero

Now

$$\left| \frac{1}{2} \sum_{\sigma=\pm 1} e^{H\sigma(\sigma_{01}+\ldots+\sigma_{02d})} \sigma \sigma_{01} \right| \leqslant \sinh 2dH \tag{13}$$

while

$$1 \leqslant e^{2dH} \frac{1}{2} \sum_{\sigma=\pm 1} e^{H\sigma(\sigma_{01}+\ldots+\sigma_{02d})}$$

Therefore

$$|<\sigma_{01}>| \leqslant e^{2dH} \sinh 2dH \tag{14}$$

which goes to zero as $H \to 0$. The argument is easily extended in the presence of matter fields and for any (abelian, or non abelian) compact group to any quantity having zero average over the group[2]. Thus explicit gauge breaking is needed to give non vanishing expectation value to non-gauge invariant quantities.

To close this section let us note that pure gauge theories are trivial in dimension 2 where the plaquette variables $g_p \equiv g_{12}\, g_{23}\, g_{34}\, g_{41}$ are independant and

$$Z = z^N \text{ where } z = \int dg\; e^{\beta\chi(g)} \; . \tag{15}$$

The free energy $\frac{1}{N}\ell n\, Z = \ell n\, z$ is analytic in β and no transition occurs. A table for some group of interest is given in section 5. In the abelian case it is trivial to see that

$$W_C^{\chi'} = \left[\frac{\int dg\; e^{\beta\chi(g)}\; \chi'(g)}{\int e^{\beta\chi(g)}} \right]^S , \tag{16}$$

where S is the area enclosed by C and therefore confinement is always valid. Eventhough it is less obvious the same property also holds in the non abelian case.

4. HAMILTONIAN FORMULATION

Up to now we have presented the Lagrangian formulation of lattice field theory. For practical purposes, in particular to investigate the particle spectrum it might be useful to introduce a Hamiltonian formulation closely related to the transfer matrix of statistical mechanics[6]. I shall use here as an example the U(1) theory where link variables are angles θ_{ij} with $\theta_{ji} = -\theta_{ij}$ mod 2π. Let us select on a $d = D + 1$ dimensional lattice a direction labelled as the time axis. A gauge transformation allows us to set all vertical variables (along the time axis) equal to zero. By horizontal we now mean directions, links, plaquette perpendicular to the time axis. Then (minus) the action takes the form

$$J = \beta \sum_t \sum_\ell \cos[\theta_{ij}(t+1) - \theta_{ij}(t)] + \beta \sum_t \sum_p \cos\theta_p(t) \ , \qquad (1)$$

where the links and plaquettes are horizontal and labelled by a fixed time. If the lattice extends over T finite time intervals (taken to be unity) then we want to identify the partition function as a trace

$$Z = \int \prod_{ij,t} \frac{d\theta_{ij}(t)}{2\pi} \ e^S = \mathrm{Tr}(U_{T,T-1} \cdots U_{21} \ U_{10}), \qquad (2)$$

where $U_{t+1,t}$ stands for a transfer matrix from time t to t+1. This is nothing but going in the reverse direction from the one followed to derive the path integral formalism. In the limit where the unit of time will be thought as being infinitesimal we will identity $U_{t+1,t}$ with the exponential of the Hamiltonian

$$U_{t+1,t} = e^{-H} \qquad (3)$$

Note that this method transforms a theory dealing with classical quantities into one involving operators in a D dimensional space. Needless to stress that the quantum mechanical operator H is distinct from the energy functional of configurations in the statistical interpretation. Indeed the latter is up to a coefficient the classical action.

Since U is an linear operator it acts in a vector space. The latter in generated by vectors indexed by angles attached to links of the spatial lattice (assumed for the moment to be limited in a finite volume)

$$|\theta\rangle = \prod_{(ij)} |\theta_{ij}\rangle \ . \qquad (4)$$

To be precise the product runs over oriented links (an arbitrary orientation is attached to each link) and when discussing the reverse orientation we have to change the sign of θ_{ij}. The label θ then stands for the collection of θ_{ij}. Up to irrelevant factors of 2π the trace of $\langle\theta'|U|\theta\rangle$ will therefore mean

$$\int_0^{2\pi} \prod_{(ij)} d\theta_{ij} \ \langle\theta|U|\theta\rangle \ .$$

A possible choice for $U_{t+1,t}$ is therefore

$$\langle\theta'|U|\theta\rangle = e^{\frac{\beta}{2} \sum_p \cos\theta_p + \cos\theta'_p + \beta \sum_\ell \cos(\theta'_\ell - \theta_\ell)} \ . \qquad (5)$$

When the time slice (taken as unity here) is very small only θ' close to θ are relevant. We replace $\cos(\theta_\ell'-\theta_\ell)$ by $1-\frac{1}{2}(\theta_\ell'-\theta_\ell)^2$. Therefore

$$<\theta'|U|\theta> \sim cst \times e^{-\frac{\beta}{2}\sum_\ell(\theta_\ell'-\theta_\ell)^2 + \beta\sum_p \cos\theta_p} . \qquad (6)$$

We recognize that H is given as

$$H = \lambda_1 \sum_x \sum_{k=1}^D E_k^2(x) - \lambda_2 \sum_p \cos\theta_p , \qquad (7)$$

where λ_1 and λ_2 are constants and for a given (spatial) link $x, x+an_k$, E_k is the conjugate variable to the angle $\theta_{x,x+an_k}$ i.e. $\frac{1}{i}\frac{\partial}{\partial\theta_{x,x+an_k}}$, taking therefore integer values. It coincides of course with angular momentum.

As a parenthetical remark we note that for a single angle

$$<\theta'|e^{-tE^2}|\theta> = \sum_{-\infty}^{+\infty} e^{in(\theta-\theta')-tn^2} = \frac{1}{\sqrt{4\pi t}}\sum_{-\infty}^{+\infty} e^{-\frac{(\theta-\theta'+2\pi pn)^2}{4t}} \qquad (8)$$

is one of the Jacobi θ-functions. The equality of the two representation is the Poisson summation formula. The last expressions may be used to formulate models of the Villain type.

In (7) E^2 is of course the electric term while $-\sum_p \cos\theta_p$ represents (up to a constant) the magnetic energy. The above derivation extends to non-abelian cases and can be generalized to include matter fields. It assumes a discretization of space alone, time being considered as a continuous variable. One could of course proceed directly to derive the Hamiltonian formalism.

Under a gauge transformation a state $|\theta> \equiv \prod_{(ij)} |\theta_{ij}>$ is modified into $|^g\theta> = \prod_{(ij)} |\theta_{ij}+\theta_i-\theta_j>$. We might wish to construct invariant linear combinations in particular for the ground state. If all θ_i except one, say θ_0, vanishes the generator of the above transformation will obviously be proportional to $\sum_{i(0)} E_{i,0}$ where E_{ij} is that component of E pertaining to the link(ij) and $E_{ij}=-E_{ji}$, $i(0)$ denote the neighbors of the site zero. The above quantity is the obvious candidate for a lattice generalization of the divergence of E. With this notation we see that a gauge invariant state like the ground state $|0>$ should obey Gauss's law

$$\text{div } E|0> = 0 . \qquad (9)$$

Points at which div $E \neq 0$ can be considered as carring an external charge.

Generalizations of this Hamiltonian formalism exist in the case of discrete groups. What is not straightforward is the treatment of the "electric" term. As an example in a Z_2 gauge theory a natural candidate for an Hamiltonian uses the Pauli matrices $\sigma^{(\alpha)}$ and

$$H = \lambda_1 \sum_{\ell} \sigma_{ij}^{(1)} - \lambda_2 \sum_{p} \sigma_{12}^{(3)} \sigma_{23}^{(3)} \sigma_{34}^{(3)} \sigma_{41}^{(3)} \tag{10}$$

in close analogy with the transfer matrix for the Ising model.

5. STRONG COUPLING EXPANSION

A virtue of the lattice theory is that it allows a large coupling (small β) expansion analogous to the large temperature one performed for statistical models. It is not necessary to enter here in a number of technical and rather tedious details to obtain these expansions [5]. The principle is simple. One expands the exponential of the action in power series in β, ignores boundary effects take the logarithm, divide by N the number of lattice sites and obtains therefore the free energy F. The last step can be avoided by using the existence of the thermodynamic limit. Since (with periodic boundary conditions say) the coefficient of any power of β is a polynomial in N (vanishing when N = 0) the logarithm is simply obtained by retaining terms linear in N. This prescription is not equivalent to the familar one in the case of the standard continuous perturbation theory where one only retains connected diagrams. This is due to the finite lattice size which prevents free interpenetration of configuration. Some amount of topological conting methods and of group theory are clearly required to go beyond the first trivial terms.

For the groups Z_2, Z_3, $U_4(1)$, $SU_2(2)$ and SU(3) we display the exact form of the action, the known result for d = 2 and the first terms of the expansion in arbitrary dim > 2. Note that the common factor d(d-1) arises from the ratio of number of plaquette $\frac{Nd(d-1)}{2}$ to the number of sites N.

Z_2

$$Z = e^{NF} = \frac{1}{2^{Nd}} \sum_{\sigma_{ij}=\pm 1} e^{\beta \sum_p (\sigma_{12}\sigma_{23}\sigma_{34}\sigma_{41})}$$

$$d = 2 \qquad F = \ln(\operatorname{ch} \beta)$$

Z_3

$$Z = e^{NF} = \frac{1}{3^{Nd}} \sum_{p_{ij}=0,1,2} e^{\beta \sum_p \cos[\frac{2\pi}{3}(p_{12}+p_{23}+p_{34}+p_{41})]}$$

$$d = 2 \quad F = \ell n \frac{1}{3} (e^{\beta} + 2e^{-2\beta})$$

$U(1)$

$$Z = e^{NF} = \int \prod_{(ij)} \frac{d\theta_{ij}}{2\pi} e^{\beta \sum_p \cos(\theta_{12}+\theta_{23}+\theta_{34}+\theta_{41})}$$

$$d = 2 \quad F = \ell n\left(\int_0^{2\pi} \frac{d\theta}{2\pi} e^{\beta \cos \theta}\right) \equiv \ell n \, I_0(\beta)$$

$SU(2)$

$$Z = e^{NF} = \int \prod_{(ij)} dU_{ij} \, e^{\beta \chi(U_{12} U_{23} U_{34} U_{41})} \tag{1}$$

χ = trace in the J = 1/2 representation

$$d = 2 \quad F = \ell n \int_0^{2\pi} \frac{d\theta}{2\pi} \sin^2\theta \, e^{2\beta \cos \theta} = \ell n\left(\frac{I_1(2\beta)}{\beta}\right)$$

SU(3) same as as above with

$$\chi = (\chi_3 + \chi_{\bar{3}})$$

$$F = \ell n\left\{ \frac{\int d\varphi_1 d\varphi_2 |\Delta(\varphi)|^2 \, e^{\beta(\chi_3(\varphi)+\chi_{\bar{3}}(\varphi))}}{\int d\varphi_1 d\varphi_2 |\Delta(\varphi)|^2} \right\} = \beta^2 + \frac{\beta^2}{3} + \ldots$$

where the group element is diagonal with elements $(e^{i\varphi_1}, e^{i\varphi_2}, e^{i\varphi_3})$, $\varphi_3 = -\varphi_1 - \varphi_2$ and $\Delta(\varphi) = \prod_{i<j}\left(e^{i\varphi_i} - e^{i\varphi_j}\right)$

Finally I_0 and I_1 are the modified Bessel functions.

For Z_2,

$$\frac{F}{d(d-1)} =$$

$$\frac{1}{4}\beta^2 - \frac{1}{24}\beta^4 + \left(\frac{1}{6}d - \frac{29}{90}\right)\beta^6 + \left(-\frac{1}{3}d + \frac{3343}{5040}\right)\beta^8$$

$$+\left(d^2 - \frac{184}{45}d + \frac{118\,471}{28\,350}\right)\beta^{10} + \left(-\frac{8}{3}d^2 + \frac{121\,153}{11\,340}d - \frac{20\,022\,781}{1\,871\,100}\right)\beta^{12}$$

$$+\left(10\,d^3 - \frac{208}{3}d^2 + \frac{935\,561}{5\,670}d - \frac{5\,647\,451\,354}{42\,567\,525}\right)\beta^{14}$$

$$+\left(-26\,d^3 + \frac{129\,161}{840}d^2 - \frac{376\,639\,121}{1\,247\,400}d + \frac{4\,021\,634\,721\,191}{20\,432\,412\,000}\right)\beta^{16} + O(\beta^{18}).$$

For Z_3,

$$\frac{F}{d(d-1)} =$$

$$\frac{1}{8}\beta^2 + \frac{1}{48}\beta^3 - \frac{1}{128}\beta^4 - \frac{1}{256}\beta^5 + \left(\frac{1}{192}d - \frac{51}{5120}\right)\beta^6 + \left(\frac{1}{128}d - \frac{153}{10\,240}\right)\beta^7$$

$$+\left(\frac{1}{1024}d - \frac{2\,187}{1\,146\,880}\right)\beta^8 + \left(-\frac{35}{6\,144}d + \frac{46\,597}{4\,128\,768}\right)\beta^9$$

$$+\left(\frac{1}{512}d^2 - \frac{1\,017}{81\,920}d + \frac{779\,381}{45\,875\,200}\right)\beta^{10} + \left(\frac{3}{512}d^2 - \frac{4\,047}{163\,840}d + \frac{2\,383\,531}{91\,750\,400}\right)\beta^{11}$$

$$+\left(\frac{137}{24\,576}d^2 - \frac{52\,709}{2\,293\,760}d + \frac{191\,096\,159}{8\,074\,035\,200}\right)\beta^{12}$$

$$+\left(-\frac{15}{8\,192}d^2 + \frac{57}{8\,192}d - \frac{3\,041\,827}{461\,373\,440}\right)\beta^{13}$$

$$+\left(\frac{5}{4\,096}d^3 - \frac{533}{32\,768}d^2 + \frac{3\,826\,173}{73\,400\,320}d - \frac{287\,774\,341\,033}{5\,877\,897\,625\,600}\right)\beta^{14}$$

$$+\left(\frac{17}{3\,072}d^3 - \frac{7\,027}{163\,840}d^2 + \frac{34\,256\,119}{314\,572\,800}d - \frac{15\,959\,874\,120\,733}{176\,336\,928\,768\,000}\right)$$

$$+\left(\frac{603}{65\,536}d^3 - \frac{4\,553\,361}{73\,400\,320}d^2 + \frac{18\,321\,594\,271}{129\,184\,563\,200}d - \frac{205\,245\,882\,159\,867}{1\,880\,927\,240\,192\,000}\right)\beta^{16} + O(\beta^{17}).$$

For U(1) (2)

$$\frac{F}{d(d-1)} =$$

$$\frac{1}{8}\beta^2 - \frac{1}{128}\beta^4 + \left(\frac{1}{192}d - \frac{11}{1\,152}\right)\beta^6 + \left(-\frac{1}{256}d + \frac{757}{98\,304}\right)\beta^8$$

$$+\left(\frac{1}{512}d^2 - \frac{85}{12\,288}d + \frac{2\,473}{409\,600}\right)\beta^{10} + \left(-\frac{29}{12\,288}d^2 + \frac{2\,467}{262\,144}d - \frac{1\,992\,533}{212\,336\,640}\right)\beta^{12}$$

$$+\left(\frac{5}{4\,096}d^3 + \frac{237}{32\,768}d^2 + \frac{178\,003}{11\,796\,480}d - \frac{38\,197\,099}{3\,468\,165\,120}\right)\beta^{14}$$

$$+\left(-\frac{15}{8\,192}d^3 + \frac{1\,485}{131\,072}d^2 - \frac{53\,956\,913}{2\,264\,924\,160}d + \frac{11\,483\,169\,709}{676\,457\,349\,120}\right)\beta^{16} + O(\beta^{18}).$$

For SU(2),

$$\frac{F}{d(d-1)} =$$

$$\frac{1}{4}\beta^2 - \frac{1}{48}\beta^4 + \left(\frac{1}{96}d - \frac{5}{288}\right)\beta^6 + \left(-\frac{1}{96}d + \frac{29}{1\,440}\right)\beta^8$$

$$+\left(\frac{1}{256}d^2 - \frac{49}{4\,608}d + \frac{1\,001}{172\,800}\right)\beta^{10} + \left(-\frac{7}{1\,024}d^2 + \frac{32\,131}{1\,244\,160}d - \frac{211\,991}{8\,709\,120}\right)\beta^{12}$$

$$+\left(\frac{5}{2\,048}d^3 - \frac{43}{4\,096}d^2 + \frac{5\,341}{368\,640}d - \frac{264\,497}{40\,642\,560}\right)\beta^{14}$$

$$+\left(-\frac{47}{8\,192}d^3 + \frac{7\,030\,933}{212\,336\,640}d^2 - \frac{97\,100\,911}{1\,486\,356\,480}d + \frac{1\,474\,972\,157}{33\,443\,020\,800}\right)\beta^{16} + 0(\beta^{18}).$$

For SU(3),

$$\frac{F}{d(d-1)} =$$

$$\frac{1}{2}\beta^2 + \frac{1}{6}\beta^3 - \frac{1}{24}\beta^5 + \left(\frac{1}{243}d - \frac{113}{3\,888}\right)\beta^6 + \left(\frac{1}{81}d - \frac{133}{6\,480}\right)\beta^7 + \left(\frac{5}{324}d - \frac{1\,069}{51\,840}\right)\beta^8$$

$$+\left(\frac{5}{972}d - \frac{509}{77\,760}\right)\beta^9 + \left(\frac{2}{6\,561}d^2 - \frac{157}{11\,664}d + \frac{490\,757}{20\,995\,200}\right)\beta^{10} + \left(\frac{4}{2187}d^2 - \frac{59}{2160}d + \frac{435\,299}{9\,797\,760}\right)\beta^{11}$$

$$+\left(\frac{1\,775}{354\,294}d^2 - \frac{218\,824\,907}{7\,255\,941\,120}d + \frac{1\,682\,010\,779}{42\,326\,323\,200}\right)$$

$$+\left(\frac{440}{59\,049}d^2 - \frac{13\,919\,677}{604\,661\,760}d + \frac{7\,603\,159}{440\,899\,200}\right)\beta^{13}$$

$$+\left(\frac{20}{531\,441}d^3 + \frac{8\,377}{2\,125\,764}d^2 - \frac{12\,469\,727}{5\,441\,955\,840}d - \frac{14\,239\,256\,399}{1\,333\,279\,180\,800}\right)\beta^{14}$$

$$+\left(\frac{514}{1\,594\,323}d^3 - \frac{69\,331}{7\,971\,615}d^2 + \frac{106\,962\,409}{2\,821\,754\,880}d - \frac{3\,474\,317\,893}{79\,361\,856\,000}\right)\beta^{15}$$

$$+\left(\frac{2\,323}{1\,594\,323}d^3 - \frac{5\,838\,272\,899}{220\,399\,211\,520}d^2 + \frac{10\,597\,782\,658\,021}{123\,423\,558\,451\,200}d - \frac{6\,402\,970\,751\,747}{82\,282\,372\,300\,800}\right)\beta^{16}$$

$$+ 0(\beta^{17}).$$

A limited analysis of these series can be done for the location of a possible singularity. Indeed if a specific gauge model undergoes a first order transition the high temperature branch might have an analytic continuation through the singular point corresponding to a metastable state (supercooling or superheating). Only if the singularity is of higher order is this elementary direct method likely to yield an interesting result.

Of course high temperature expansions are not limited to the free energy. One can also study the behavior of the Wilson loop average. As an example consider the Z_2 gauge group and a curve C equal to a square of side L. Then W decreases for large L according to the area law and one finds

$$\lim_{L^2\to\infty} \frac{\ln W}{L^2} = \ln \tanh \beta + 2(d-2)[\tanh \beta^4 + \tanh \beta^6 + (9d-22) \tanh \beta^8 + \ldots] \quad (3)$$

For d = 2 this reduces to a trivial result $W = (\mathrm{th}\ \beta)^{L^2}$ and we know that no transition occurs. In higher dimensions we can look for a location of a first zero when d = 3 and 4, with this limited amont of information one obtains respectively β=0.691 and 0.574 while transitions are known to occur at β = 0.761 and 0.441 (see the following section). Since we have taken here series to a rather low order it is not unlikely that such expansions might be useful to locate the transition.

Much work has been and is being carried on these high temperature expansions in both Lagrangian and Hamiltonian formalism that we have no space to report here.

6. DUALITY TRANSFORMATIONS

The possibility of relating the small and large coupling regimes of systems with abelian symmetries offers a mean to provide some exact results (for very unphysical models) which may serve as testing ground for approximation methods. I shall limit myself here to the simple example of a Z_2 pure gauge group although interesting applications have been discussed for compact QED (i.e. the abelian model with group U(1) coupled to matter (by M. Peskin[11] and others) and the ideas extended to the non abelian case by t'Hooft, Mandelstam, Polyakov...

The original example is of course the the Kramers Wannier duality for the two dimensional Ising model later extended by Wegner to gauge groups[5] [8] [9]. I should also note here a beautiful application of similar ideas by Jose, Kadanoff, Kirkpatrick and Nelson[10] to the 2-dimensional x, y model.

An interpretation of this idea as follows. Assume that a system undergoes a transition to an ordered state. There will be configurations which depart from the ordered structure with various types of defects. Some may be of the nature of spin waves, the analogs of perturbative particle excitations, some others may have a stability insured by topological reasons. The idea is therefore to study both the order and disorder variables and their interactions. At low enough coupling one might expect that the defects have a small interaction and therefore behave as an (almost) free gas.

Take for instance a coupled Z_2 gauge model

$$Z = 2^{-N(d+1)} \sum_{\sigma_{ij}=\pm 1,\ \sigma_i=\pm 1} \exp S$$

$$S = \beta_\ell \sum_\ell \sigma_i\, \sigma_{ij}\, \sigma_j + \beta_p \sum_p \sigma_{ij}\, \sigma_{jk}\, \sigma_{kl}\, \sigma_{\ell i} \ . \tag{1}$$

A duality transformation will associate to the original (hypercubical) lattice a dual one with its lattice nodes at the center of the hypercubes of the original one. More generally a δ-dimensional manifold in the original lattice will be put in correspondence with a $(d-\delta)$-dimensional one on the dual lattice.

Specifically let us first set $d = 2$ for the model described by eq.(1) use a gauge transformation $\sigma_{ij} \to \sigma_i\, \sigma_{ij}\, \sigma_j$, to eliminate the "matter" variables σ_i and insert the elementary (character) expansion

$$e^{\beta\sigma} = \cosh \beta + \sigma \sinh \beta = \cosh\beta[1 + \beta \tanh \beta] \ , \tag{2}$$

valid when $\sigma = \pm 1$. Then

$$Z = 2^{-2N}(\cosh^2\beta_\ell \cosh \beta_p)^N \sum_{\sigma_{ij}=\pm 1} \prod_\ell (1 + \tanh \beta_\ell\, \sigma_{ij}) \prod_p (1 + \tanh \beta_p \sigma_{ij}\sigma_{jk}\sigma_{k\ell}\sigma_{\ell i}) \tag{3}$$

We expand the products and sum over $\sigma_{ij} = \pm 1$. Non vanishing terms are in one to one correspondence with configurations of P plaquettes bordered by L links which belong to an odd number (here one) of these P plaquettes. These contribute a term $(\tanh \beta_\ell)^L (\tanh \beta_p)^P$. To each of these selected plaquettes we assign a "spin" $\tilde{\sigma}_i = -1$ at the dual site and set $\tilde{\sigma}_i = +1$ otherwise, in such a way that $P = \frac{1}{2} \Sigma_i (1-\tilde{\sigma}_i)$, $L = \frac{1}{2} \Sigma_\ell (1-\tilde{\sigma}_i\tilde{\sigma}_j)$ with the last sum running over nearest neighbours of the dual lattice. Consequently

$$Z = \frac{1}{4}(\cosh^2\beta_\ell \cosh \beta_p)^N \sum_{\tilde{\sigma}_i=\pm 1} \exp\left\{ \frac{1}{2} \ln \tanh \beta_p \sum_i (1-\tilde{\sigma}_i) + \frac{1}{2} \ln \tanh \beta_\ell \sum_\ell (1-\tilde{\sigma}_i\tilde{\sigma}_j)\right\}. \tag{4}$$

We recognize the partition function of the two dimensional Ising model in an external field $h = -\frac{1}{2} \ln \tanh \beta_p$ and inverse temperature $\tilde{\beta} = -\frac{1}{2} \ln \tanh \beta_\ell$. In particular the pure gauge field is trivial, as we know, with $F(0,\beta_p) = \ln \cosh \beta_p$.

The same reasoning in dimension three where a plaquette is dual to a link shows that the coupled model is self dual with

$$F(\beta_\ell,\beta_p) = \frac{3}{2} \ln[\sinh 2\beta_\ell \sinh 2\beta_p] + F\left(-\frac{1}{2} \ln \tanh \beta_p, -\frac{1}{2} \ln \tanh \beta_\ell\right) . \tag{5}$$

Note the interchange of the two indices ℓ and p on the two sides of this equation.

In four dimensions the coupled model is dual to a model defined as

$$\exp N\tilde{F}(\tilde{\beta}_\ell,\tilde{\beta}) = 2^{-6N} \sum_{\tilde{\sigma}_{ijk\ell}=\pm 1} \exp \tilde{\beta}_\ell \sum_p \tilde{\sigma}_{ijk\ell} + \tilde{\beta}_p \sum_c \Pi \, \tilde{\sigma}_{ijk\ell} , \quad (6)$$

with variables $\tilde{\sigma}_{ijk\ell}$ attached to plaquettes and interacting through products over the 6 faces of 3-dimensional cubes. The precise relation is

$$\tilde{\beta}_\ell = -\frac{1}{2} \ell n \tanh \beta_p \qquad \tilde{\beta}_p = -\frac{1}{2} \ell n \tanh \beta_\ell$$

$$F(\beta_\ell,\beta_p) = 2\ell n(\frac{1}{2} \sinh 2\beta_\ell) + 3\ell n(2 \sinh 2\beta_p) + \tilde{F}(\tilde{\beta}_\ell \, \tilde{\beta}_p). \quad (7)$$

In the case of a pure gauge field we have to study the limit of these relations when $\beta_\ell \to 0$ or $\tilde{\beta}_p \to \infty$. For d = 2 we find a trivial result with no transition. When d = 3 the limit $\tilde{\beta}_p \to \infty$ entails that any product of dual variables along a plaquette $\tilde{\sigma}^p_{ij}\tilde{\sigma}_{jk}\tilde{\sigma}_{k\ell}\tilde{\sigma}_{\ell i} = 1$. This is a cohomology condition which on a cubic lattice yields $\tilde{\sigma}_{ij} = \tilde{\sigma}_i\tilde{\sigma}_j$ with $\tilde{\sigma}_i = \pm 1$. As a consequence the dual of the Z_2 gauge system is an Ising model

$$F(\beta_\ell{=}0,\beta_p) = -\frac{1}{2} \ell n \, 2 + \frac{3}{2} \ell n \sinh 2\beta_p + F_I(-\frac{1}{2} \ell n \tanh \beta_p) \, (8)$$

This system is known to exhibit a second order transition. Finally the same limit $\beta_\ell \to 0$ in dimension 4 reduces the relation (7) to a self duality property (very much as the pure Ising model in dimension two is self dual). Consequently, if the system exhibits a unique transition (and this is in principle a big if) it can only occur at the fixed point $\beta_c = -\frac{1}{2} \ell n \tanh \beta_c$ or $\sinh 2\beta_c = 1$.

Collecting the results we have the following table for the Z_2 gauge group

d	2	3	4	∞
β_c	∞	0.7613	0.4407	$\sim \frac{cst}{d}$
order		2	1	1

(8)

The value quoted for β_c in dimension 3 follows from known numerical values for the Ising model which also yields the order of the transition. We shall return to the situation in large dimension later.

It is possible to extend the duality transformation to average values of observables. Consider for instance the Wilson loop in dimension three. Then by a dual transformation

$$W_c = e^{-(\mathfrak{F}-\mathfrak{F}')} , \qquad (9)$$

where $\mathfrak{F}$ is the total free energy (i.e. including the volume factor) of the dual Ising model (at $\tilde{\beta}_\ell = -\frac{1}{2} \ln \tanh \beta_p$) and ' is equal the same quantity when one reverses all the signs of the Ising coupling along a set of links of the dual lattice crossing a surface bounded by C. In other words $-\ln W$ has the interpretation of a surface tension. At low enough temperature for the Ising model i.e. for **small** β_p we expect this quantity to vary with the area enclosed by C. This means than Wilson's confinement criterion is fulfilled. In this direction we obtained with Balian and Drouffe the following statement. For $\beta < \beta_c$, where β_c is such that $\tanh \beta_c \geqslant \frac{1}{2d-3}$ there exist two positive constants a_1 and a_2 such that

$$a_2 \leqslant -\frac{\ln W}{|S|} \leqslant a_1 . \qquad (10)$$

Similarly a high β_p expansion shows that $-\ln W$ goes like the perimeter of the curve C.

Let us close this section with a comment of the extension of these dual transformation to other groups. Consider for instance in a pure gauge model the expansion of a plaquette contribution in terms of irreducible characters $\chi^{(r)}$

$$e^{\beta\chi(g_p)} = \sum_{r \geqslant 0} \beta_{(r)} \chi^{(r)}(g_p) , \qquad (11)$$

with $\beta_{(r)}$ functions of $\beta (\beta_{(r)} \underset{\beta \to 0}{\to} \delta_{r,0})$ and $\frac{\beta(r)}{\beta(0)}$ the generalization of $\tanh \beta$ for Z_2. When dealing with the product of these quantities we find a sum of terms with each plaquette carring a representation indexed by (r) and a factor $\prod_p \beta_{(r_p)}$. These plaquettes may be considered as points, links or plaquettes ... of the dual lattice according to the dimension d = 2,3,4...

The integral over the link variables introduces constraints on the r_p. The product of representations attached to the plaquettes meeting a given link has to contain the identity representation in which case it yields a generalized Clebsch Gordan coefficient. For an abelian group, the set of representation is itself an abelian group and this constraint can be solved, yielding a dual model involving the dual group (as a global or local symmetry) arising from the amount of arbitrariness in this solution. This explains the successes encountered and also the limitations of duality. In the non abelian case the dual of a group is not a group and apparently one does not know how to solve for these cohomological constraints.

7. PHASE TRANSITIONS

The possibility of applying the lattice dynamics to the QCD particle models relies on the existence of a phase transition to a confining regime with long range correlations. According to popular belief this transition should occur for zero coupling (or temperature) in dimension four in the case of non abelian groups due to asymptotic freedom. In other words dimension 4 would be the lower critical dimension for non abelian gauge fields.

We have just seen examples of transitions for the Z_2 gauge group. For models involving a global symmetry a familiar tool of investigation is the mean field theory. It yields a first framework suggesting a transition (a second order one for the examples which interest us here like the Ising, xy or non linear σ-model). The study of fluctuations enables one to determine the dimension (upper critical) when the critical indices depart from there mean field value.

The naive transposition of these ideas to gauge fields is not possible due to the lack of a local order parameter. One way out is to eliminate gauge invariance by integrating over some surabundant gauge degrees of freedom. For instance we could set equal to one all vertical g_{ij}'s. But this procedure leaves a (time independent) gauge invariance and suffers from its arbitrariness. As a first hint we can however transform the mean field idea into a mathematical inequality (Peirl's inequality) and study the corresponding result. This goes as follows let us write

$$Z = \langle e^S \rangle , \tag{1}$$

where the average is understood as the integral over the link variables (recall that we normalized the measure for each link). Now

$$Z = \langle e^{\sum_\ell \chi(g_{ij}h_{ji}) + S - \sum_\ell \chi(g_{ij}h_{ji})} \rangle , \tag{2}$$

where h is for each link an element of a linear vector space in which G is naturally inbedded. For instance in the case of U(1), each g may be thought as a (real) two dimensional unit vector and h will be a two dimensional vector. For SU(2) we may write $g = u_0 + i\,\vec{u}.\vec{\sigma}$ with $(u_0,\vec{u})$ a unit real 4-dimensional vector so that we can set $h = h_0 + i\,\vec{h}.\vec{\sigma}$ with $(h_0,\vec{h})$ a 4-dimensional vector. ... Then the character χ used in (2) is the trace in the defining representation of the (linear) group.

From the convexity of the exponential function if follows that $\langle e^A \rangle \geqslant e^{\langle A \rangle}$. Thus

$$Z \geqslant \langle e^{\sum_\ell \chi(g_{ij}h_{ji})} \rangle \exp \frac{\langle e^{\sum_\ell \chi(g_{ij}h_{ji})} (S - \sum_\ell \chi(g_{ij}h_{ji}) \rangle}{\langle e^{\sum_\ell \chi(g_{ij}h_{ji})} \rangle} \tag{3}$$

Let us use this inequality for a uniform $h_{ij} \equiv h$. Taking the logarithm and dividing by N the number of sites (number of links Nd, number of plaquettes $N\frac{d(d-1)}{2}$) we find

$$F \geqslant d\left[u(h) - h\frac{d}{dh}u(h)\right] + \frac{d(d-1)}{2}\frac{<e^{\chi(h(g_{12}+g_{23}+g_{34}+g_{41}))}S_p>}{<e^{\chi(h[g_{12}+\ldots+g_{41}])}>} \quad (4)$$

where

$$<e^{\chi(gh)}> = e^{u(h)} \quad (5)$$

and S_p is the contribution of one plaquette to the action . The last average can generally be expressed in terms of u(h). In the cases of U(1) or SU(2) pure gauge fields where u depends only on the length of the vector h and the character used in the action is the basic one it is equal to $u'(h)^4$.

The r.h.s. of eq.(4) can obviously be maximized with respect to h. Therefore

$$F \geqslant \sup_h F(h)$$
$$F(h) = \frac{\beta}{2}d(d-1)\,u'(h)^4 + d(u(h) - hu'(h)). \quad (6)$$

We recognize the standard maximazing principle of -U+TS in thermodynamics where the first term would correspond to energy, the second to entropy.

The analog of the r.h.s. of (6) for systems with a global symmetry is the mean field approximation.

The maximum of F(h) can be studied for fixed β. For $\beta \leqslant \beta_c$ where β_c is of order d^{-1}, the maximum is obtained for h = 0, F(0)=0. For larger values of β a secondary maximum exceeds the one at the origin and yields positive F(h(β))(Figure 2). The resulting curve has a discontinuous derivative at β_c which on face value would suggest a first order transition. Systematic (but non variational) corrections can be made to the above estimates. If they are compared with the known value of β_c in dimension 4 for the Z_2 theory they seem to converge to the exact value in spite of the suspicion one may have on the procedure.

More elaborate tools than the above inequalities may be used to derive exact results. For instance in a series of work de Angelis, de Falco, Guerra and Marra have shown the unicity of the gauge invariant phase at small enough β in the thermodynamic limit[12]. Recent work by Mack and Petkova, Frohlich and Korthals Altes indicate a relationships between the non abelian gauge theory of SU(N) and the one pertaining to its center Z_N. Namely when the second confines so does the first according to Wilson's

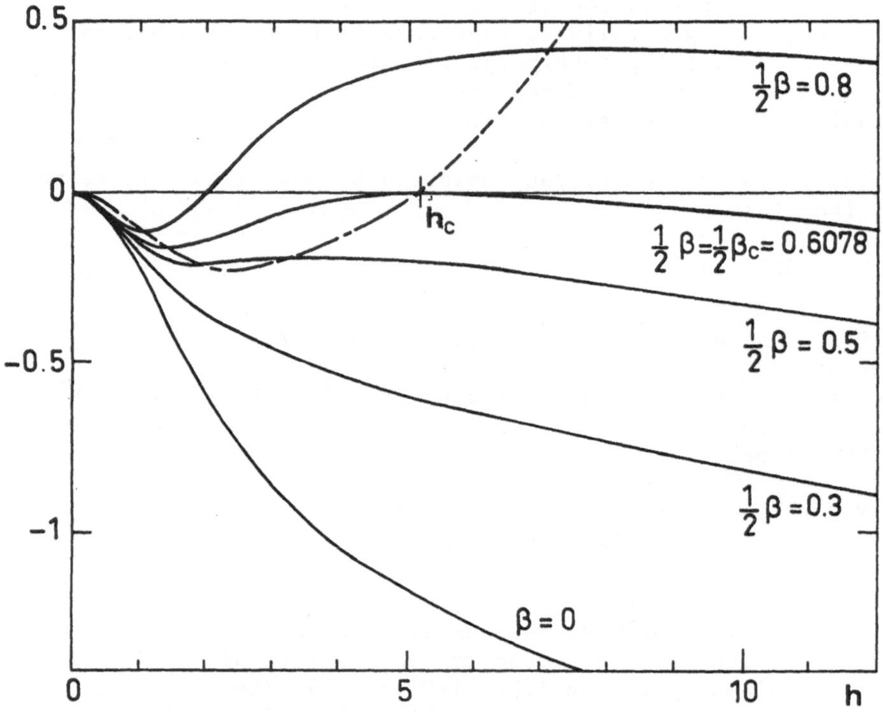

Fig. 2. Trial function F(h) for various values of β U (1) with d = 4.

criterion[13]. These work make use of a generalization of the Griffiths Kelly Sherman and Ginibre inequalities obtained in the case of continuous groups by Messager, Miracle-Sole and Pfister.

We have already quoted the numerical studies of Creutz, Jacobs and Rebbi[4] which support a first order transition for the Z_2 gauge group in four dimensions[18]. They have also carried calculations for other abelian and non abelian case. In the latter case they seem to agree with the absence of a transition in four dimension and a transition for d $\geqslant 5$. In the case of Z_n groups in four dimensions and for n large enough (larger than four or perhaps five) they find three phases separated by what seems higher order transitions. The small and very large β phases are analogous to the similar ones for the Z_2 gauge field. An intermediate one would be of the type of a U(1) small coupling phase with presumably Goldstone (photons) long range excitations.

Except for details the overall picture seems to fit the observations by Migdal[15] of a correspondence of the phenomena between globally symmetric models in dimension close to two and those for gauge fields close to four. These are summarized in the following table

d = 2 systems with global symmetry

Z_2 $\beta_c \neq 0$ self dual unique second order transition

Z_3 $\beta_c \neq 0$ unique transition

Z_n $n \geqslant 4$ (or 5) 3 phases

U(1) (or xy model) unique $\beta_c \neq 0$. No long range order in the low temperature phase. Second order transition with critical indices from the vortex theory

$O_n (n \geqslant 3)$ or $SU_n (n \geqslant 2)$ no transition, asymptotic freedom.

d = 4 systems with local symmetry

Z_2 $\beta_c \neq 0$ self dual unique first order transition

Z_3 $\beta_c \neq 0$ unique transition

Z_n $n \geqslant 4$ (or 5) 3 phases

U(1) (free compact electrodynamics) unique $\beta_c \neq 0$, should be analogous to free QED in the small coupling region. Behavior not known on the other side of the transition below β_c in the critical regime. Perhaps analogous to a form of non linear electrodynamics.

$O_n(n\geqslant 3)$ or $SU_n(n\geqslant 2)$ no transition, asymptotic freedom.

8. LARGE COORDINATION NUMBER

To conclude this brief survey I would like to present some attempts at formulating substitutes to the mean field theory valid when the coordination number (the number of neighbours of a given lattice point) is large.

The method proposed by Drouffe, Parisi and Sourlas[16] is to resum the dominant terms of the high temperature series when the dimension d is very large. To start with one can look at a model where mean field ideas apply. Take for instance the Ising model in large dimension d.

The susceptibility defined as

$$\chi(\beta) = \frac{1}{Z} \sum_k \sum_{\sigma_i=\pm 1} \sigma_0\sigma_k \, e^{\beta \sum_\ell \sigma_i\sigma_j} \tag{1}$$

can be written

$$\chi = \sum \chi_n \tanh \beta^n \tag{2}$$

where χ_n is essentially for large d the number of open paths on the lattice with fixed origin and length n. For large n (and d) this behaves as $(2d)^n$ thus

$$\chi \sim \frac{1}{1 - 2d \tanh \beta} \tag{3}$$

which indicates a transition at $\beta_c \sim \frac{1}{2d}$ where the susceptibility becomes infinite.

For gauge theories we observe on the expansions for the free energy given in section 5 that the polynomials in d which appear as coefficients of the various powers of β have a degree which increases by one unit every four powers of β. This can be traced as follows.

For β small we may retain in the Fourier series of e^S in irreducible characters only the first two terms. Higher terms start with higher powers of β. So all gauge theories behave for small β in essence like the Z_2 gauge theory with tanh β replaced by $\beta_{(1)}/\beta_{(0)}(\beta)$ of order β for small β. The free energy is expanded in terms of closed surfaces made of plaquettes, each one carrying a factor proportional to β. When the power in β is increased by four units, there appears the possibility of deriving from a given diagram a new one with in extra bump corresponding to sticking a cube in 2d - 4 directions.

Drouffe Parisi and Sourlas give arguments to the effect that the leading diagrams to a given order in β are trees of cubes which they succeed to count. This yields an expansion in terms of a parameter $(2d\ \beta^4)$. The dominant contribution has a singularity at a point $\beta \sim \frac{1}{d^{1/4}}$ (i.e. much further away than the critical point of the Peirl's inequality) where the plaquette-plaquette correlation length becomes infinite.

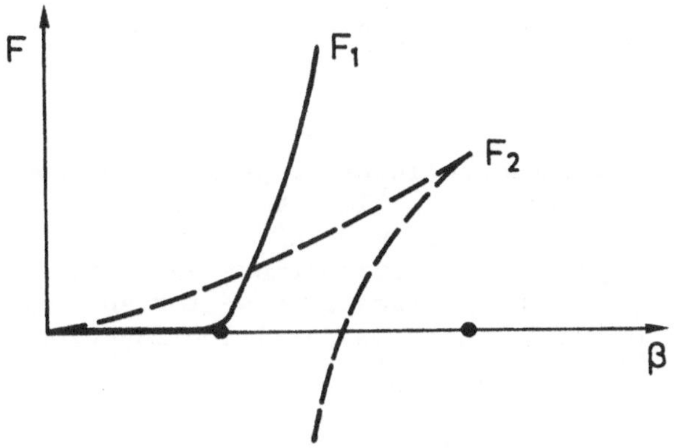

Figure 3.

Sketch of the free energy: F_1 is the lower bound from Peirls inequality, F_2 is the result of resuming the dominant diagrams in the large d limit.

This phenomenon, might be interpreted as the indication that a metastable phase exists below a first order transition point. The true domain of validity of the approximation shrinks to zero as $d \to \infty$.

The formula obtained for F_2 reads

$$F_2 = \frac{d^{3/2}\ n^2}{12\sqrt{2}}\ y^{3/2}(1-3y)\ [1 + O(d^{-1/4})] \ , \tag{4}$$

with n = dimension of the first non trivial representation occuring in the Fourier series of the action, and

$$y(1-y)^4 = 2d \left(\frac{\beta_{(1)}}{n\beta_{(0)}}\right) \tag{5}$$

Thus for Z_2, $n = 1$ and $\frac{\beta_{(1)}}{n\beta_{(0)}} = \tanh\beta$.

Corrections to the leading approximation do not modify qualitatively the above puzzling results.

To conclude let me mention yet an other model studied in collaboration with J.B. Zuber to understand the nature of the gauge theory for large coordination number. The inspiration comes from the Sherrington Kirkpartrick model of spin glasses[3]. Let us eliminate all reference to topology by assuming that all pair of sites of the lattice with N nodes are neighbors. To simplify matters, assume a Z_2 model where the link variables are of the form $\sigma_{ij} = \sigma_{ji} = \pm 1$ $(i \neq j)$. Thus we have an action proportional to $\mathrm{tr}\,\sigma^4$ and

$$Z = \frac{1}{2^{\frac{N(N-1)}{2}}} \sum_{\sigma_{ij}=\pm 1} e^{\frac{\beta}{4N}\mathrm{tr}\,\sigma^4} \tag{6}$$

The factor $\frac{1}{N}$ is inserted to insure a proper "thermodynamic" limit which in his case reads $F = \lim_{N\to\infty} \frac{1}{N^2} \ln Z$.

The above expression is furthermore approximated by replacing the condition $\sigma_{ij} = \pm 1$ by trace $\sigma^2 = N^2$ for an arbitrary symmetric matrix σ. Thus our model uses as partition function

$$Z = \frac{\int_{\mathrm{tr}\sigma^2=N^2} \prod_{i\leqslant j} d\sigma_{ij}\, e^{\frac{\beta}{4N}\mathrm{tr}\,\sigma^4}}{\int_{\mathrm{tr}\sigma^2=N^2} \prod_{i\leqslant j} d\sigma_{ij}} \tag{7}$$

The study of the $N \to \infty$ limit of this expression involves the same mathematics as the one developped in the context of the planar approximation[17]. One rewrites the above expression in terms of the eigenvalues λ_k of σ as

$$Z = \int_{\Sigma\lambda_i^2=N^2} \prod d\lambda_i \prod_{i<j} |\lambda_i-\lambda_j|\, e^{\frac{\beta}{4N}\Sigma\lambda_i^4} \Big/ \text{same with } \beta = 0 \tag{8}$$

and finds a normalized saddle point distribution of eigenvalues $u(\frac{\lambda}{\sqrt{N}})$ given by

$$u(x) = \frac{4}{\Pi a^4}\left[a^2-2 - 2(a^2-4)\frac{x^2}{a^2}\right]\sqrt{a^2 - x^2} \qquad x^2 \leqslant a^2 \tag{9}$$

$$a^2 - 4 = \frac{a^6}{8}\beta$$

This yields a positive distribution for $0 \leq \beta \leq \beta_c = \frac{2}{27}$ for which a^2 varies between 4 and 6.

Between these values the shape of u given by the semi-circle law for $\beta = 0$ flattens with square root singularities at the edges until for β_c they turn into singularities of the type $(x-a)^{3/2}$. Therefore the analog of the Wilson loop parameter

$$\lim_{p\to\infty} \frac{1}{N^{\frac{p}{2}-1}} \langle \mathrm{tr}\, M^p \rangle = \lim_{p\to\infty} \int_{-a}^{+a} x^p\, u(x)\, dx \qquad (10)$$

which picks the end point singular behavior has a discontinuous character as β crosses β_c.

Work is under progress to analyse the behavior of this model beyond β_c to ascertain the nature of the transition.

REFERENCES

[1] K.G. Wilson, Phys. Rev. D10, 2445 (1974).

[2] S. Elitzur, Phys. Rev. D12, 3978 (1975).

[3] D. Sherrington, S. Kirkpartick, Phys. Rev. Lett. 35, 1792 (1975)
D.J. Thouless, P.N. Anderson, R.G. Palmer, Phil. Mag. 35, 593 (1977).

[4] M. Creutz, L. Jacobs, C. Rebbi, Phys. Rev. Lett. 42, 1390 (1979) and BNL 26307 (1979).

[5] R. Balian, J.M. Drouffe, C. Itzykson, Phys. Rev. D10, 3376 (1974), D11, 2098 (1975), D11, 2104 (1975), D19, 2514 (1979).
J.M. Drouffe, C. Itzykson, Phys. Rep. 38C, 133 (1978).

[6] J. Kogut, L. Susskind, Phys. Rev. D11, 395 (1975).

[7] J. Villain, Journal de Physique 36, 581 (1975).

[8] F.J. Wegner, Journ. Math. Phys. 12, 2259 (1971).

[9] J.M. Drouffe, Phys. Rev. D18, 1174 (1978).

[10] J.V. José, L.P. Kadanoff, S. Kirkpatrick, D.R. Nelson, Phys. Rev. B16, 1217 (1977).

[11] M.E. Peskin, Ann. of Phys. (NY) 113, 122 (1978).

[12] G.F. de Angelis, D. de Falco, F. Guerra, R. Marra in Facts and Prospects of Gauge Theories ed. by P. Urban, Springer (1978).

[13] G. Mack and V.B. Petkova, Preprint Hamburg (1978).
J. Frohlich, Preprint IHES (1979).
Korthals Altes

[14] A. Messager, S. Miracle-Sole, C. Pfister, Comm. Math. Phys. 58, 19 (1978).

[15] A.A. Migdal, Zh E.T.F. 69, 810, 1477 (1975).

[16] J.M. Drouffe, G. Parisi, N. Sourlas, Preprint Dph-T 104-79 (1979).

[17] E. Brezin, C. Itzykson, G. Parisi, J.B. Zuber, Comm. Math. Phys. 59, 35 (1978).

[18] It is interesting to quote the remarkable agreement between the Monte Carlo work and the analytical estimates obtained from strong coupling expansions both with an accuracy of a few percent at the present stage. Take for instance the self-dual 4-dimensional Z_2 model with critical coupling β_c given by $\tanh \beta_c = \sqrt{2}-1$. From section 5 the average plaquette action defined as $E = \frac{1}{6}\frac{dF}{d\beta}$ admits an expansion
$E = t + 4t^5 - 4t^7 + 60t^9 + 56t^{11} + 1088t^{13} - 878t^{15} + \ldots$
with $t = \tanh \beta$. From duality it follows that
$E(\beta_c+\varepsilon) + E(\beta_c-\varepsilon) = \sqrt{2}$ while the above series yields
$E(\beta_c-\varepsilon) = 0.4819$. The predicted discontinuity at the critical point is therefore $\Delta E = E(\beta_c+\varepsilon) - E(\beta_c-\varepsilon) = 0.4504$ which agrees with the "observed" value reported in [4].

CLASSICAL GAUGE THEORIES AND THEIR QUANTUM ROLE

Arthur Jaffe*

Harvard University

Cambridge, Mass. 02138

The classical Euclidean variational equations $\delta\mathscr{A} = 0$, where $\mathscr{A}$ is the action functional, play two important roles. First, in dimension $d \leq 3$ they provide time independent solutions to classical four-dimensional physics. In that way, for example, the Ginzburg-Landau model has proved important in the theory of superconductivity. The second role of Euclidean stationary points of the action is to provide a starting point for classical approximations to quantum fields. We make some remarks on these two topics; for classical gauge theories we restrict attention to the Higgs model.

1. GINZBURG-LANDAU THEORY

The Ginzburg-Landau equations [1] couple the $d = 2$ Euclidean electromagnetic field to a ϕ^4 complex (charged) scalar theory. In other words they are the same as the Abelian Higgs equations, arising from the action

$$\mathscr{A}(a,\phi) = \frac{1}{2}\int\left\{F^2 + |D_a\phi|^2 + \frac{\lambda}{4}(|\phi|^2 - 1)^2\right\} d^2x \quad . \tag{1}$$

Here $F = da$ and $D\phi = D_a\phi = d\phi - ia\phi$. The variational equations are (with Euclidean coordinate indices)

$$\begin{aligned} \partial^\mu F_{\mu\nu} &= J_\nu \\ D_\mu D^\mu \phi &= -\lambda\phi(|\phi|^2 - 1) \end{aligned} \tag{2}$$

*Supported in part by the National Science Foundation under Grant PHY79-16812.

where the current J_ν is defined by $J_\nu = \mathrm{Im}(\bar{\phi} D_\nu \phi)$. The field $F = F_{01}$ we sometimes denote by B (magnetic field). Solutions to these $d = 2$ Euclidean equations are special solutions to $d = 4$ (Minkowski) equations which are time and x_3 independent. The vortex number (first Chern number) is

$$n = \frac{1}{2\pi} \int_{R^2} B\, d^2x = \frac{1}{2\pi} \oint a \quad . \tag{3}$$

Assuming sufficient regularity for integration by parts to be valid, the line integral of A in (3) can be taken on a circle S_1 at ∞.

In the case of the pure electromagnetic field, $n \neq 0$ requires infinite actions, e.g. $a = n d\theta$ where θ is the polar angle about the vortex. For the Higgs model, however, finite action n-vortex solutions for the case $\lambda = 1$ have recently been shown by C. Taubes to exist [2], and we discuss his results below.

The interpretation of the Ginzburg-Landau vortices arises in the theory of superconductivity. The assumptions $\mathcal{A} < \infty$, $\delta\mathcal{A} = 0$, lead to exponential decay of F at infinity, and to integer n [2]. It follows that the total flux is quantized, and after inserting dimensional constants

$$\text{Flux} = \int B\, d^2x = n\left(\frac{ch}{2e}\right) = n(2 \times 10^{-7} \text{ gauss cm}^2) \tag{4}$$

where c = velocity of light, h = Planck's constant and e is the electron charge. The basic flux unit is associated with a charge 2e, because of the "Cooper pairs" of electrons which occur in a superconductor. Thus each vortex is interpreted as a flux tube carrying one magnetic flux unit 2×10^{-7} gauss cm^2. Experimentally, a triangular lattice of vortices is observed in type II superconducting films, illustrated in Figure 1.

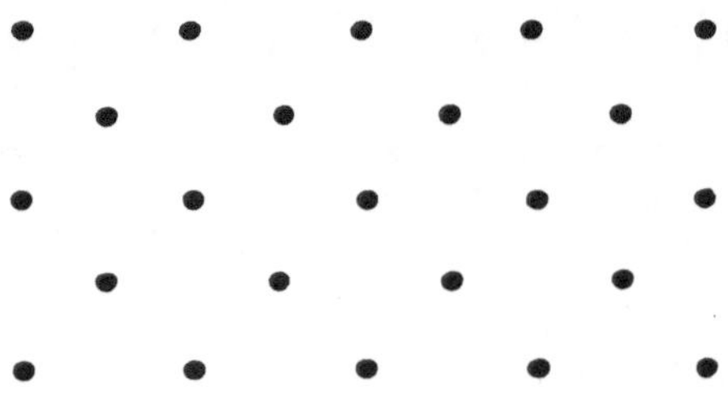

Fig. 1. Triangular lattice of vortices giving location of magnetic flux tubes (perpendicular to film).

A type II superconductor is characterized by the diagram in Fig. 2 of induced field B versus applied field H. For $H_{C_1} < H < H_{C_2}$, the induction B increases with H due to the appearance of vortices. For $H < H_{C_1}$ no vortices are present and the induced field is zero (Meissner effect). For $H > H_{C_2}$ the vortices cannot be distinguished from one another because of overlap and the conductor is normal. The interval (H_{C_1}, H_{C_2}) is said to be a "mixed" state, in which normal flux tubes are superimposed on a bulk superconductor without flux.

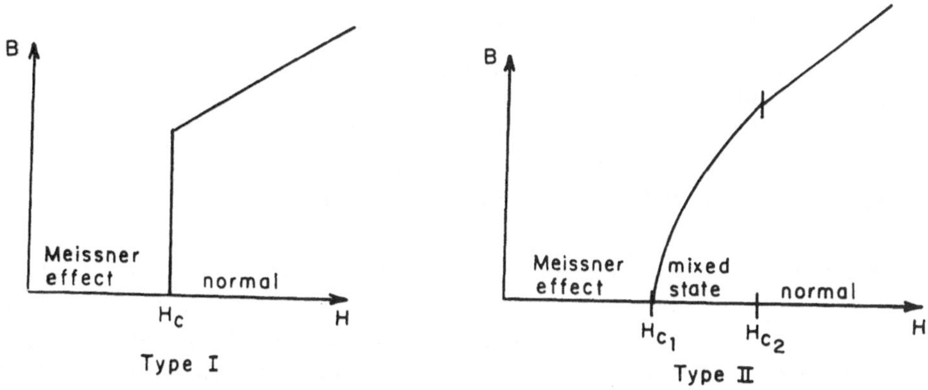

Fig. 2. Magnetic induction vs. applied field in superconductors.

The theoretical work on which this picture is based involves the phenomenological parameter λ. The equations for $\lambda < 1$ are supposed to describe type I superconductors (no flux tubes observed and B(H) discontinuous). For these superconductors $B(H) = 0$ for $H < H_C$ and $B(H) > B_C > 0$ for $H > H_C$. For the parameter $\lambda > 1$, the equations describe type II superconductors. The case $\lambda = 1$ is borderline, and in fact has a special theoretical significance.

In case $\lambda >> 1$, Abrikosov performed an approximate variational calculation, which gave evidence that the configuration (a, ϕ) of minimum action and fixed boundary conditions has a triangular lattice of vortices [1]. Recently, a numerical calculation by Jacobs and Rebbi [3] determined a vortex-vortex potential for the action $\mathcal{A}$. They found an attractive potential for $\lambda < 1$, zero potential for $\lambda = 1$ and a repulsive potential for $\lambda > 1$. One expects on the basis of this numerical model that for $\lambda < 1$ multi-vortex configurations are energetically unstable and the minimum of $\mathcal{A}$ can have at most one singular point. Thus it is energetically favorable to exclude the magnetic flux. On the other hand, for $\lambda > 1$ multivortex configurations could be stabilized by the boundary conditions and in this case a lattice of vortices (e.g. the Abrikosov picture) is expected. For the case $\lambda = 1$ the vortices (like instantons for pure Yang-Mills theory) do not interact. Thus we expect that global multivortex solutions can be found.

2. MULTIVORTEX SOLUTIONS FOR $\lambda = 1$

In case $\lambda = 1$, the action $\mathscr{A}$ can be written as $\pm\pi n$ plus a sum of squares [4],

$$\mathscr{A} = \frac{1}{2}\int\left\{F \pm \frac{1}{2}\,(|\phi|^2 - 1)\right\}^2 dx + \frac{1}{4}\int |\mathscr{D}_\pm\phi|^2\, dx \pm \pi n \quad . \tag{5}$$

Here $\mathscr{D}_\pm = D_\mp i{*}D$ is a complex covariant derivative. It follows from (5) that any critical points of $\mathscr{A}$, i.e. a solution to (2), which also minimizes $\mathscr{A}$, satisfies the first order equations

$$\mathscr{D}_+\phi = 0, \qquad F = \frac{1}{2}\,(1 - |\phi|^2), \qquad \text{if} \quad n > 0$$

$$\mathscr{D}_-\phi = 0, \qquad F = -\frac{1}{2}(1 - |\phi|^2), \qquad \text{if} \quad n < 0 \quad , \tag{6}$$

and vanishes for $n = 0$. For these minima $\mathscr{A} = \pi|n|$, so the vortices have zero interaction energy. In the following it is convenient to use the complex coordinate $z \equiv x_1 + ix_2$ to parameterize the plane.

THEOREM 1. [2]. Given $n \geq 0$ integer, there exists a finite action solution to the equations (6) with the following properties:

(i) The solution is globally C^∞.

(ii) The zeros of ϕ are a set of points $\{\alpha_j\}$, $j = 1,2,\ldots,n$, and as $z \to \alpha_j$,

$$\phi(z) \sim (z-\alpha_j)^{n_j}$$

where n_j is the multiplicity of α_j in the set $\{\alpha_j\}$. Hence $n_j > 0$.

(iii) Given $\{\alpha_j\}$, the solution is unique.

Furthermore: (a) Both $|\phi| - 1$ and $|D\phi|$ vanish as $|z| \to \infty$,

$$n = \frac{1}{2\pi}\int F\, d^2x = \sum_{\text{distinct } \alpha_j} n_j = \pi^{-1}\mathscr{A} \quad . \tag{7}$$

(b) Any solution to (6) with $\mathscr{A} = \pi n$ is gauge equivalent to one of the above.

(c) For $n \leq 0$ integer the above statements hold, but with $\mathscr{A} = -\pi n$, $\overline{\phi(z)} \sim (z-\alpha_j)^{n_j}$, and $n = -\Sigma n_j = (2\pi)^{-1}\int F$. Here $n_j > 0$.

We call the case $n > 0$ "vortices" and $n < 0$ "antivortices." The theorem shows that an n-vortex solution which minimizes $\mathscr{A}$ can be regarded as having the vortices or antivortices located at the points $x = x_j = \{\mathrm{Re}\alpha_j; \mathrm{Im}\alpha_j\}$, $j = 1,2,\ldots,n$. There are no vortex-antivortex solutions which minimize $\mathscr{A}$, since for given n the set of solutions is parameterized by the locations of the vortices $(n > 0)$ or the antivortices $(n < 0)$.

Consider, for simplicity, the case $n \geq 0$. Write $\phi = \exp(u/2 + iv)$, for u real and $v = \Sigma_{j=1}^{n} \theta_j$, θ_j the polar angles measured from the points α_j. Then the equations (6) can be reduced to solving

$$\Delta u = e^u - 1 \qquad \text{where} \quad \Delta = \partial_\mu \partial^\mu \quad . \tag{8}$$

Before the work of Taubes, deVega and Schnaposnik [5] had proved existence with $\alpha_1 = \alpha_2 = \cdots = \alpha_n$. In that case (8) reduces to an ordinary differential equation. Furthermore E. Weinberg [6] had used index theorem methods to show count 2n zero modes near each such solution. The method of Taubes is to regard (8) as a variational problem (with infinite action). As in studying meron equations [7], subtraction of an explicit divergent term in the action leads to a modified renormalized action whose minimum gives the solution to (8).

3. GENERAL SOLUTIONS FOR $\lambda = 1$

In the preceding section we characterized the multivortex solutions to the first order equations (6) obtained by minimizing $\mathscr{A}$. We ask what are the solutions to the second order equations (2) with $\lambda = 1$, $\mathscr{A} < \infty$? Of course the multivortex solutions satisfy these equations, but in general we expect others. A remarkable result of Taubes shows that more general solutions do not exist. (In the special case that the solution is close in an appropriate Banach space to a vortex solution, the methods of [9] are also presumably applicable.)

THEOREM 2. Every weak solution to the equations (2) with $\mathscr{A} < \infty$ has

(i) $n = 1/2\pi \int F \, d^2x$ integer,

(ii) $\mathscr{A} = \pi|n|$,

and is gauge equivalent to a multivortex solution given in Theorem 1. In particular, no vortex-antivortex solutions exist.

The proof of the theorem is a clever application of the maximum principle for the Laplacian [2]. It would be very interesting to generalize these results to the nonabelian case and to $d=3$ (monopoles). What solutions exist for $\lambda > 1$?

We remark that these results, and those characterizing multi-vortex solutions are similar in some respects to the characterization of finite action critical points of the σ-model action by n-instanton or n-anti-instanton solutions [8]. As above, the instantons are parameterized by positions $\{\alpha_j\}$; also $\mathscr{A} = n\mathscr{A}_1$, where $\mathscr{A}_1$ is the one-instanton action. A deeper connection between these models is not known.

Furthermore, the abelian-Higgs model on hyperbolic space is equivalent to an SU(2), pure Yang-Mills$_4$ theory with O(3) spatial symmetry [10]. The first order equations correspond to the self-duality equations $F = \pm *F$, i.e. Yang-Mills instantons. Taubes results generalize to hyperbolic space, and show that every O(3)-symmetric, finite action solution to the SU(2) Yang-Mills equations $D{*}F = 0$, must also satisfy $F = \pm *F$. Thus every such solution is n-instantons (or n-anti-instantons) on a line. No instanton-anti-instanton solutions with this symmetry exist. People conjecture this result is true without the assumption of O(3) symmetry.

4. QUANTIZATION

Why should classical solutions be of any importance for quantization? After all, the quantization of an action functional* $\mathscr{A}(\phi)$ can be viewed as the definition of a path space measure $d\mu(\phi)$ (functional integral) of the form

$$Z^{-1}e^{-\mathscr{A}(\phi)}\,[d\phi] \tag{9}$$

Here $[d\phi]$ denotes the formal product Lebesgue measure $d\phi(x)$ over space-time points x, and Z is a normalizing factor. Normally we establish the existence of $d\mu$ by approximating (9), e.g. by replacing space-time by a lattice, and then removing the approximation by e.g. taking the limit of zero lattice spacing. The measure $d\mu(\phi)$ is an integral over all classical field configurations, and (as in the case of Wiener measure where $\mathscr{A}(\phi) = \int \phi'(t)^2\,dt$) is concentrated on paths $\phi(\cdot)$ for which $\mathscr{A}(\phi)$ is not defined (i.e. $\mathscr{A}(\phi) = \infty$). Generically, the integral $d\mu(\phi)$ is concentrated on some space of distribution-valued classical fields. For a detailed elaboration of these ideas, see [11].

---

*For simplicity, we let ϕ denote here the set of all fields under consideration.

The classical solutions studied above, on the other hand, have finite action $\mathscr{A} < \infty$. Even without this restriction, the special ϕ's which are critical points of $\mathscr{A}$, i.e. solutions $\delta \mathscr{A}(\phi) = 0$,

$$\int_{\text{all } \phi} d\mu(\phi) = 1, \qquad \int_{\text{finite action } \phi} d\mu(\phi) = 0 \ . \quad (10)$$

So why are solutions useful?

The answer to this question lies in the details of $d\mu(\phi)$. So far, the only gauge theory for which $d\mu(\phi)$ has been constructed is the Higgs_2 model [12]. It is an interesting open problem to establish the spectral properties (e.g. mass gap) related to the vortex configurations in the Higgs model discussed above. But so far, the role of classical solutions has not been analyzed in this model.

The role of classical solutions, and more generally the "semi-classical approximation," has been understood in $P(\phi)_2$ models [13], so let me explain it in that case. Let us begin with a picture of our space of ϕ's, namely the classical configuration space of the field. For $P(\phi)_2$ models we take this to be $\mathscr{S}'(R^2)$, the space of all tempered distributions. This space contains both finite and infinite action ϕ's, in fact as explained above the measure of the finite action ϕ's is zero.

In the space $\mathscr{S}'$, let us choose one *finite action* configuration $\phi_{c\ell}$, where $c\ell$ denotes "classical." Around $\phi_{c\ell}$ we denote $\mathscr{D}$ its "domain of influence," cf. Figure 3. For all $\phi \in \mathscr{D}$, we expand the action $\mathscr{A}(\phi)$ in a Taylor series about the point $\phi_{c\ell}$, namely

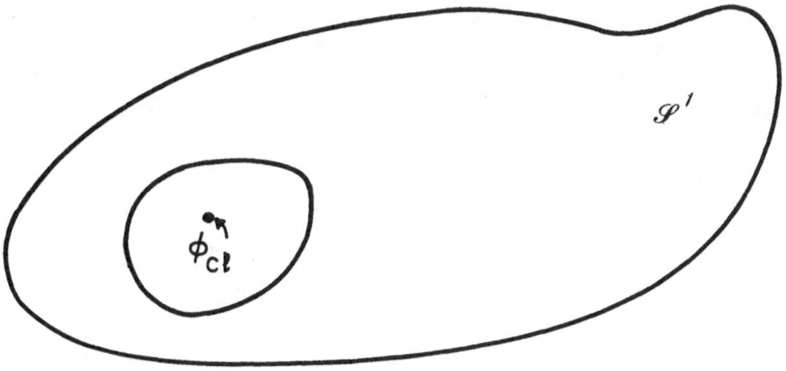

Figure 3. Schematic picture $\mathscr{S}'$, the space of ϕ's. $\mathscr{D}$ is the domain of influence of a particular classical configuration $\phi_{c\ell}$.

$$\mathscr{A}(\phi) = \mathscr{A}(\phi_{c\ell}) + Q(\phi-\phi_{c\ell}) + R(\phi-\phi_{c\ell}) \quad . \tag{11}$$

Here Q contains the quadratic terms in $\phi-\phi_{c\ell}$ and R denotes the remainder, $\mathscr{A}(\phi) - \mathscr{A}(\phi_{c\ell}) - Q$. We let $\mathscr{A}_{c\ell} \equiv \mathscr{A}(\phi_{c\ell})$ be the classical action of $\phi_{c\ell}$, which can be evaluated. Then

$$\exp(-\mathscr{A}(\phi)) = \exp(-\mathscr{A}_{c\ell})\exp(-Q)\exp(-R) \tag{12}$$

is a product of three factors:

$$(\text{classical}) \times (\text{Gaussian}) \times (\text{Remainder}) \quad . \tag{13}$$

Since the classical factor a constant which can be evaluated, and the Gaussian factor has moments which can be calculated, the success of using the expansion (12) rests on whether the remainder factor $\exp(-R)$ is close to 1 for all ϕ in $\mathscr{D}$. Whether or not this is true depends both on the choice of $\phi_{c\ell}$ and of $\mathscr{D}$. If this is the case, we neglect R and write

$$d\mu(\phi) \sim e^{-\mathscr{A}_{c\ell}} \, d\mu_{\text{Gaussian}}(\phi-\phi_{c\ell}) \quad . \tag{14}$$

Of course equation (14) is approximately valid only for $\phi \in \mathscr{D}$, and to give a complete argument we need to estimate the error term $\exp(-R)$. (In fact for some applications we may be interested in keeping the leading terms in the power series in R, namely $\Sigma_{j=0}^{N} R^j/j!$, and these provide corrections to the semiclassical approximation, cf. [13].)

5. PARTITION OF UNITY

In the best of all worlds, it will be possible to choose an (infinite) set of $\phi_{c\ell}^{(i)}$'s, labelled by an index i, such that our function space $\mathscr{S}'$ is divided up as in Figure 4. The domains $\mathscr{D}_i$ of dependence of $\phi_{c\ell}^{(i)}$ are chosen such that

$$\mathscr{S}' = \bigcup_i \mathscr{D}_i \tag{15}$$

In each $\mathscr{D}_i$ we have a small error term $R^{(i)}$. Then it is clear that we want to expand every integral over $d\mu(\phi)$ into a sum of integrals over the $\mathscr{D}_i$

$$d\mu(\phi) = \sum_i d\mu_i(\phi) = \sum_i \chi_i \, d\mu(\phi) \tag{16}$$

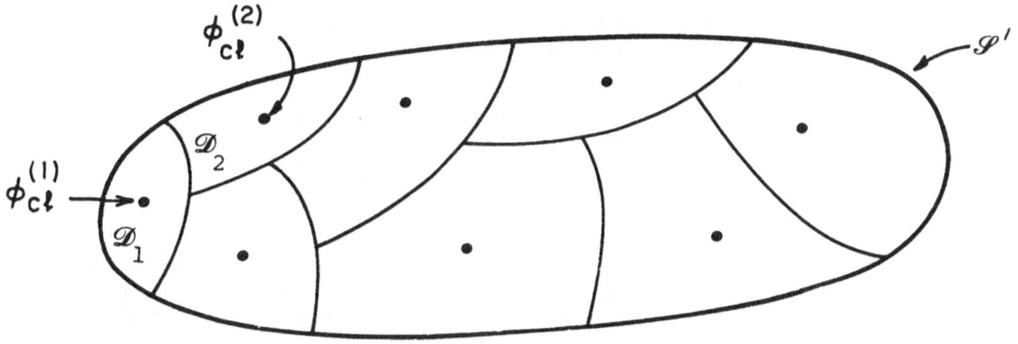

Figure 4. Dividing $\mathscr{S}'$ into domains $\mathscr{D}_i$ close to $\phi_{c\ell}^{(i)}$.

where

$$\chi_i = \chi_i(\phi) = \begin{cases} 1 & \phi \in \mathscr{D}_i \\ 0 & \text{otherwise} \end{cases} .$$

This is a partition of unity of function space, in this case $\mathscr{S}'$, into domains $\mathscr{D}_i$. On each $\mathscr{D}_i$ we use the expansion (12) about the appropriate point $\phi_{c\ell}^{(i)}$, and the problem is to show the choice of $\phi_{c\ell}^{(i)}$, $\mathscr{D}_i$ exists such that the $R^{(i)}$ have a small effect. The usefullness of the classical solutions, i.e. $\delta\mathscr{A} = 0$, is classical ϕ can motivate the choice of the $\phi_{c\ell}^{(i)}$. We may include among the $\phi_{c\ell}^{(i)}$ all minima of $\mathscr{A}$, i.e. all classical solutions for the above examples. But to analyze the errors it is generally necessary to also include configurations $\phi_{c\ell}^{(i)}(x)$ which are not solutions, but which are close to different classical solutions in different regions of space. In that case, the factors

$$e^{-\mathscr{A}(\phi_{c\ell}^{(i)})} \tag{17}$$

arising from the classical term in (12) can be interpreted, as i varies, as yielding a statistical mechanics of the classical solutions, interpreted locally. (This is the "instanton gas," and instanton-antiinstanton configurations should presumably be included, even if they are not solutions.) To be more concrete, we must study an example.

6. THE $\lambda\phi^4 - \phi^2$ MODEL

In order to understand the classical approximation, we study in detail the $d=2$ model [13]

$$\mathscr{A}(\phi) = \frac{1}{2}\cdot\int :(\nabla\phi)^2: dx + \int :\left\{\lambda\phi^4 - \phi^2 + \frac{1}{4\lambda}\right\}: dx \tag{18}$$

$$= T(\nabla\phi) + :V(\phi): \quad .$$

The basic idea is to insert (18) in (9) and to expand $\mathscr{A}$ about a point $\phi_{c\ell}$, i.e. a classical (given) approximation. Here $\phi_{c\ell}$ is not necessarily a solution. Then with $M^2 = V''(\phi_{c\ell})$, the classical mass squared, and $\phi = \phi_{c\ell} + \psi$,

$$\mathscr{A}(\phi) = \mathscr{A}(\phi_{c\ell} + \psi) = \mathscr{A}(\phi_{c\ell}) \frac{1}{2} \int : \left\{(\nabla\psi)^2 + \frac{1}{2} M^2\psi^2\right\}: dx$$
$$+ R(\psi) \quad , \tag{19}$$

and

$$d\mu(\phi) = e^{-\mathscr{A}(\phi_{c\ell}) - R(\psi)} d\psi_M \quad . \tag{20}$$

Here $d\psi_M$ is the mass M free field measure, i.e. a Gaussian with covariance $(-\Delta + M^2)^{-1}$. Then $R(\psi)$ is the remainder, consisting of linear terms and terms of degree 3 or 4 in ψ. (If $\phi_{c\ell}$ is a solution, the linear terms vanish.)

The goal of a good approximation is to pick $\phi_{c\ell}$ so that $R(\psi)$ is small. The potential $V(\phi)$ has the form shown in Figure 5. The only classical solutions are

$$\phi_+(x) = (2\lambda)^{-1/2} , \qquad \phi_-(x) = -(2\lambda)^{-1/2} \quad .$$

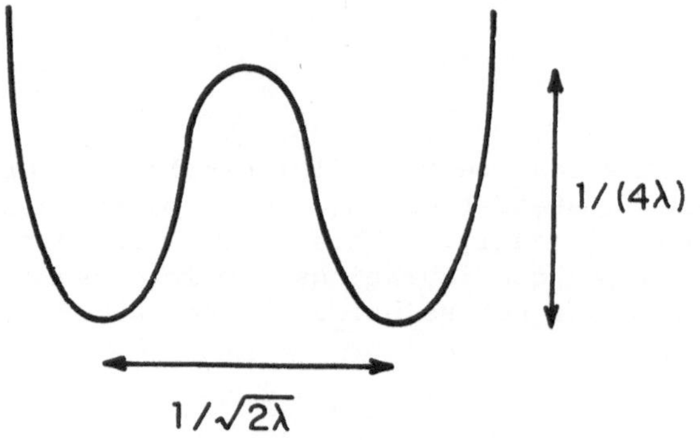

Figure 5. The $\lambda\phi^4 - \phi^2$ potential, $\lambda \ll 1$.

However these two (constant) configurations are insufficient to give a convergent expansion of the form (16), i.e., to show that the remainder $R_{\pm}$ is small for $\mathcal{D}_{\pm}$ which cover $\mathcal{S}'$. It is necessary [13] to consider in addition a set of $\mathcal{D}_i$'s defined by whether in each unit square Δ of space-time, the average ϕ, $\int_{\Delta} \phi(x)\, dx$, is positive (near ϕ_+) or negative (near ϕ_-). Thus we obtain an expansion in "phase islands" labelled by the two phases namely the classical solutions $\phi = \phi_+$, $\phi = \phi_-$, cf. Figure 6.

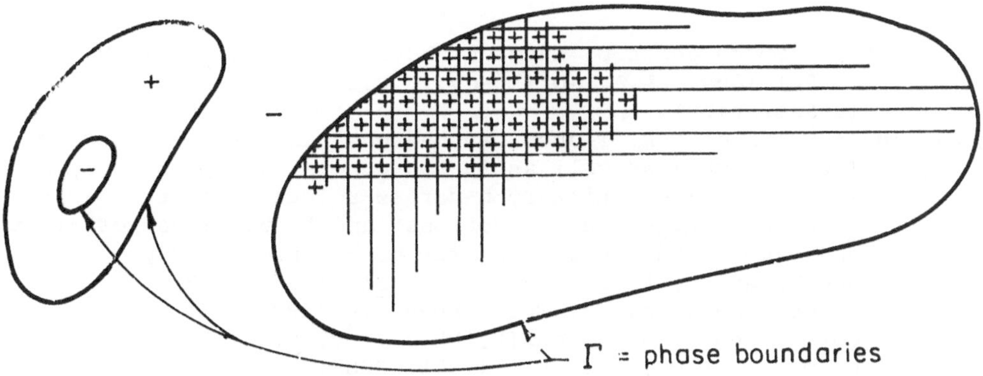

Figure 6. Phase domains (or islands) for the $\lambda\phi^4 - \phi^2$ model. Here a grid of unit squares is superimposed on space-time. A square Δ is + or − according to whether sgn $\int_{\Delta} \phi(x)\, dx$ is positive or negative. The boundaries Γ separating + from − regions (i.e., phase islands) index the domains $\mathcal{D} = \mathcal{D}_{\Gamma}$. This illustration shows small + islands in a sea of −.

For this choice of $\mathcal{D}_i = \mathcal{D}_{\Gamma}$, the classical action in (19) is

$$\mathcal{A}(\phi_{c\ell}) \sim \frac{1}{\lambda} |\Gamma|$$

where $|\Gamma|$ is the length of the boundary of the phase islands Γ in $\phi_{c\ell}$. Thus the classical action is similar to the energy of an Ising model with phase islands bounded by Γ. Equation (14) then becomes

$$d\mu(\phi) \sim d\mu_{\text{ISING}} \otimes d\psi_{\text{GAUSSIAN}}$$

and (16) becomes

$$d\mu \sim \sum_{\Gamma} d\mu^{\Gamma}_{\text{Ising}} \otimes d\psi_{\text{Gaussian}} \quad .$$

The analysis of phase transitions of the $\lambda\phi^4 - \phi^2$ model is contained in the Ising factor which suppresses tunnelling. The estimates to show that R can be neglected are worked out in detail in [13]. An extension to the two-phase Yukawa_2 models has been analyzed by Balaban and Gawedzki [14]. As mentioned above, it would be very interesting to develop such an expansion for the Higgs_2 model.

REFERENCES

1. V.L. Ginzburg and L.D. Landau, Zh. Eksp. Theor. Fiz. 20 (1950) 1064. A.A. Abrikosov, J.E.T.P. 5 (1957) 1174 (translation).
2. The results of C.H. Taubes can be found in his Harvard Ph.D. Thesis (1980) as well as in two papers to appear in Comm. Math. Phys.: "Arbitrary n-vortex solutions to the first order Ginzburg-Landau equations" and "Equivalence of first and second order equations for gauge theories."
3. L. Jacobs and C. Rebbi, Phys. Rev. B19 (1979) 4486.
4. E. Bogomol'nyi, Sov. J. Nucl. Phys. 24 (1976) 449.
5. H. deVega and F. Schnaposnik, Phys. Rev. D14 (1976) 1100.
6. E. Weinberg, Phys. Rev. D19 (1979) 3008.
7. T. Jonsson, O. McBryan, F. Zivilli, J. Hubbard, Comm. Math. Phys. 68 (1979) 259.
8. G. Woo, J. Math. Phys. 18 (1977) 1264. W. Garber, S. Ruijsenaars, E. Seiler and D. Burns, Ann. Phys. 119 (1979) 305.
9. R. Flume, Phys. Lett. 76B (1978) 593.
10. E. Witten, Phys. Rev. Lett. 38 (1977) 121. D. Forcas and N. Manton, Commun. Math. Phys., to appear.
11. J. Glimm and A. Jaffe, 1976 Cargèse Lectures, and References there, as well as their forthcoming book.
12. D. Brydges, J. Fröhlich and E. Seiler, Ann. Phys. 121 (1979) 227, and Comm. Math. Phys. to appear.
13. J. Glimm, A. Jaffe and T. Spencer, Ann. Phys. 101 (1976) 610 and 101 (1976) 631.
14. T. Balaban and K. Gawedski, preprint.

THE COUPLING CONSTANT IN A ϕ^4 FIELD THEORY

James Glimm*[1] and Arthur Jaffe+[2]

*The Rockefeller University, New York, NY 10021

+Harvard University, Cambridge, MA 02138

Abstract

Under simple hypothesis, the following statements are proved in a ϕ^4 field theory: (1) The coupling constant is continuous in the parameters, e.g. as the scaling limit is approached or as ultraviolet cutoffs are removed. (2) Zero coupling corresponds to a free field or a generalized free field. (3) Different types of amputation in the definition of the coupling constant differ only by a finite nonzero factor.

* * * * *

In this note, we show by elementary methods that the ϕ^4 coupling constant is a continuous function of the parameters under assumptions which should be appropriate for the critical point scaling limit or removal of the ultraviolet cutoff. In particular the limit theory is nontrivial (not a generalized free field) if and only if the coupling constants of the approximating theories are bounded away from zero. Our analysis applies also to the Ising model, at its critical point.

The motivation for this paper is the statement of Baker [1] that high temperature series expansions for the d=3 Ising model

---

1. Supported in part by the National Science Foundation under Grant PHY 78-08066.

2. Supported in part by the National Science Foundation under Grant PHY 79-16812.

give small but significant violations of hyperscaling: $d\nu + \gamma \neq 2\Delta$. If these numbers are significant, we agree with Baker that the critical point defines a trivial field theory.

We work with a coupling constant defined off mass shell (at p=0), in order to be able to perform the analysis. Particle amputation multiplies an external leg of a n-point function by

$$m^2 - p^2 \Big|_{p=0} = m^{+2}$$

while propagator amputation multiplies by

$$\left(Z^{-1} G^{(2)\sim}(p)\right)^{-1} \Big|_{p=0} = Z/\int G^{(2)}(x-y)\, d(x-y)$$

where

$$G^{(2)} = \langle\phi(x)\ \phi(y)\rangle - \langle\phi\rangle^2$$

is the truncated two point function of the ϕ^4 field theory. Also let $\chi = \int G^{(2)}\, d(x-y)$. The the coupling constant is

$$(1) \qquad 0 \leq g_\ell = -Z^{-2} m^{d+4} \int G^{(4)}(x_1,\ldots,x_4)\, dx_1 \ldots dx_3 \left(\frac{Z}{m^2\chi}\right)^\ell$$

In g_ℓ , ℓ legs have propagator amputation and $4-\ell$ legs have particle amputation. The bound

$$(2) \qquad Z/(m^2\chi) \leq 1$$

follows from the definition of field strength renormalization, so that

$$(3) \qquad g \equiv g_0 \geq g_1 \geq \cdots \geq g_4 \ .$$

It was first shown in [2] that

$$(4) \qquad g_2 = -m^d \int G^{(4)}/\chi^2 \leq \text{const.}$$

Since g_2 has the critical exponents,

(5) $$g_2 \sim \tau^{-2\Delta + \gamma + d\nu} \quad ,$$

this leads to the inequality

(6) $$-2\Delta + \gamma + d\nu \geq 0 \quad .$$

The subsequent consideration of (4)-(6) by other authors is summarized in [1].

In this paper we work only with g_0 . The finiteness of

(7) $$g_\ell / g_{\ell+1} = Z^{-1} m^2 \chi$$

follows from conventional scaling ideas, which we now assume. In particular we assume

$$0 \leq Z^{-1} m^{-d+2} G^{(2)}(x/m) \leq \begin{cases} O(1)\ x^{-(d-1)/2} e^{-|x|} , & |x| \geq 1 \\ O(1)\ x^{-d+2-\eta} , & |x| \leq 1 \end{cases}$$

We take $\eta < 2$ and the constants $O(1)$ above to be universal constants. Then (7) is bounded uniformly in all parameters. This gives us

Theorem 1. Assume (8) with universal constants. Then with the hypothesis of [2], g and all g_ℓ are bounded, by a universal constant. Either the g_ℓ are all nonzero or they are all zero. In the latter case, and only then, the theory is a generalized free field.

The final statement follows from [3].

Now we turn to the continuity of g under a limit process. We assume that m is bounded away from zero in the limit process. This allows a (positive mass) scaling limit at the critical point, and removal of ultraviolet cutoffs, for noncritical couplings. By scaling, we assume m is fixed at 1.

To illustrate the idea, we see that $g_\ell / g_{\ell+1}$ is convergent by the Lebesgue bounded convergence theorem. Thus is is sufficient to consider g_0 .

Theorem 2. Assume (8) with universal constants and the hypothesis of [2]. The each g_ℓ is continuous in a limit process for which the mass m is fixed and not zero, and for which the two and four point functions $G^{(2)}(x)$ and $G^{(4)}(x_1,\ldots,x_4)$ are pointwise convergent a. e.

Proof. By Lebowitz' and Griffiths' inequalities,

$$0 \leq -G^{(4)}(x_1,\ldots,x_4) \leq G^{(2)}(x_{i_1} - x_{i_2})G^{(2)}(x_{i_3} - x_{i_4})$$

for any permutation $\{i_1,\ldots,i_4\}$ of $\{1,\ldots,4\}$. After a permutation and translation of variables, we can suppose that $i_\nu = \nu$, $x_1 = 0$, and $|x_2 - x_3| \leq |x_3 - x_4|$. Also we may take $m = 1$. Then with $F(x)$ denoting the right hand member of (8),

$$\int F(x_1 - x_2)\ F(x_3 - x_4)\ dx_2\, dx_3\, dx_4$$

$$\leq \int_{|x_2 - x_3| \leq |x_3 - x_4|} F(x_3 - x_4)F(x_2)\ d(x_2 - x_3)\ d(x_3 - x_4)\ dx_2$$

$$\leq \int x^d\ F(x)\ dx \int F(x)\ dx < \infty\ .$$

Thus continuity follows from the Lebesgue bounded convergence theorem.

References

1. G. Baker, Phys. Rev. B, 15 p.1553 (1977).

2. J. Glimm and A. Jaffe, Ann. Inst. H. Poincaré 22 p.97 (1975).

3. C. Newman, Commun. Math. Phys. 41 p.1 (1975).

EXACT INSTANTON GASES

M. Lüscher

DESY

Hamburg

ABSTRACT

A brief summary of recent exact calculations of quantum fluctuations about multi-instanton configurations in non-linear σ-models and Yang-Mills theories is given.

INTRODUCTION

The infrared divergencies, which appear in the dilute instanton gas approximation[1] in QCD have so far been considered as "just another of the various infrared disasters of the theory to which we have no answer". A possible way out of this calamity has recently been discovered[2),3] in the analogous case of the two-dimensional CP^{n-1} models[4),13]. The suggestion is that one should drop the assumption that the instanton gas is dilute. Rather, one should work out all multi-instanton contributions exactly and sum them before taking the infinite volume limit. Physical quantities such as the free energy density of the gas or correlation functions of local fields may then well come out to be infrared finite.

A serious obstacle to this procedure is the fact that there are no stable instanton anti-instanton solutions to the Yang-Mills field equations. Such mixed configurations, however, must necessarily be taken into account to guarantee the cluster property of the approximation (cp. Sect. 4). What will be discussed here, is an amputated instanton gas involving pure multi-instanton solutions only.

To set up the exact instanton gas, several mathematical

problems must be solved:

(i) The multi-instanton solutions should be parametrized in an explicit and intuitively appealing way.

(ii) The determinant of the fluctuation operator about multi-instanton fields must be calculated.

(iii) The existence of the infinite volume limit of the ensuing exact instanton gas must be established.

All these steps have been carried out almost completely for the CP^{n-1} case. The resulting formulae are summarized and commented in Sect. 3. In the Yang-Mills case, no very explicit parametrization of multi-instanton solutions is known and the existence of the thermodynamic limit has only been speculated about [5]. Some progress has been made, however, in calculating determinants. More specifically, a closed expression for the determinant of the Dirac operator in an arbitrary external instanton field has been found [6]. How this has been achieved is reviewed in Sect. 2. Some conceptual difficulties with the exact instanton gas are pointed out in the final Sect. 4.

COMPUTATION OF DETERMINANTS

For the sake of definiteness we here consider the special case of the Dirac operator

$$(1) \qquad D = i\,\gamma_\mu(\partial_\mu + A_\mu)$$

where $A_\mu = -A_\mu^+$ is an arbitrary SU(n) k-instanton solution as given by the Atiyah-Hitchin-Manin-Drinfeld [7] construction. To compute the determinants of the fluctuation operators in CP^{n-1} models one can follow the general method explained below [3].

First of all we must define the determinant of the Dirac operator. A convenient volume cutoff is provided by replacing D by

$$(2) \qquad D_R = 2\,\Omega^{-1/2} D\,; \qquad \Omega \overset{\text{def}}{=} 4R^4(R^2+x^2)^{-2}$$

Formally,

$$(3) \qquad \lim_{R\to\infty} D_R = D$$

The advantage of working with D_R instead of D is that D_R has a purely discrete spectrum:

$$(4) \quad D_R \psi_n = \lambda_n \psi_n \; ; \qquad \int d^4x \, \Omega^{1/2} |\psi_n|^2 < \infty$$

This follows from the observation that if we consider x_μ to be stereographic coordinates of a sphere S^4 with radius R, then eq. (2) defines an elliptic operator on S^4, which has a discrete spectrum, of course.

A popular way to deal with the ultra-violet divergence, which appears when one naively tries to compute $\det D_R$, is the ζ-function finite part prescription. Namely, one defines a meromorphic function

$$(5) \quad \zeta(s) = \sum_n{}' (\lambda_n^2)^{-s} \qquad \text{(zero eigenvalues omitted)}$$

and sets

$$(6) \quad \Gamma_\zeta = \frac{1}{2} \ln \det{}' D_R^2 \overset{\text{def}}{=} -\frac{1}{2} \zeta'(0)$$

This makes sense, since $\zeta(s)$ is everywhere holomorphic except for simple poles at $s = 1$ and $s = 2$.

More familiar to physicists is the Pauli-Villars regularization method. Here, one introduces a set of regulator masses M_j, $j = 1,\ldots,\nu$ and signs $\varepsilon_j = \pm 1$ such that

$$(7) \quad \sum_{j=1}^{\nu} \varepsilon_j = -1$$

$$\sum_{j=1}^{\nu} \varepsilon_j M_j^{2p} = 0 \qquad (p = 1,\ldots,\nu-1)$$

Then one defines a regularized determinant of D_R:

$$(8) \quad \Gamma_{reg} \overset{\text{def}}{=} \frac{1}{2} \ln\{ \det{}' D_R^2 \prod_{j=1}^{\nu} \det(D_R^2 + M_j^2)^{\varepsilon_j} \}$$

Using the Seeley expansion [9] for the heat kernel of D_R^2, one can show that up to terms vanishing for large M_j, this equals

$$(9)\quad \Gamma_{PV} = \sum_{j=1}^{\nu} \varepsilon_j \ln M_j \{ a M_j^4 + b M_j^2 + c \} + d ,$$

where a,b,c and d are some numbers independent of M_j.

Whether the ζ-function method or the Pauli-Villars scheme is used is essentially irrelevant, since the two are closely related:

$$a = \lim_{s\to 2} (s-2)\,\zeta(s)$$

$$b = -2 \lim_{s\to 1} (s-1)\,\zeta(s)$$

(10)

$$c = 2\,\zeta(0) + \frac{k}{2}$$

$$d = -\,\zeta'(0)$$

In what follows we stick to the Pauli-Villars prescription, but one can obtain equivalent results by the ζ-function method .

In outline the calculation of Γ_{PV} proceeds as follows. One first determines the variation of Γ_{reg} with respect to the parameter of the instanton background field A :

$$(11)\quad \delta\Gamma_{reg} = \frac{1}{2}\,\mathrm{Tr}\{\delta D_R^2 [G + \sum_{j=1}^{\nu} \varepsilon_j G^{M_j}]\}$$

Here, G and G^{M_j} denote the Green's functions of D_R^2 and $D_R^2 + M_j^2$ respectively:

$$D_R^2 G = 1 - P_o \;;\; P_o\text{: zero mode projector}$$

(12)

$$(D_R^2 + M^2) G^M = 1$$

The crucial point is of course that the right hand side of eq. (11) is exactly calculable, because G is known [10] and G^M can be expanded for large M. Having computed $\delta\Gamma_{reg}$ (for large M_j) one obtains Γ_{PV} by integrating $\delta\Gamma_{reg}$ along the instanton manifold. The integration constant is fixed by computing Γ_{PV} for a special, highly symmetric instanton configuration, where all the eigenvalues of D_R are known and can be summed explicitly. Obviously, the whole calculation is quite complicated and lengthy, and it is surprising that the outcome can be written down in a few lines. To do this, we first have to introduce some more notation.

The general U(n) gauge potential A_μ with topological charge k giving rise to self-dual strengths can be represented in the form [7),10)]

(13) $$A_\mu = v^+\partial_\mu v \qquad ; \quad v^+v = 1$$

where v is an (n+2k) x n complex matrix. v is the solution matrix of a set of 2k complex linear equations

(14) $$v^+\Delta_{A'} = 0 \qquad (A' = 1,2)$$

with $\Delta_{A'}$ a spinor of (n+2k) x k matrices depending linearly on x_μ:

(15) $$\Delta_{A'} = a_{A'} + b^A x_{AA'}$$

$$x_{AA'} = x_\mu (e_\mu)_{AA'} ; \quad e_0 = 1\!\!1 , \; e_a = -i\sigma^a$$

Here, σ^a denote the Pauli matrices and spinor indices are raised and lowered following the rule

(16) $$\xi^A = \xi_B\, \epsilon^{BA} ; \quad \xi_A = \epsilon_{AB}\, \xi^B$$

$$\epsilon_{AB} = \epsilon^{AB} = -\epsilon_{BA} ; \quad \epsilon_{12} = 1$$

$a_{A'}$ and b_A are constant (n+2k) x k matrices, which parametrize the instanton solution and must be chosen such that [+)]

+) $\Delta^+_{A'}$ is the adjoint of $\Delta_{A'}$ in the spinor sense, i.e.

$$\Delta^+_1 = -\bar{\Delta}^T_2 ; \quad \Delta^+_2 = \bar{\Delta}^T_1$$

$$\Delta^{+}_{A'}\,\Delta_{B'} = -\,\epsilon_{A'B'}\,f^{-1} \qquad \text{for all } x. \tag{17}$$

To insure that the instanton solution A_μ thus constructed is non-singular, it is necessary and sufficient that the complex k x k matrix f^{-1} defined by eq. (17) is invertible for all x.

The determinant of the Dirac operator in an arbitrary background instanton field A_μ described by matrices $a_{A'}$ and b_A has been calculated along the lines explained above for any fixed world radius R [6]. In the flat space limit $R \to \infty$ the outcome reduces to

$$\begin{aligned}\lim_{R\to\infty}\left(\Gamma_{PV} - n\Gamma^{0}_{PV}\right) &= k\left\{\tfrac{2}{3}\sum_{i=1}^{\nu} \varepsilon_i \ln M_i - 4\zeta'(-1) - \tfrac{2}{3}\ln 2 + \tfrac{5}{12}\right\} \\ &+ \frac{1}{24\pi^2}\int d^4x \left\{ I_1(x) + \int_0^1 dt\; I_2(t,x)\right\}\end{aligned} \tag{18}$$

Here, Γ^{0}_{PV} is the determinant with vanishing background field, $\zeta(z)$ is Riemann's ζ-function and the integrands I_1 and I_2 are

$$\begin{aligned} I_1 = {} & \mathrm{Tr}\{ f\partial_\mu f^{-1}\, f\partial_\mu f^{-1}\, f\partial_\nu f^{-1}\, f\partial_\nu f^{-1}\} \\ & - 5\,\mathrm{Tr}\{ b^{+}_{A} b^{A} f\, b^{+}_{B} b^{B} f\} + 4k\,(1+x^2)^{-2} \end{aligned} \tag{19}$$

$$I_2 = \epsilon_{\mu\nu\rho\sigma}\,\mathrm{Tr}\{K^{-1}\partial_t K\, K^{-1}\partial_\mu K\, K^{-1}\partial_\nu K\, K^{-1}\partial_\rho K\, K^{-1}\partial_\sigma K\} \tag{20}$$

$$K(t,x) = (1-t)\,(1+x^2)\,\tfrac{1}{2}\, b^{+}_{A} b^{A} + t\, f^{-1}(x)$$

Thus, the log of the determinant of the Dirac operator can be expressed as an explicit integral of a rational function of the instanton parameter matrices $a_{A'}$ and b_A and the integration variables x_μ and t. Nevertheless, the result is disappointingly complicated so that I fear the estimation of multi-instanton effects in gauge theories will be rather difficult.

THE CP^{n-1} INSTANTON GAS

CP^{n-1} models in two Euclidean dimensions describe fields

$$z_\alpha(x); \quad \alpha = 1,\ldots,n; \quad x = (x_1,x_2); \quad |z|^2 = 1$$

of complex unit vectors. Fields z_α and z'_α related by a gauge transformation

$$z'_\alpha(x) = e^{i\Lambda(x)} z_\alpha(x)$$

are considered equivalent. The gauge invariant action and topological charge are respectively

$$S = \frac{n}{2f}\int d^2x\ \overline{D_\mu z}\cdot D_\mu z\ ; \qquad D_\mu = \partial_\mu - \bar{z}\cdot\partial_\mu z$$

(21)

$$Q = \frac{i}{2\pi}\int d^2x\ \epsilon_{\mu\nu}\ \overline{D_\mu z}\cdot D_\nu z$$

where $f > o$ denotes a coupling constant.

Instanton solutions are obtained by minimizing the action S in a fixed topological sector. For $Q = k > o$ almost all instanton configurations can be parametrized by

$$z_\alpha(x) = \frac{p_\alpha(s)}{|p(s)|}\ ; \qquad s = x_1 - i x_2$$

(22)

$$p_\alpha(s) = c_\alpha \prod_{j=1}^{k}(s - a_\alpha^j); \quad \alpha = 1,\ldots,n; \quad c_n = 1$$

There are thus $nk + n-1$ complex parameters:

$$c_\alpha \qquad (\alpha = 1,\ldots,n-1)$$

(23)

$$a_\beta^j \qquad (\beta = 1,\ldots,n\ ;\ j = 1,\ldots,k)$$

Let $\mathcal{O}$ be any observable composed from z-fields, for example

$$\mathcal{O} = \bar{z}_\alpha z_\beta(x)\, \bar{z}_\gamma(0)\, z_\delta(0)$$

The expectation value of $\mathcal{O}$ is formally defined by

$$(24) \quad \langle \mathcal{O} \rangle = Z^{-1} \int \mathcal{D}[z]\, \mathcal{O}\, e^{-S}$$

where Z is a normalization factor such that $\langle 1 \rangle = 1$. The pure instanton gas now arises from integrating (24) by the saddle point method, the saddle points being all the instanton solutions (22). Thus,

$$(25) \quad \langle \mathcal{O} \rangle_{\text{inst.}} = Z^{-1} \sum_{k=0}^{\infty} (k!)^{-n} \int \prod_j d^2\lambda_j \; J(\lambda)\, \mathcal{O}(\lambda) \times \\ \times e^{-\frac{n\pi}{f}k} \left[\det\left(\tfrac{2n}{f}\Delta(\lambda)\right)\right]^{-1}$$

Here, λ_j stands for the instanton parameters, $J(\lambda)$ is a collective coordinate Jacobian and $\mathcal{O}(\lambda)$ denotes the observable $\mathcal{O}$ evaluated for the instanton solution with parameters λ_j. The determinant of the fluctuation operator (Hessian) $\Delta(\lambda)$ must be regularized[3)+)] and can then be computed following the scheme explained in § 2. After renormalizing the coupling constant one obtains

$$(26) \quad \langle \mathcal{O} \rangle_{\text{inst.}} = Z^{-1} \sum_{k=0}^{\infty} (k!)^{-n} \, \mathfrak{z}^{nk} \int \frac{\prod_{\alpha=1}^{n-1} d^2 c_\alpha}{(\bar{c}_\gamma c_\gamma)^n} \prod_{\beta=1}^{n} \prod_{j=1}^{k} d^2 a_\beta^j \\ \times \mathcal{O}(c,a) \exp - \mathcal{U}(c,a)$$

with (cp. eq. (22))

+) In the CP^1-case, the determinant can also be calculated exploiting conformal invariance, see Ref. 2.

(27) $$z = m \frac{2n}{f} e^{-\frac{n+2}{2n}}$$

(28) $$U = \frac{n}{2\pi}\int d^2x \ln|p|^2 \partial_s \partial_{\bar{s}} \ln|p|^2 + \tfrac{1}{2}nk(\ln(\bar{c}_\gamma c_\gamma) - 1) - \sum_{\alpha=1}^{n}\left[k \ln|c_\alpha|^2 + \sum_{i<j} \ln|a^i_\alpha - a^j_\alpha|^2\right]$$

Here, m denotes the renormalization group invariant mass, which is related to the bare coupling constant by

(29) $$\frac{2\pi}{f} = - \sum_{i=1}^{\nu} \varepsilon_i \ln \frac{M_i^2}{m^2}$$

The expression (26) is still too complicated to say much about the properties of the instanton gas except in the CP^1 case, where the many body potential U simplifies:

(30) $$U = \sum_{i,j} \ln|a^i - b^j|^2 - \sum_{i<j}\{\ln|a^i - a^j|^2 + \ln|b^i - b^j|^2\}$$

($a^i = a^i_1$ and $b^i = a^i_2$). This is precisely the Coulomb interaction energy of k positive and k negative charges sitting at a^i respectively b^j. Thus, the CP^1 instanton gas is identical to the neutral Coulomb gas at temperature T = 1.

The thermodynamics of the Coulomb gas has been discussed by Fröhlich.[11] The gas is infrared finite but condenses at T = 1, i.e. the instanton density is infinite (this effect is not expected to occur in the CP^{n-1}, $n \geq 3$, and Yang-Mills instanton gases, because the interactions between instanton poles are less attractive in these cases). Some correlation functions still may make sense, however. Consider for example the (gauge invariant) spin field

$$q^a = \bar{z}_\alpha \sigma^a_{\alpha\beta} z_\beta \; ; \qquad \sigma^a : \text{Pauli matrices.} \tag{31}$$

Define the azimuthal phase

$$e^{i\phi} = \frac{q^1 + i q^2}{|q^1 + i q^2|} \tag{32}$$

From ordinary perturbation theory one would expect the two-point function of $e^{i\phi}$ decays with at most a power of the distance between the points. However, taking the instantons into account, one can show that [2) 12)]

$$\langle e^{i\phi(x)} e^{-i\phi(0)} \rangle_{\text{inst.}} \sim \exp - c\, z |x| \quad (|x| \to \infty) \tag{33}$$

where c is a constant number and z is the fugacity (27). Thus, instantons are capable to produce enough disorder such that at least some correlation functions of local fields cluster exponentially.

CONCEPTUAL QUESTIONS

Besides the purely computational difficulties, which arise when setting up the instanton gas approximation, there are also some more conceptual problems, to which I have no satisfactory answer.

a) The pure instanton gas violates parity, i.e.

$$\langle F_{\mu\nu} {}^* F_{\mu\nu} \rangle_{\text{inst.}} = \rho \neq 0 \tag{34}$$

Parity may of course be restored by adding the anti-instanton contribution. The resulting vacuum, however, would not be a pure quantum mechanical state: the cluster property would not hold. For example, one would expect that

$$\langle (F_{\mu\nu} {}^* F_{\mu\nu})(x) (F_{\rho\sigma} {}^* F_{\rho\sigma})(0) \rangle \xrightarrow[|x| \to \infty]{} \rho^2 \neq 0 \tag{35}$$

but

$$(36)\quad \langle F_{\mu\nu}{}^{*}F_{\mu\nu}\rangle = \frac{1}{2}\{\langle F_{\mu\nu}{}^{*}F_{\mu\nu}\rangle_{\text{inst.}} + \langle F_{\mu\nu}{}^{*}F_{\mu\nu}\rangle_{\text{anti-inst.}}\} = 0$$

What apparently is lacking are mixed instanton anti-instanton configurations. How to exactly incorporate those, in particular dense ones, is not known.

b) The scale size integral appearing in the single instanton contribution in the CP^{n-1} model is proportional to

$$(37)\quad \int_0^\infty d\lambda\, \lambda^{n-3}$$

For n = 2 this is not only divergent for large scale sizes λ but also at λ = o.+) The condensation of the pure instanton gas in the CP^1 case can be traced back to this ultraviolet divergence. In particular it leads to

$$(38)\quad \langle i\epsilon_{\mu\nu}\overline{D_\mu z}\cdot D_\nu z\rangle_{\text{inst.}} = \infty ,$$

which is unacceptable. If the model is renormalizable, all UV divergencies should ultimately be absorbed into appropriate coupling constant and wave function renormalizations. How the divergence (38) is to be removed is unclear, but it is conceivable that it disappears, when problem a) is solved.

c) One of the motivations to study CP^{n-1} models was to check the reliability of the instanton gas approximation by comparing it with the 1/n-expansion [13]. No such comparison has been possible yet, because the pure CP^{n-1} instanton gas is rather complicated and problem a) should presumably be solved first. To show that it is conceivable that instanton contributions do neither rise nor fall exponentially, when $n \to \infty$, consider the free energy F of the pure instanton gas neglecting the many body potential U (cp. eq. (26)). Thus,

+) A UV divergent scale size integral also occurs in QCD if there are sufficiently many (but not too many) flavors.

$$F = \lim_{V\to\infty} \frac{1}{V} \ln Z$$

(39)

$$Z = \sum_{k=0}^{\infty} (k!)^{-n} z^{nk} \int \frac{\Pi\, d^2c_\alpha}{(\bar{c}_\gamma c_\gamma)^n} \int_V \Pi\, d^2 a_\beta^j$$

which gives

$$F = nz = \frac{2m}{f\sqrt{e}} n^2 + O(n) \qquad (40)$$

REFERENCES

1. e.g. C. Callan, R. Dashen and D. Gross, Phys. Rev. D17 (1978) 2717
2. V.A. Fateev, I.V. Frolov and A.S. Schwarz, Nucl. Phys. B154 (1979) 1
3. G. Berg and M. Lüscher, Comm. Math. Phys. 69 (1979) 57
4. H. Eichenherr, Nucl. Phys. B146 (1978) 215
   V. Golo and A. Perelomov, Phys. Lett. 79B (1978) 112
5. A.A. Belavin, V.A. Fateev, A.S. Schwarz and Yu.S. Tyupkin, Phys. Lett. 83B (1979) 317
6. B. Berg and M. Lüscher, to appear in Nucl. Phys.
7. M.F. Atiyah, N.J. Hitchin, V.G. Drinfeld and Yu.I. Manin, Phys. Lett. 65A (1978) 185
   V.G. Drinfeld and Yu.I. Manin, Comm. Math. Phys. 63 (1978) 177
8. E. Corrigan, P. Goddard, H. Osborn and S. Templeton, Preprint Calt-68-726 (1979)
9. e.g. P.B. Gilkey, "The Index Theorem and the Heat Equation", Publish or Perish, Boston 1974
10. E.F. Corrigan, D.B. Fairlie, P. Goddard and S. Templeton, Nucl. Phys. B140 (1978) 31
    N.H. Christ, E.J. Weinberg and N.K. Stanton, Phys. Rev. D18 (1978) 2013
11. J. Fröhlich, Commun. Math. Phys. 47 (1976) 233
12. G. Lazarides, to be published
13. A. D'Adda, P. Di Vecchia and M. Lüscher, Nucl. Phys. B146 (1978) 63

PROPERTIES OF LATTICE GAUGE THEORY MODELS

AT LOW TEMPERATURES

Gerhard Mack

II. Institut für Theoretische Physik

Universität Hamburg, Germany

INTRODUCTION

In quark confinement physics, the center of the gauge group plays a crucial role.[1] This can be seen from a proper formulation of the problem. One would like to explain

(1) Quark confinement: There are no physical states with the flavor quantum numbers of a quark = all physical particles have integral baryon number etc.,

(2) Saturation of forces: The known physical hadrons are made of three quarks, or a quark and an antiquark, but not six or nine quarks etc..

The center $\Gamma = Z(3)$ of the gauge group $G = SU(3)$ of quantum chromodynamics consists of matrices γ^n, $n = 0, 1, 2$, with

$$\gamma = \exp 2\pi i\lambda_8/\sqrt{3} = e^{2\pi i/3}\mathbf{1} \tag{1.1}$$

Quark and gluon field transform under γ according to

$$q(x) \longrightarrow e^{2\pi i/3}q(x) \quad ; \quad A_\mu(x) \longrightarrow A_\mu(x) \tag{1.2}$$

The U(1) symmetry group generated by baryon number B consists of elements $\exp i\varphi B$ with $0 \leq \varphi < 6\pi$. They act on quark and gluon fields according to

$$q(x) \longrightarrow e^{i\varphi/3}q(x) \quad ; \quad A_\mu(x) \longrightarrow A_\mu(x) \ . \tag{1.3}$$

By comparing with (1.2) we see that

$$\gamma = e^{2\pi i B} \quad , \tag{1.4}$$

since the action of both sides on quark and gluon fields, and therefore also on all physical states, is identical. In conclusion, the combined symmetry group is

$$(SU(3)_{local} \times U(1)_{global})/Z(3)_{global} \tag{1.5}$$

and quark confinement in quantum chromodynamics as formulated above means that all physical states transform trivially under global (x-independent) transformations in the center of the gauge group.

It is instructive to imagine a world with quarks that transform as octets, decuplets, or any other representation of the gauge group SU(3) that restricts to a trivial representation of the center. Such quarks would come free, i.e. physical states with their flavor quantum numbers would exist. From the work of Kogut and Susskind[2] we know that this is true even in the strong coupling limit of a lattice theory, where quark triplets are confined[3] (cp. Glimms lectures at this school[4]). The reason is simple. A string between such hypothetical quarks can break by creation of gluon pairs.

In these lectures we will study pure Yang Mills theories on a Euclidean lattice in three and four dimensions. Often we will take G = SU(2) as our gauge group. Its center $\Gamma = Z(2)$. The Euclidean lattice formulation makes methods of classical statistical mechanics applicable. Choosing an SU(2) gauge group brings about some technical simplifications. Omission of dynamical quarks is a possibly serious mutilation though. It has not been proven that dynamical quarks cannot play an important role in their own confinement. There are models[5] in which charged scalar fields help in stabilizing the confining "high temperature phase" of the model. One of them will be considered briefly in the next section. In quantum chromodynamics one could think of Cooper pair formation to get scalars[6,7]. But saturation of forces is left unexplained in such a scheme. One hopes therefore that gluons, which are charged in a nonabelian theory, do the stabilizing job themselves. One hopes also that understanding of the phase structure and universality and stability properties of a pure Yang Mills theory will provide the tools to justify perturbative treatment of dynamical quarks, at least if they are not massless.

In view of what was said earlier, the idea emerges naturally to look at the SU(N) theory as a kind of Z(N) gauge theory. Such "reduction of the gauge group" will be an important theme in these lectures. A Z(N) theory of quark confinement provides a natural qualitative explanation of saturation of forces because, roughly speaking, only objects with nonzero N-ality can be confined by Z(N) gauge quanta[8]. It is not so obvious at first sight how fluctuations of Z(N) variables may continue to be important and confine static quarks in the continuum limit, though. (Indeed, in the standard Z(2) model on a lattice these variables freeze and this leads to a phase transition to a nonconfining phase.) An answer to this question will be proposed in section 10. It will be seen there that the values of the relevant Z(2) variables in a SU(2) theory are determined by topological properties of a SO(3) gauge field in this limit. (SO(3) = SU(2)/Z(2)).

2. A HIGGS MODEL WITH PERSISTENT CONFINEMENT OF THE (SCALAR) QUARKS

On a lattice, quarks which transform nontrivially under the center of the gauge group are confined at high temperatures, i.e. for small values of $\beta = 4/g^2$, g = bare coupling constant. One hopes for a smooth transition with persistent confinement to a continuum limit. The continuum limit is supposed to be reached when $\beta \to \infty$. There exists a model[6] in four dimensions in which this hope can be shown to materialize. It is one of the models that were investigated by Fradkin and Shenker[5] and others[3,9].

The model differs from quantum chromodynamics (QCD) in that the matter fields are scalar rather that fermion fields, and there is a buildt-in mass scale (other than the lattice spacing). The gauge group is SU(2). In addition there is a global SU(2) symmetry which we call isospin. It takes the place of the baryon U(1) in QCD. The combined symmetry group is

$$(SU(2)_{local} \times SU(2)_{global})/Z(2)_{global} \tag{2.1}$$

in place of (1.5). The scalar quarks transform nontrivially under the center Z(2) of the gauge group. They carry isospin $\frac{1}{2}$. By quark confinement we mean again absence of physical states with the flavor quantum numbers of the scalar quarks. This will be true if all physical states have integral isospin. This is again equivalent to the requirement that all physical states transform trivially under the center of the gauge group.

The model possesses a continuum limit in which quark confinement in this sense persists. It describes an isotriplet of free massive vector mesons.

The Euclidean action L(U, φ) of the model is a function of string bit variables $U(b) \in SU(2)$ which are attached to links b of the 4-dimensional lattice, and of a doublet of complex scalar fields

$$\varphi(x) = \begin{pmatrix}\varphi_1(x)\\ \varphi_2(x)\end{pmatrix} \quad ; \quad |\varphi_1(x)|^2 + |\varphi_2(x)|^2 = 1 \tag{2.2}$$

It splits in the usual way,

$$L(U, \varphi) = L_G(U) + L_M(U, \varphi) \quad . \tag{2.3}$$

The kinetic term L_G for the gauge field has the usual form[3] for an SU(2) lattice gauge theory. The boundary $\dot{p} = \partial p$ of a plaquette consists of four links $b_1 \ldots b_4$, and we write

$$U(\dot{p}) = U(b_4) \ldots U(b_1) \tag{2.4}$$

In this notation (tr = trace)

$$L_G(U) = \frac{\beta}{2} \sum_p \mathrm{tr}(\, U(\dot{p}) - 1\,) \tag{2.5}$$

Sum is over all unoriented plaquettes of the lattice.

The matter part L_M takes its most convenient form when expressed in terms of SU(2) matrices

$$\Phi(x) = \begin{pmatrix} \varphi_1(x) & -\bar{\varphi}_2(x) \\ \varphi_2(x) & \bar{\varphi}_1(x) \end{pmatrix} \tag{2.6}$$

In this notation

$$L_M(U, \varphi) = \varkappa \sum \mathrm{tr}(\Phi(x)^* U(xy) \Phi(y) - 1) \quad . \tag{2.7}$$

The parameter $\varkappa > 0$ introduces a mass scale into the theory.

Under a gauge transformation $V_1(x)$,

$$\Phi(x) \longrightarrow V_1(x)\Phi(x) \tag{2.8a}$$

$$U(xy) \longrightarrow V_1(x) U(xy) V_1(y)^{-1} \tag{2.8b}$$

(2.8a) says that the matter field $\varphi(x)$ transforms as a doublet under gauge transformations.

From Eq.(2.7) we see that the model has in addition another global SU(2) symmetry. We call it isospin. Isospin rotations act according to

$$\Phi(x) \longrightarrow \Phi(x) V_2^{-1} \tag{2.9a}$$

$$U(xy) \longrightarrow U(xy) \tag{2.9b}$$

for $V_2 \in SU(2)$ independent of x.

If $V_2 = -1$ then it commutes with all SU(2) matrices. Its action on Φ, U agrees therefore with the action of a global gauge transformation $V_1(x) \equiv -1$ in the center Z(2) of the gauge group. The combined symmetry group is therefore as indicated in (2.1).

To visualize the action of the symmetry transformations, it is convenient to regard $\varphi(x)$ as a real unit 4-vector whose components are real and imaginary parts of $\varphi_1(x)$ and $\varphi_2(x)$. Both the gauge transformations (2.8a) and the isospin rotations (2.9a) act as rotations on this 4-vector. They are different unless $V_1 = V_2 = -1$. The rotations are specified by the isomorphism

$$(SU(2) \times SU(2))/Z(2) \simeq SO(4) \tag{2.10}$$

4-vectors transform according to the representation $(\frac{1}{2},\frac{1}{2})$ of SO(4). It restricts to the real representation $(\frac{1}{2})\oplus(\frac{1}{2})$ on each of the two SU(2) ideals. Thus our matter field $\varphi(x)$ carries isospin $\frac{1}{2}$. The gluon field carries no isospin according to (2.9b).

It was shown by Fradkin and Shenker that there exist values $\beta_o > 0$ and $k_o < \infty$ such that the model admits convergent cluster expansions whenever either

$$0 \leqslant \beta < \beta_o \quad \text{or} \quad \frac{\varkappa}{\beta} > k_o \; . \tag{2.11}$$

The two areas overlap. Convergence of the cluster expansions ensures analyticity of the free energy and correlation functions, and uniqueness and invariance of the ground state under isospin rotations. For small β and $\varkappa$ one can also find a complete set of eigenstates of the transfer matrix (Hamiltonian) which retain finite energy in the infinite volume limit. They all carry integral isospin, therefore quarks are confined in the sense of our definition. It is reasonable to believe (and ought to be proven) that this remains valid throughout the domain (2.11).

The cluster expansions are obtained[5] by first transforming to gauge invariant variables W(b) given by

$$W(xy) = \Phi(x)^* U(xy)\Phi(y) \quad , \quad \text{whence} \quad trU(\dot{p}) = trW(\dot{p}) \; . \tag{2.12}$$

Subsequently one writes

$$\exp\frac{\beta}{2}tr(W(\dot{p}) - 1) = 1 + f_p(W) \tag{2.13}$$

and expands in products of f's. Results are expressed in terms of cluster integrals A(P),

$$A(P) = \int \prod_b d\nu(W(b)) \prod_{p\in P} f_p(W) \tag{2.14}$$

$$d\nu(W) = z^{-1} dW \exp\varkappa trW \quad ; \quad z = \int dW\, e^{\varkappa tr\, W}$$

dW is normalized Haar measure on SU(2). P are connected collections of plaquettes.

Given a function A of P, one defines the ξ-norms for $\xi > 1$ by[10,11]

$$\|A\|_\xi = \max_b \frac{1}{\xi} \left(1 + \sum_{\substack{P \\ b\epsilon P}} |A(P)|\, \xi^{|P|}\right) \tag{2.15}$$

b ϵP means b ϵ $\dot{p}$ for a plaquette p in P . $|P|$ is the number of plauettes in P. It is well known[11] that cluster expansions for free energy, correlation functions etc. converge if there is $\xi > 1$ such that $\|A\|_\xi < 1$. This is fulfilled in the range (2.11) of parameters; ξ can be chosen to depend only on $k = \varkappa/\beta$. To prove this, one estimates A(P) by use of the inequality

$$\mathrm{tr}\,(W(\dot{p}) - 1) \geq 4 \sum_{b\epsilon\dot{p}} \mathrm{tr}\,(W(b) - 1) \tag{2.16}$$

Summation is over all four links b in the boundary $\dot{p}$ of a plaquette p.

A continuum limit of the model can be approached in two steps. First one lets

$$\beta \rightarrow \infty \;, \; \varkappa \rightarrow \infty \qquad \text{with} \quad \frac{\varkappa}{\beta} = k \quad \text{fixed and} > k_o \;. \tag{2.17}$$

In this way one stays inside the domain (2.11) of validity of the cluster expansions. For fixed $k > k_o$, their convergence is uniform in β. In this limit one obtains a free field theory on the lattice, with a mass determined by k. Convergence to a free field theory can be proven by showing that

$$\|A(P) - A_o(P)\|_\xi \rightarrow 0 \quad \text{as} \quad \beta \rightarrow \infty \;. \tag{2.18}$$

where $A_o(P)$ are the cluster integrals for a free field theory. This implies convergence of the free energy and of all correlation functions.

Subsequently one lets $k \rightarrow 0$, so that the correlation length in units of the lattice spacing becomes infinite. Such a continuum limit of a free field theory presents no problems. It suffices to inspect the two point functions. As a result one obtains a theory of free massive vector mesons. They are described by a triplet $B^c{}_\mu(x)$ of vector fields which are related to the W-variables on the lattice by

$$W(b) = \exp -igaB(b) \quad ; \quad B(b) = \sum_c B^c{}_\mu(x)\, \frac{\tau^c}{2} \tag{2.19}$$

τ^c are Pauli matrices, $g = (4/\beta)^{1/2}$, a = 1 in units of lattice spacing, and b is the link which leaves x in the μ-direction. It follows that these fields carry isospin one since they transform under isospin rotations according to

$$B^c{}_\mu(x) \longrightarrow R(V_2)^c{}_d B^d{}_\mu(x) \tag{2.20}$$

$R(V_2)$ is the SO(3) rotation associated with the element V_2 of SU(2). In conclusion, there are no physical states with the flavor quantum numbers of the quarks in the continuum limit - they remain confined.

3. DESCRIPTION OF SU(2) MODELS

From now on we restrict our attention to pure Yang Mills theories without matter fields, with a gauge group G that possesses a nontrivial center Γ. The action depends on variables $U(b) \in G$ that are attached to links b, $U(b) \longrightarrow U(b)^{-1}$ under reversal of the direction of the link b. If C is a path consisting of links $b_1 \ldots b_n$ one defines the parallel transporter U(C) along C by

$$U(C) = U(b_n) \ldots U(b_1) \tag{3.1}$$

In the special case (2.4), $C = \dot{p}$ is the boundary of a plaquette p. The action will be taken to be of the form

$$L(U) = \sum_p \mathcal{L}(U(\dot{p})) \tag{3.2}$$

Sum is over all unoriented plaquettes. $\mathcal{L}$ is supposed to be bounded above, real, and it must satisfy $\mathcal{L}(V) = \mathcal{L}(V^{-1}) = \mathcal{L}(V_1 V V_1^{-1})$ to ensure gauge invariance.

In the standard SU(2) model of Wilson[3]

$$\mathcal{L}(V) = \frac{\beta}{2} \operatorname{tr} (V - 1) \tag{3.3}$$

and the path measure is given by

$$d\mu(U) = \frac{1}{Z} e^{L(U)} \prod_b dU(b) \quad . \tag{3.4}$$

dU(b) is normalized Haar measure on SU(2).

We will also consider a modified model proposed by Petkova and the author[12]. It has the same Lagrangean (3.3), but the admissible configurations U are restricted by the constraint

$$\prod_{p \in \dot{c}} \operatorname{tr} U(\dot{p}) > 0 \qquad \text{for every elementary cube c.} \tag{3.5}$$

Product is over the six plaquettes in the boundary $\dot{c}$ of a 3-dimensional cube. As a result, the path measure takes the form

$$d\mu(U) = \frac{1}{Z} e^{L(U)} \prod_c \Theta\Big(\prod_{p \in \dot{c}} \operatorname{tr} U(\dot{p})\Big) \prod_b dU(b) \tag{3.6}$$

For either model, the expectation value of an observable F(U) is given by

$$\langle F \rangle = \int d\mu(U)\, F(U) \tag{3.7a}$$

We are particularly interested in the expectation value of the Wilson loop observable

$$\langle \mathrm{tr}\, U(C) \rangle = \int d\mu(U)\, \mathrm{tr}\, U(C) \tag{3.7b}$$

where C is a rectangular path enclosing an area of L·T plaquettes. The partition function Z is always defined by the requirement that $\langle 1 \rangle = 1$.

In the modified model, it is most convenient to impose cyclic boundary conditions on the cosets $\bar{U}(b) = U(b)\Gamma \in SO(3)$ only.

The modified model has formally the same continuum limit as the standard model. In the continuum limit $\beta \to \infty$ one has in either model

$$U(\dot{p}) \to 1 \qquad \text{as} \qquad \beta \to \infty \tag{3.8}$$

It follows that constraint (3.5) is almost always fulfilled in the standard model for any particular cube c, in the limit. The limit behavior (3.8) is to be understood in the following probabilistic sense.

Let P be any collection of n distinct plaquettes $p_1 \dots p_n$. Then the probability $p(\xi_1,\dots,\xi_n)$ that $\mathrm{tr}(U(\dot{p}_j)-1) \leq -\xi_j$ for all $j=1\dots n$ is bounded by

$$p(\xi_1,\dots,\xi_n) \leq \prod_j D(\beta, \xi_j) \tag{3.9}$$

$$D(\beta, \xi) = \mathrm{const}\cdot\beta\, e^{-\beta\xi/48} \to 0 \quad \text{as} \quad \beta \to \infty \quad \text{if } \xi \neq 0.$$

This implies (3.8) because $\mathrm{tr}(U(\dot{p})-1) \leq 0$, and $= 0$ only if $U(\dot{p}) = 1$. Inequalities (3.8) derive from chessboard estimates[13,12]. The factor 1/48 in the exponent is probably far from optimal.

In a later section, we shall want to compare the SU(2) models with the standard Z(2) gauge theory model[14,15]. It has variables $\sigma(b) = \pm 1$ and action

$$L(\sigma) = \sum_p \beta(\sigma(\dot{p}) - 1) \tag{3.10}$$

It is obtained from the SU(2) models with the same value of ß by restricting variables U(b) to the center of the gauge group. The constraint (3.5) is then automatically fulfilled.

4. THE 'T HOOFT DISORDER PARAMETER

In a pure Yang Mills theory on a lattice, the quantum field theoretic Hilbert space of physical states consists of wave functions $\Psi(U)$. They depend on variables $U(b)$ that are attached to links b in the (Euclidean) time t=0 hyperplane Σ.The scalar product of two such wave functions Ψ_1 and Ψ_2 is of the form

$$(\Psi_1,\Psi_2) = \int d\rho(U)\, \bar{\Psi}_1(U)\, \Psi_2(U) \tag{4.1a}$$

One may take $d\rho(U) = \Pi dU(b)$ (product over all links b in Σ). For our purposes it is however more convenient to extract a common multiplicative factor from all wave functions so that

$$d\rho(U) = \frac{1}{Z} \prod_{b\in\Sigma} dU(b) \exp \sum_{p\in\Sigma} \mathcal{L}(U(\dot{p})) \tag{4.1b}$$

The vacuum state is given by

$$\Omega(U) = \int \prod_{b>0} dU(b) \exp \sum_{p>0} \mathcal{L}(U(\dot{p})) \tag{4.2}$$

$b > 0$ resp. $p > 0$ are all links resp. plaquettes in the half space $t > 0$, excluding those in Σ.Note that $U(\dot{p})$ may involve variables $U(b)$ with $b\in\Sigma$, cp. (2.4). These are not integrated over, instead Ω depends on them.

Observables $F(U)$ which depend only on variables $U(b)$ with $b\in\Sigma$ act on wave functions Ψ as multiplication operators. One has in this case

$$\langle F \rangle = (\Omega, F\Omega) \tag{4.3}$$

They form a complete set of commuting operators.

The formulae given above are appropriate for the standard SU(2) model, etc. For the modified model, step functions must be included which restrict integration to configurations which satisfy constraint (3.5).

Let S be any set of links b in the time t=0 hyperplane Σ,and σ an element of the center Γ of the gauge group G. The 't Hooft operator $B_\sigma(S)$ is defined by its action on wave functions[16]

$$(B_\sigma(S)\Psi)(U) = \Psi(U_\sigma) \tag{4.4}$$

with

$$U_\sigma(b) = \begin{cases} U(b)\sigma^{-1} & \text{if } b \in S \\ U(b) & \text{otherwise} \end{cases}$$

For $\Gamma = Z(2)$ there is only one nontrivial element $\sigma = -1$ of Γ , and we write $B(S)$ in place of $B_{-1}(S)$ in this case. The 't Hooft disorder parameter is the expectation value of $B_\sigma(S)$,

$$\langle B_\sigma(S) \rangle = (\Omega, B_\sigma(S)\Omega) \quad . \tag{4.5}$$

It is convenient to use the coboundary operator $\hat{\partial}$ (= boundary operator on the dual lattice)[12,17]. Let us restrict attention to the $\nu-1$ dimensional lattice Σ for a moment, and suppose that $\Gamma = Z(2)$ from now on. If S is a set of links in Σ then $\hat{\partial}S$ consists of those plaquettes p in Σ which have an odd number of links $b \epsilon S$ in their boundaries. One is mainly interested in S, $\hat{\partial}S$ of the form shown in figure 1. Because of gauge invariance, $\langle B_\sigma(S) \rangle$ turns out to depend on S only through $\hat{\partial}S$.

As a consequence of its definition, the 't Hooft operator satisfies the following commutation relations with the multiplication operator tr U(C) for closed loops C in Σ ('t Hooft algebra[16])

$$B(S) \operatorname{tr} U(C) = \xi \operatorname{tr} U(C)\, B(S) \quad , \quad \xi = \pm 1 \quad \text{if} \quad \Gamma = Z(2) \; .$$

Let C be the boundary of a surface Ξ in Σ. Then $\xi = -1$ if Ξ contains an odd number of plaquettes in $\hat{\partial}S$, and $\xi = +1$ otherwise. ξ counts thus how many times (mod 2) $\hat{\partial}S$ winds around C.

It follows from the definitions (4.4), (4.5) and the explicit formula (4.2) for the vacuum wave function that[12]

$$\langle B(S) \rangle = \langle \exp \sum_{b \epsilon S} \left\{ \mathcal{L}(-U(\dot{p}_b)) - \mathcal{L}(U(\dot{p}_b)) \right\} \rangle \geq 0 \; . \tag{4.6}$$

p_b is the plaquette which protrudes from link $b \epsilon \Sigma$ in positive time direction.

A duality transformation reveals[12,42] that $\langle B(S) \rangle$ may be interpreted as expectation value of a Wilson loop operator for a small static Z(2) monopole (of size 1 lattice spacing cubed) in $\nu=4$ dimensions (cp. section 8). The loop is $\hat{\partial}S$, this is indeed a closed loop on the dual lattice of Σ (see figure 1). It will be seen in section 10 that both the modified and the standard SU(2) model may be interpreted as Z(2) gauge theories without dynamical Z(2) monopoles. Therefore the asymptotic behaviour of $\langle B(S) \rangle$ for large loops $\hat{\partial}S$ will determine whether small static Z(2) monopoles are confined or not.

All the formulae of this section are also valid for a theory with gauge group $G = \Gamma = Z(2)$, in particular for the standard Z(2) model of section 3.

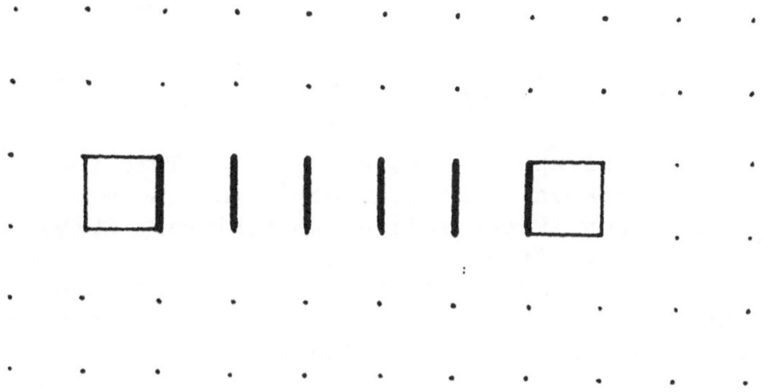

Figure 1a

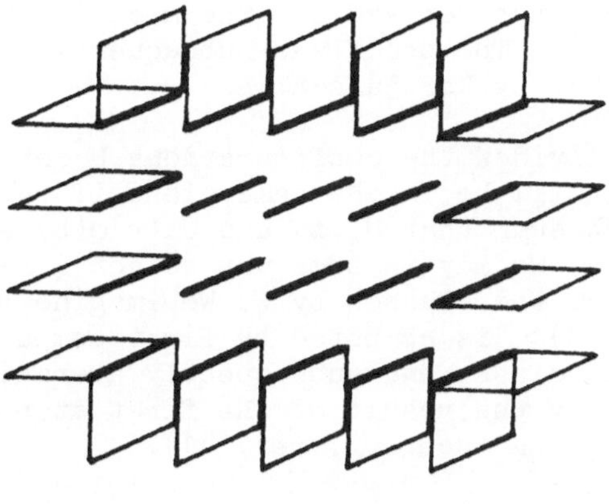

Figure 1b

Fig.1. Argument S of the 't Hooft operator B(S) in the time zero hyperplane Σ. Sets S of links (heavy lines), and plaquettes in $\hat{\partial}S$ (squares).
(a) for a $\nu-1 = 2$ dimensional lattice Σ.
(b) for a $\nu-1 = 3$ dimensional lattice Σ.

5. INEQUALITIES RELATING THEORIES WITH GAUGE GROUPS SU(2) AND Z(2)

We consider the SU(2) models described in section 3 and compare them with the standard Z(2) model, with the same value of β in the two Lagrangeans (3.3) and (3.10). The expectation values of the Wilson loop observables in these SU(2) and Z(2) models are related by the inequality

$$|\langle \mathrm{tr}\, U(C) \rangle_{SU(2)}| \leq 2 \langle \sigma(C) \rangle_{Z(2)} \quad . \tag{5.1}$$

The factor of 2 comes about because tr $1_2 = 2$. This inequality was first derived by Petkova and the author[12] for the modified SU(2) model. It was subsequently generalized to arbitrary SU(N) models by Fröhlich[18]. It follows from (5.1) that the Wilson loop expectation value shows an area law decay in the SU(2) theory whenever the same is true in the Z(2) theory. According to Wilson[3], an area law decay implies that static quarks are confined.

There is also an inequality for the 't Hooft disorder parameter. It goes in the opposite direction.

$$\langle B(S) \rangle_{SU(2)} \geq \langle B(S) \rangle_{Z(2)} \geq 0 \tag{5.2}$$

This inequality was also first derived for the modified SU(2) model by Petkova and the author[12]. The proof was subsequently generalized to the standard SU(2) model by Korthals Altes[19].

To prove (5.1) one divides the configurations U into classes[20]. Configurations U_1 and U_2 will be in the same class if there exist $\sigma(b) = \pm 1$ for each link b such that $U_1(b)$ and $U_2(b)\sigma(b)$ are related by a gauge transformation. We may in some way select a representative W in each class, and label the classes by W. We imagine that the expectation value $\langle \mathrm{tr}\, U(C) \rangle$ is computed by first computing the average $\langle \mathrm{tr}\, U(C) \rangle_W$ in each class, and subsequently averaging over W also. In probability theory the result of the first step is called a <u>conditional expectation value</u> (see e.g. ref. 41).

The conditional expectation values $\langle \cdot \rangle_W$ are computed by averaging over the variables σ which relate configurations in the same class. (The average over the gauge group is trivial.) Explicitly,

$$\langle \mathrm{tr}\, U(C) \rangle_W = \mathrm{tr}\, W(C) \left\{ Z_W^{-1} \int \prod_b d\sigma(b)\ \sigma(C)\ \exp \sum_p K_p(W)\sigma(\dot{p}) \right\} \tag{5.3}$$

with

$$K_p(W) = \frac{\beta}{2}\, \mathrm{tr}\, W(\dot{p}) \tag{5.4}$$

$d\sigma$ is normalized Haar measure on Z(2), viz.

$$\int d\sigma \; (\ldots) = \frac{1}{2} \sum_{\sigma=\pm 1} (\ldots) \quad , \tag{5.5}$$

and

$$Z_W = \int \prod_b d\sigma(b) \; \exp \sum_p K_p(W) \sigma(\dot{p}) \quad . \tag{5.6}$$

The expression in { } in (5.3) is expectation value of the Wilson loop in a Z(2) gauge theory with space time dependent coupling constants $K_p(W)$. Because of Eq.(5.4) they are bounded by

$$|K_p(W)| \leq \beta \tag{5.7}$$

In the modified SU(2) model one can choose W so that $K_p(W) \geq 0$. It follows then from the second Griffiths inequality[21] that the expectation value $\langle \sigma(C) \rangle$ is decreased when one replaces coupling constants K_p by ß. As a result

$$|\langle \text{tr } U(C) \rangle_W| \leq 2 \langle \sigma(C) \rangle_{Z(2)} \qquad \text{for all } W \; . \tag{5.8}$$

There are generalized Griffiths inequalities[22] which assert that (5.8) follows from (5.7) even if K_p are not necessarily nonnegative. The result generalizes therefore to the standard SU(2) model[18,19].

Finally one can average over the classes to obtain $\langle \text{tr } U(C) \rangle$. Inequalities (5.8) imply (5.1) because

$$|\langle \text{tr } U(C) \rangle| \leq \sup_W |\langle \text{tr } U(C) \rangle_W| \tag{5.9}$$

The other inequalities (5.2) were derived in ref. 12 by first performing a duality transformation on the Z(2) variables σ, followed by a similar argument as above. The direction of the inequality (5.2) is the reverse of that in (5.1) because the duality transformation takes small coupling constants into large ones and vice versa.

6. PHASE TRANSITION IN THE MODIFIED SU(2) MODEL

It was shown by Petkova and the author[12] that the <u>modified SU(2) model</u> described in section 3 possesses a high temperature phase (ß small) and a low temperature phase (ß large) that are distinguished by a qualitatively different behavior of the 't Hooft disorder parameter. This is true both in $\nu=3$ and 4 dimensions.

Let S be a set of $|S|$ links in the time zero hyperplane as shown in figure 1, and denote by $|\hat{\partial}S|$ the number of plaquettes in its coboundary. One finds that

$$\langle B(S) \rangle \geq \text{const} \cdot e^{-\alpha|\hat{\partial}S|} \quad \text{for small } \beta \tag{6.1}$$

whereas

$$\langle B(S) \rangle \leq \text{const} \cdot e^{-\alpha'|S|} \quad \text{for large } \beta \, . \tag{6.2}$$

In four dimensions, (6.1) implies a perimeter law, whereas (6.2) is an area law decay. It follows according to the discussion at the end of section 4 that small Z(2) monopoles are confined at low temperatures β^{-1}. This does however not imply monopole confinement in the sense in which this term is used in 't Hooft's lectures at this school[23], and one cannot conclude that quarks are not confined in the low temperature phase of our modified model.

The bound (6.1) follows from inequality (5.2) and known properties of the standard Z(2) lattice gauge theory model. (In four dimensions, a perimeter law decay (6.1) in the Z(2) model for small β is equivalent, by virtue of a duality transformation, to a perimeter law decay of the Wilson loop expectation value at large β - i.e. the nonconfinement of static quarks at low temperatures.)[14,15,26,34]

The result (6.2) is derived in several steps.

<u>1st step</u>. One regards the model as a Z(2) gauge theory model with fluctuating coupling constants K_p in the same manner as described in section 5. That is, one expresses $\langle B(S) \rangle$ in terms of conditional expectation values $\langle B(S) \rangle_W$. Starting from Eq. (4.6), one finds in place of Eq. (5.3)

$$\langle B(S) \rangle_W = Z_W^{-1} \int \Pi \, d\sigma(b) \left\{ \exp -2 \sum_{b \in S} K_{p_b}(W)\sigma(\dot{p}_b) \right\} \cdot \exp \sum_p K_p(W)\sigma(\dot{p}) \tag{6.3}$$

Z_W and $K_p(W)$ are the same as in section 5. It is crucial for the following second step that the representatives W of the classes can be chosen so that always

$$K_p(W) \geq 0 \qquad \text{(in the modified model)} \tag{6.4}$$

This follows from validity of the constraint (3.5) in the modified model.

It follows from standard properties of conditional expectation values that

$$\begin{aligned} \langle B(S) \rangle &= \int d\mu(U) \, \langle B(S) \rangle_{W(U)} \\ &= Z^{-1} \int \prod_b dU(b) \; Z_{W(U)} \langle B(S) \rangle_{W(U)} \quad . \end{aligned} \tag{6.5}$$

μ is the path measure defined in section 3, and W(U) is the representative of the class of configurations to which U belongs (cp. section 5).

2nd step. We observe that expression (6.3) is an expectation value in a Z(2) gauge theory with space time dependent coupling constants K_p. One performs a Kramers Wannier duality transformation[15]. In four dimensions, one obtains another Z(2) gauge theory with new coupling constants $\hat{K}_p$ that are related to the old one by

$$\hat{K}_p(W) = \frac{1}{2} \ln \coth K_p(W) \geq 0 .$$

These new coupling constants come out real because the old coupling constants were positive by (6.4). If it were otherwise, the new model would not be statistical mechanics. The new model has variables $\omega(c) = \pm 1$ attached to cubes c (= links of the dual lattice). It is convenient to use the notation

$$\int d\omega \, (\ldots) = \sum_{\omega=\pm 1} (\ldots)$$

Expression (6.3) becomes

$$\langle B(S) \rangle_W = \hat{Z}_W^{-1} \int \prod_c d\omega(c) \; \omega(C_S) \exp \sum_p K_p(W)\omega(\hat{\partial} p) \qquad (6.6)$$

$\hat{\partial}p$ consists of the cubes c that have p in their boundary $\dot{c}$, and

$$\omega(\hat{\partial} p) = \prod_{\substack{c \\ p \in \dot{c}}} \omega(c) \quad , \quad \text{while} \quad \omega(C_S) = \prod_{p \in \hat{\partial} S} \omega(c_p) \; . \qquad (6.7)$$

c_p is the cube protruding from plaquette p in positive time direction. $\hat{Z}_W$ is the partition function of the new model with coupling constants $\hat{K}_p$. It differs from Z_W by a certain W-dependent factor. C_S is a closed loop on the dual lattice, and expression (6.6) is therefore expectation value of a Z(2) Wilson loop in the new Z(2) model.

Remark. Under the duality transformation, Z(2) monopoles go into Z(2) quarks and vice versa[42,43]. The Wilson loop interpretation of Eq. (6.6) explains thus the remark about monopoles at the end of section 4, at least for the modified model.

3rd step. Inequalities (3.9) and expression (5.4) for $K_p(W)$ tell us that $K_p(W)$ tend to be large, of order ß, for large ß. It follows that $\hat{K}_p(W)$ are small, except for a set of classes (labelled by W) of small measure. This suggests to use high temperature expansions. Because of the exceptional W's, a little care is needed, though. One writes down a cluster expansion for (6.3) on a finite lattice where it is a finite sum (and converges therefore). Then one inserts into Eq. (6.5). One uses chessboard estimates to derive estimates on the individual terms of the expansion which results by integrating term by term. They are uniform in the size of the lattice. They show that the expansion continues to converge in the infinite volume limit.

4th step. One identifies the leading term of the expansion. Because of the Wilson loop interpretation of expression (6.6), the result (6.2) follows from the standard proof of the area law decay of the Wilson loop expectation value at high temperatures[3].

7. THIN VORTICES

For the sake of comparison, let us first recall some properties of the two dimensional Ising ferromagnet. Its spin variables $\sigma(x) = \pm 1$ are attached to vertices x of the lattice. A link b of the lattice has a boundary $\dot{b}$ which consists of two vertices x and y, and we write

$$\sigma(\dot{b}) = \sigma(x)\sigma(y) \tag{7.1}$$

in analogy with (2.4). Links b with $\sigma(\dot{b}) = -1$ form closed paths on the dual lattice. They separate regions with spins up from regions with spins down. These domain walls are the famous Peierls contours[24]. A configuration σ is determined by its Peierls contours up to a global spin rotation $\sigma \rightarrow -\sigma$. The two point correlation function $\langle \sigma(x)\sigma(z) \rangle$ is determined by the probability distribution of Peierls contours that wind around either x or z. Let p_N be the probability that there are N of them. Then

$$\langle \sigma(x)\sigma(z) \rangle = \sum_N (-)^N p_N \quad . \tag{7.2}$$

See figure 2a. Absence of spontaneous magnetization and exponential falloff of the correlation function obtains if long Peierls contours are abundant. This is true at high temperatures. Conversely, at low temperatures β^{-1}, long Peierls contours are very rare since they cost energy proportional to their length, and the entropy S, which is also proportional to length, cannot make up for this since its contribution to the free energy $F = E - TS$ of a contour is suppressed by a factor $T = \beta^{-1}$. Short contours winding around x cannot see the other point z in $\langle \sigma(x)\sigma(z) \rangle$. They can therefore not produce a falloff as $z \rightarrow \infty$.

The situation in the standard Z(2) lattice gauge theory model in $\nu = 3$ and 4 dimensions is quite analogous. The variables $\sigma(b) = \pm 1$ are now attached to links b. A plaquette p of the lattice has a boundary $\dot{p}$ which consists of four links $b_1 \ldots b_4$, and we write

$$\sigma(\dot{p}) = \sigma(b_4)\ldots\sigma(b_1) \tag{7.3}$$

in place of (7.1). As a consequence of this definition, for every cube c

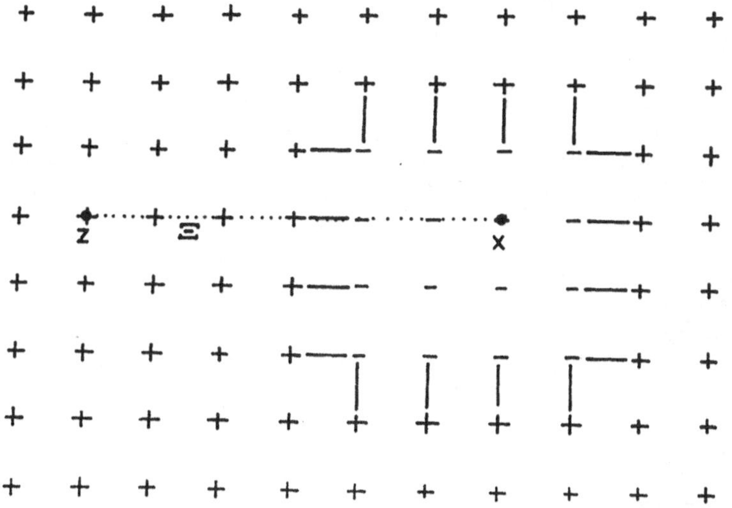

Figure 2a

Fig. 2a. Peierls contour in a two dimensional Ising ferromagnet. The contour consists of the black links. It winds around x or z if it contains an odd number of links in Ξ. (Ξ is any line from x to z.) Formula (7.2) obtains because $\sigma(z)\sigma(x) = \prod_{b\in\Xi}\sigma(\dot{b})$; $\sigma(\dot{b}) = -1$ on the contour by definition.

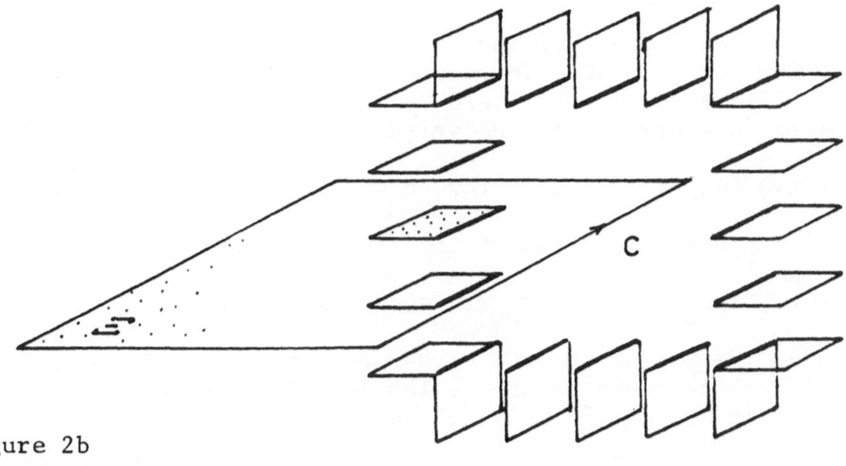

Figure 2b

Fig. 2b. Vortex in a Z(2) lattice gauge theory. Drawing for three dimensions. The vortex consists of the plaquettes shown as squares in the figure. It winds around C if it contains an odd number of plaquettes in Ξ. (Ξ is any surface with boundary C). Formula (7.5) obtains because $\sigma(C) = \prod_{p\in\Xi}\sigma(\dot{p})$; $\sigma(\dot{p}) = -1$ on the vortex by definition.

$$\prod_{p\in\dot{c}} \sigma(\dot{p}) = 1 \tag{7.4}$$

Plaquettes p with $\sigma(\dot{p}) = -1$ form closed paths ($\nu=3$) resp. surfaces ($\nu=4$) T on the dual lattice. See figure 2b. They are called vortices. A configuration σ is determined by its vortices up to a gauge transformation.

Let C be a rectangular closed path consisting of links $b_1 \ldots b_n$, and $\sigma(C) = \sigma(b_n)\ldots\sigma(b_1)$. The Wilson loop expectation value $\langle \sigma(C) \rangle$ is determined by the probability distribution of vortices that wind around the path C. Let p_N be the probability that there are N of them. Then

$$\langle \sigma(C) \rangle = \sum_N (-)^N p_N \; . \tag{7.5}$$

At high temperatures, large vortices are abundant and lead to an area law decay of $\langle \sigma(C) \rangle$. At low temperatures, however, large vortices are very rare, for the same reasons as discussed after Eq. (7.2) for the Ising ferromagnet[25]. Therefore the only vortices that are relevant to (7.5) wind tightly around the path C. They are only able to produce a perimeter law decay

$$\langle \sigma(C) \rangle \gtrsim \text{const}\cdot e^{-\alpha|C|} \qquad \text{for large } \beta. \tag{7.6}$$

A proof of (7.6) along these lines has been worked out by Göpfert[26].

In order to get more familiar looking formulae it is convenient to use additive notation. One introduces vector potentials $A_\mu(x)$ which take their values in the field $\mathbb{F}_2 = \{0, 1\}$, etc. Let e_μ be the lattice vector in μ-direction, and write

$$\sigma(b) = \exp i\pi A_\mu(x) \qquad , \qquad \text{for} \quad b = (x, x+e_\mu) \, . \tag{7.7a}$$

Then $\sigma(\dot{p})$ is expressed in terms of the field strength,

$$\sigma(\dot{p}) = \exp i\pi F_{\mu\nu}(x) \qquad , \; F_{\mu\nu}(x) = \Delta_\mu A_\nu(x) - \Delta_\nu A_\mu(x) \tag{7.7b}$$

if $p=p_{\mu\nu}(x)$ is the plaquette with corner points $x, x+e_\mu, x+e_\mu+e_\nu, x+e_\nu$. Δ_μ is the difference operator on the lattice. In this notation, (7.4) takes the familiar form (2nd Maxwell equation)

$$\Delta_\mu F_{\nu\lambda} + \Delta_\lambda F_{\mu\nu} + \Delta_\nu F_{\lambda\mu} = 0 \, . \tag{7.8}$$

Consider now the intersection $\hat{\partial}S$ of a vortex with the time zero plane Σ in four dimensions. (It is closed in the dual lattice of Σ and is therefore indeed coboundary $\hat{\partial}S$ of some set S of links in Σ, compare figures 1a, 2b for the three dimensional case). By definition,

$\hat{\partial}S$ consists of spacelike plaquettes $p = p_{ij}(x)$ with $F_{ij}(x) = 1$. There is thus a quantum of magnetic flux passing through p, and $\hat{\partial}S$ is a magnetic flux loop.

A vortex is (Euclidean) world sheet of such a flux loop. From Eq. (4.4) one sees that the 't Hooft operator (for the Z(2) theory) creates a magnetic flux loop on $\hat{\partial}S$, since $U_{-1}(\dot{p}) = -U(\dot{p})$ for $p \in \hat{\partial}S$. Because of the conservation law (7.4) or (7.8) it cannot just disappear but must evolve in time into a vortex. At low temperatures large vortices are very rare, this leads to an area law decay of $\langle B(S) \rangle$ in the standard Z(2) model at low temperatures. The vortices in a Z(2) theory have a thickness of only one lattice spacing. They will therefore be called <u>thin</u> vortices.

Let us now turn to SU(2) theories. Thin vortices also exist in such theories[37]. We may (for instance) define a thin vortex to consist of a coclosed set T of plaquettes p such that sign tr $U(\dot{p}) = -1$ for every p in T. In the modified model, the set of <u>all</u> plaquettes p on the lattice with sign tr $U(\dot{p}) = -1$ is always coclosed (= closed on the dual lattice). In the standard model this is not so.

The 't Hooft operator creates a thin magnetic flux loop on $\hat{\partial}S$. In the modified model, this magnetic flux loop must evolve into a thin vortex much as in the standard Z(2) model, due to the constraint (3.5) which replaces (7.4). But large thin vortices are still very rare at low temperatures because of the large cost of energy associated with them. This leads to the area law decay (6.2) of the 't Hooft disorder parameter in the modified model.

In the standard model, the conservation law (3.5) need not hold. Therefore the flux loop created by the 't Hooft operator need not evolve into a thin vortex. This does not mean that the excitation created by the 't Hooft operator may disappear from one instant of time to the next. It still has to evolve into a more general type of vortex, with a "vortex soul" that is coclosed. This can be seen from the formalism of section 10 below. But in the standard model, a thin flux loop may spread out as time evolves. As a consequence, the scarcity of <u>thin</u> vorticies at low temperatures need not necessarily lead to an area law of the 't Hooft disorder parameter in this case.

In contrast with the Ising model, the domain walls in ferromagnets with continuous symmetry can <u>spread</u>. The spins may rotate gently across such a domain wall so that no two neighboring ones make a large angle. In two dimensions there is no spontaneous magnetization because a domain wall which surrounds a cluster of n aligned spins can lower its free energy to a value independent of n by spreading. This physical explanation has been known for a long time[27]. A mathematical proof that it works has been given by Dobrushin and Shlosman[28]

This suggests that one should consider thick vortices in gauge theories with continuous gauge group. In such a thick vortex, $U(\dot{p}) \approx 1$ is possible for all plaquettes p. In the next section we will derive a sufficient condition for confinement of static quarks by condensation of such thick vortices.

8. SUFFICIENT CONDITION FOR CONFINEMENT OF STATIC QUARKS

Let D be a unitary representation of the gauge group G and $\chi(\cdot) = \mathrm{tr}\, D(\cdot)$ the corresponding character. If C is a rectangular path enclosing T·L plaquettes, with $T \gg L$, then

$$\langle \chi(U(C)) \rangle \sim \mathrm{const} \cdot e^{-TV(L)} \quad \text{for} \quad T \gg L \,, \tag{8.1}$$

in the limit of large T. According to Wilson[3], static quarks which transform according to representation D of G will be confined if $V(L) \to \infty$ as $L \to \infty$. In this section we will discuss an inequality for V(L). A sufficient condition for confinement of static quarks follows from it. This result was first obtained by Petkova and the author.[29] A similar condition was independently proposed by 't Hooft[23,30]. The condition is valid for pure Yang Mills theories. It generalizes to theories with fundamental scalar fields, provided these scalar fields transform trivially under the center Γ of the gauge group G (triplets, quintuplets etc., if G = SU(2)).

The representation D of G determines a character ω of a 1-dimensional representation of Γ.

$$D(U\gamma) = D(U)\omega(\gamma) \;; \tag{8.2}$$

$\omega(\gamma)$ are complex numbers of modulus 1.

Let us choose sublattices Λ_i of our total lattice Λ which wind around the path C as shown in figure 3. They will be called vortex containers. Different vortex containers are allowed to touch each other, but they may not intersect each other or the path C. We imagine that variables U(b) are fixed for the links b on the boundary $\partial\Lambda_i$ of Λ_i. We consider the partition functions $Z(\Lambda_i, U)$ as functions of these boundary conditions. In particular we consider their change when a singular gauge transformation[31,32] $U \to U_\sigma$ is applied to these boundary conditions ($\sigma \in \Gamma$).

A singular gauge transformation by σ acts as follows. One chooses a set T_i of links in $\partial\Lambda_i$ which is coclosed in $\partial\Lambda_i$ and winds once around C as shown in figure 3. Then

$$U(b) \to U_\sigma(b) = \begin{cases} U(b)\sigma^{-1} & \text{if } b \in T_i \\ U(b) & \text{otherwise} \end{cases} \tag{8.3}$$

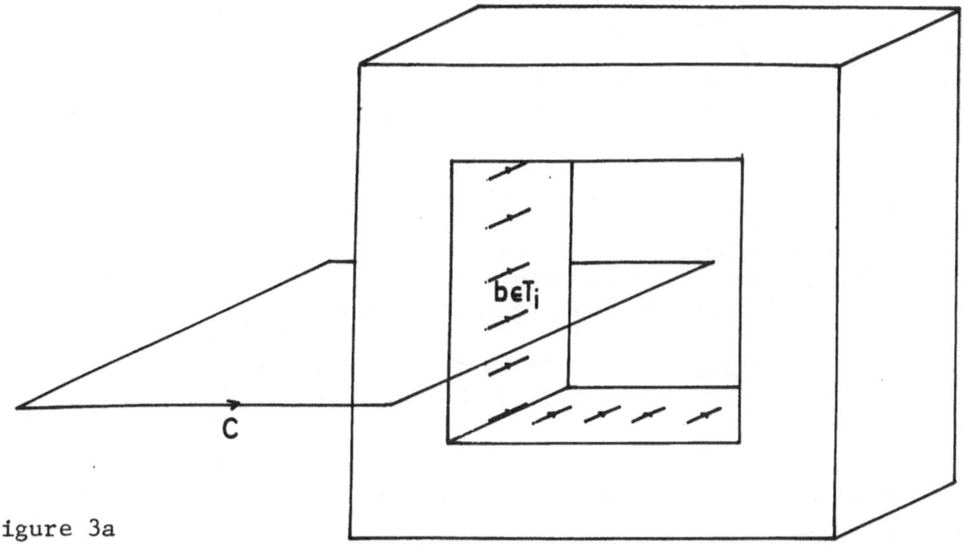

Figure 3a

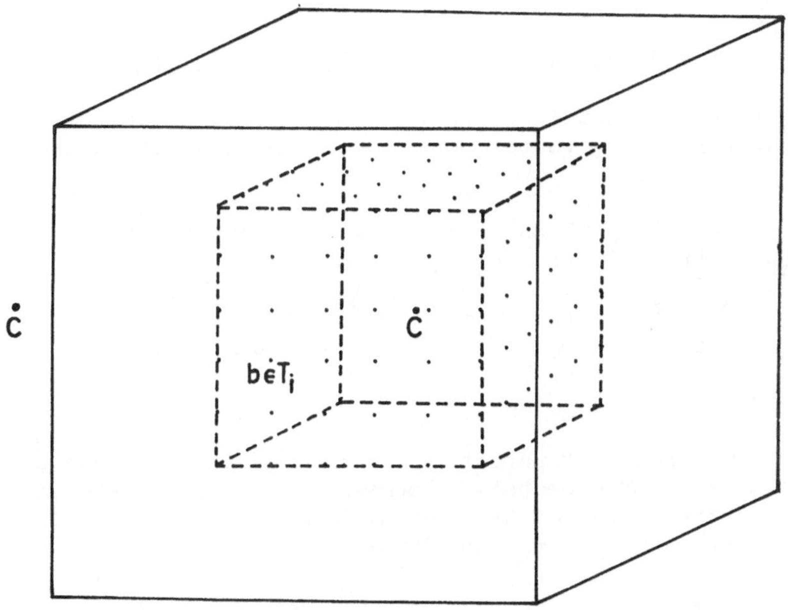

Figure 3b

Fig. 3. Vortex container Λ_i winding around the Wilson loop C, and the links b ε T_i that appear in definition (8.3) of a singular gauge transformation; (a) in 3 dimensions, (a),(b) in 4 dimensions (intersections with the hyperplanes $x^4 = 0$, resp. $x^1 = 0$).

The precise choice of T_i is immaterial since different choices lead to transformations that differ only by ordinary gauge transformations. A singular gauge transformation on $\partial\Lambda_i$ agrees with some ordinary gauge transformation on each topologically trivial piece of $\partial\Lambda_i$, but not on all of it.

In ν dimensions, vortex containers have the topology of $I \times I \times S^{\nu-2}$ (I = interval, S^n = unit sphere in n+1 dimensions), therefore $\partial\Lambda_i$ has the topology of $S^1 \times S^{\nu-2}$. (This differs from 't Hooft's choice in ≥ 4 dimensions.)[23]. T_i has the topology of $S^{\nu-2}$ (when considered as a cell complex in the dual lattice of $\partial\Lambda_i$).

Our inequalities are as follows. Consider the following normalized probability distributions on Γ, ($d\sigma$ = normalized Haar measure on Γ)

$$p_{\Lambda_i,U}(\sigma) = Z(\Lambda_i, U_\sigma) / \int d\sigma' \, Z(\Lambda_i, U_{\sigma'}) \tag{8.4}$$

Let $\hat{p}$ be their Fourier transforms on Γ,

$$\hat{p}_{\Lambda_i,U}(\omega) = \int d\sigma \, \omega(\sigma) \, p_{\Lambda_i,U}(\sigma) \quad . \tag{8.5}$$

Then

$$|\langle \chi(U(C)) \rangle| \leq \chi(1) \prod_i \left\{ \max_U |\hat{p}_{\Lambda_i,U}(\omega)| \right\} \tag{8.6}$$

ω is the character of Γ defined by (8.2). Product is over all the nonintersecting vortex containers that are fitted around C as described above.

Let us spell out this result in more detail for G = SU(2), D(U) = tr U. In this case

$$\hat{p}_{\Lambda_i,U}(\omega) = \left(1 - \frac{Z(\Lambda_i, U_\sigma)}{Z(\Lambda_i, U)}\right)\left(1 + \frac{Z(\Lambda_i, U_\sigma)}{Z(\Lambda_i, U)}\right)^{-1} \tag{8.7}$$

with $\sigma = -1$.

The ratio of partition functions in (8.7) can also be expressed as a ratio of expectation values of a certain observable $\tilde{B}(S_i)$ in the two Gibbs states of the system in Λ_i which are specified by the two boundary conditions U and U_σ on Λ_i.

$$\ln\left[Z(\Lambda_i, U_\sigma)/Z(\Lambda_i, U)\right] = -\frac{1}{2} \ln\left\{\langle \tilde{B}(S\ ) \rangle_{\Lambda_i,U_\sigma} / \langle \tilde{B}(S\ ) \rangle_{\Lambda_i,U}\right\} \tag{8.8}$$

S is a set of plaquettes which is such that it could be location of a hypothetical thin vortex that winds around C inside Λ_i, and

$$\langle \tilde{B}(S) \rangle = \langle \exp \sum_{p \in S} \{ \mathcal{L}(-U(\dot{p})) - \mathcal{L}(U(\dot{p})) \} \rangle \tag{8.9}$$

Eq. (8.8) follows from the following relation (8.8') and $(U_\sigma)_\sigma = U$.

$$Z(\Lambda_i, U_\sigma) = Z(\Lambda_i, U) \langle B(S) \rangle_{\Lambda_i, U} \tag{8.8'}$$

Let us now illustrate the possible uses of inequality (8.6) with an example. Suppose that the vortex container Λ_i has thickness d_i and extension $|T_i|$ (= number of links in T_i ≈ number of plaquettes in S). We must expect that the change of free energy ln Z will be proportional $|T_i|$. S can be so chosen that it has distance ≈ d/2 from the boundary $\partial\Lambda_i$. Suppose that the influence of the boundary conditions on the expectation value $\langle B(S) \rangle$ decays exponentially with this distance so that $|\ln Z(\Lambda_i, U_\sigma)/Z(\Lambda_i, U)| \leq |T_i| \exp -md/2$. We may choose vortex containers of suitable thickness , so that this is less than $\xi < 1$ for all of them. By counting the number of them that can be fitted around C we find from inequality (8.6) and (8.1) that $V(L) \geqslant \text{const}\cdot L\,(\ln L)^{-2}$, i.e. an at least almost linear increase with L which ensures confinement.

We will call

$$\frac{1}{|T_i|} \ln Z(\Lambda_i, U_\sigma)/Z(\Lambda_i, U) = \beta\cdot\text{vortex free energy} \tag{8.10}$$

per unit area ($\nu=4$) resp. length ($\nu=3$) of the vortex. It depends on boundary conditions U on $\partial\Lambda_i$. Our results show that <u>confinement is true if the vortex free energy decreases sufficiently fast with the thickness d</u> (cross section) of the vortex container.

Exactly the opposite behavior corresponds to what 't Hooft calls a <u>Higgs phase</u>. There

$$\frac{1}{|T_i|} \ln p_{\Lambda i, U}(\sigma) = \frac{1}{|T_i|} \ln\left[Z(\Lambda_i, U_\sigma)/\int d\sigma' Z(\Lambda_i, U_{\sigma'})\right] \not\to 0$$

$$\text{when } d \to \infty \;, \text{ with } T_i \gg d \;, \tag{8.11}$$

for some boundary conditions U, for instance U(b) = 1. The limit $f(\sigma)$ is a gauge theory analog of the <u>surface tension</u>[46] in ferromagnets. The limit $|T_i| \to \infty$ is taken first. (This eliminates the "translational" entropy associated with the penetration point of the vortex soul (cp. section 10) in some fixed surface Ξ across Λ_i).

Münster has shown[33] that lattice <u>Higgs models</u> with sufficiently many fundamental scalars (with trivial transformation law under Γ) <u>possess such a Higgs phase</u>. Göpfert has shown[26] that the <u>low temperature phase of the standard Z(2) model is</u> also <u>a Higgs phase</u> in this sense. Proofs are based on a Peierls argument. We shall show in the next section that pure Abelian Yang Mills theories with continuous gauge group have no Higgs phase. All this holds in $\nu \geq 3$ dimensions.

We emphasize that 't Hoofts notion of Higgs phase is much more restrictive than what used to be called a "Higgs mechanism". In particular, it is customary to speak of a Higgs mechanism also when the fundamental scalars do not transform trivially under any nontrivial subgroup of Γ. The model of section 2 is an example.

We wish to emphasize also that the inequality (8.6) and the resulting sufficient condition for confinement as well as 't Hoofts variant of it are results of a kinematical nature. They do not depend on the details of the dynamics of a pure Yang Mills theory but are also valid for theories with fundamental scalars.

The proof of inequality (8.6) uses only very basic properties of the theories - nearest neighbor interaction and gauge invariance. We sketch the proof for a theory of the general form (3.2), (3.4) of the action and path measure. The result is also valid for our modified SU(2) model.

Let Λ^c be the complement of the interior of all vortex containers Λ_i. Then $\Lambda^c \cap \Lambda_i = \partial\Lambda_i$. From Eq. (3.7) and the definition of the partition functions $Z(\Lambda_i, U)$ one obtains

$$\langle \chi(U(C)) \rangle = \frac{1}{Z} \int \prod_{b \epsilon \Lambda^c} dU(b)\, \chi(U(C)) \exp \sum_{p \epsilon \Lambda^c} \mathcal{L}(U(\dot{p})) \prod_i Z(\Lambda_i, U) \tag{8.12}$$

$Z(\Lambda_i, U)$ depends on the restriction of the configuration U to $\partial\Lambda_i$.

One performs a variable substitution $U(b) \longrightarrow U(b)\sigma(b)$, $\sigma(b) \epsilon \Gamma$, with the following properties.

i) For all plaquettes $p \epsilon \Lambda^c$, $U(\dot{p})$ stays invariant.

ii) On $\partial\Lambda_i$, the transformation is gauge equivalent to a singular gauge transformation (8.3),

$$U(b) \longrightarrow U(b)_{\gamma_i} = \begin{cases} U(b)\gamma_i^{-1} & \text{if } b \epsilon T_i \\ U(b) & \text{otherwise.} \end{cases}$$

iii) On the Wilson loop

$$U(C) \longrightarrow U(C) \prod_i \gamma_i$$

The transformation is determined by the values of the γ_i's. One averages over all γ_i's, using normalized Haar measure on Γ(e.g. (5. 5) if $\Gamma = Z(2)$). Since the partition functions are gauge invariant functions of U, this gives

$$\langle \chi(U(C)) \rangle = \frac{1}{Z} \int \Pi \, dU(b) \, \chi(U(C)) \exp \Sigma \, \mathcal{L}(U(\dot{p})) \cdot$$
$$\cdot \prod_i \int \left\{ d\gamma_i \, \omega(\gamma_i) \, Z(\Lambda_i, U_{\gamma_i}) \right\}$$
$$= \frac{1}{Z} \int \prod_{b \epsilon \Lambda^c} dU(b) \left[\exp \sum_{p \epsilon \Lambda^c} \mathcal{L}(U(\dot{p})) \right] \chi(U(C))$$
$$\cdot \prod_i \left\{ \hat{p}_{\Lambda_i, U}(\omega) \int d\sigma_i \, Z(\Lambda_i, U_{\sigma_i}) \right\} \quad (8.13)$$

In the first equation we have used that $\chi(U(C)\Pi\gamma_i) = \chi(U(C))\Pi\omega(\gamma_i)$. In the second equation we have inserted the definitions (8.4) and (8.5).

By the same procedure one finds that

$$1 = \frac{1}{Z} \int \prod_{b \epsilon \Lambda^c} dU(b) \exp \sum_{p \epsilon \Lambda^c} \mathcal{L}(U(\dot{p})) \prod_i \left\{ \int d\sigma_i \, Z(\Lambda_i, U_{\sigma_i}) \right\} \quad (8.14)$$

Since $|\chi(U(C))| \leq \chi(1)$, inequality (8.6) follows from (8.13) and (8.14).

9. COMPUTATIONS OF VORTEX FREE ENERGIES

High Temperature Expansions.

For small values of ß, the vortex free energy can be computed by high temperature expansion in the standard models. A set P of plaquettes will be called a polymer[11] if it is connected, i.e. cannot be split into two sets P_1 and P_2 such that no plaquette in P_1 touches a plaquette in P_2. For polymers P one defines activities

$$A(P) = \int \Pi \, dU(b) \prod_{p \epsilon P} (e^{\mathcal{L}(U(\dot{p}))} - 1) \quad (9.1)$$

We are interested in a finite lattice Λ_i with prescribed boundary conditions U(b) for links b in $\partial\Lambda_i$. A(P) will depend on these boundary conditions if P touches the boundary. We will write A(P, U) to exhibit this dependence.

The free energy possesses a high temperature expansion of the form[11,34]

$$\ln Z(\Lambda_i, U) = \sum_Q a(Q) \prod_{P\varepsilon Q} A(P, U) \tag{9.2}$$

Summation is over all linked clusters Q of polymers P. They are collections of not necessarily distinct polymers which are linked in the following sense. Draw a graph with a vertex for each polymer, and a link between two vertices if the two polymers share a plaquette, or a plaquette in one of them touches a plaquette in the other. This graph must be connected. a(Q) are combinatorial factors. a(Q) = 1 if Q consists of a single polymer.

From (9.2) one obtains an expansion for the vortex free enery, viz. $\ln Z(\Lambda_i, U) - \ln Z(\Lambda_i, U_\sigma)$. All those terms in the sum drop out for which $A(P, U_\sigma) = A(P, U)$ for all polymers P in Q. Since A(P, U) is a gauge invariant function of U, this equality will hold unless the intersection of at least one polymer P in Q with $\partial\Lambda_i$ is a topologically nontrivial part of $\partial\Lambda_i$ - i.e. it should contain a closed path C which winds around Λ_i. The requirement that $A(P, U_\sigma)$ or $A(P, U) \neq 0$ leads to the further requirement that P must contain all plaquettes in a whole surface Ξ which cuts through Λ_i and has C as its boundary. But for such P, $A(P, U) = O(\beta^{|\Xi|})$.

For the standard SU(2) and Z(2) models one finds to leading orders in β in ν=3 and 4 dimensions (σ=-1)

$$\frac{1}{|T|} \ln Z(\Lambda_i, U_\sigma)/Z(\Lambda_i, U) = -4c(U)\, e^{-\alpha|\Xi|} \tag{9.3}$$

with $$|c(U)| \leq 2 .$$

$|T|$ is the extension of the vortex container, and $|\Xi|$ its cross section. $|\Xi| = d^2$ if the container has thickness d. The exponent

$$\alpha = -\ln \frac{\beta}{4} + \frac{2}{3}\left(\frac{\beta}{4}\right)^2 + \ldots \qquad \text{for} \quad G = SU(2) \tag{9.4a}$$

$$\alpha = -\ln \beta + \frac{1}{3}\beta^2 + \ldots \qquad \text{for} \quad G = Z(2) \tag{9.4b}$$

and

$$c(U) = \frac{1}{|T|} \sum_\Xi \operatorname{tr} U(\partial\Xi) + \ldots \tag{9.5}$$

Summation in (9.5) is over all surfaces Ξ of minimal area $|\Xi|$ that cut through Λ_i. There are $|T|$ of them for containers of regular shape. Some higher order corrections to α have been computed by Münster[35]. See also Yaffe[36] (Appendix). In order β^4 a difference appears between 3 and 4 dimensions.

We see that our sufficient condition for confinement is fulfilled at high temperatures $ß^{-1}$. Eq. (9.3) is an area law. It is interesting that a less rapid falloff, e.g. a perimeter law, would still suffice to produce confinement; see the discussion in the last section.

Low Temperature Behavior

For a vortex container of fixed cross section $|\Xi|$, the limiting behavior of the vortex free energy as $ß \to \infty$ can be found. It is a fact that

$$\lim_{ß\to\infty} \left(\int d\mu \, e^{ßS} \right)^{1/ß} = \|e^S\|_\infty = \text{ess. sup. } e^S \tag{9.6}$$

for any normalized measure μ. For the path measures of section 3 and $S = L/ß$ this gives

$$\frac{1}{ß} \ln Z(\Lambda_i, U) \longrightarrow \max_{U'} L(U') \tag{9.7}$$

The maximum is with respect to all configurations U' in Λ_i which agree with U on the boundary $\partial\Lambda_i$.

We restrict our attention to special boundary conditions U. L assumes its absolute maximum for U' = 1. This takes care of boundary conditions U = 1 (i.e. U(b) = 1 for all links b in $\partial\Lambda_i$).

Consider now boundary conditions 1_σ ($1_\sigma(b) = \sigma$ if $b \in T_i$, and = 1 otherwise). For abelian theories and large containers, the maximum in (9.7) is easy to determine. It corresponds to a uniform distribution of magnetic flux, if G is continuous. One finds

$$-\frac{1}{ß|T|} \ln Z(\Lambda_i, U_\sigma)/Z(\Lambda_i, U) \longrightarrow \begin{cases} 2 & \text{if } G = Z(2) \quad (9.8a) \\ \dfrac{\text{const}}{|\Xi|} \begin{cases} \text{if } G = U(1) \\ \text{or } G = SU(2) \end{cases} & (9.8b) \end{cases}$$

σ can be any element of the center Γ of G. The proportionality contant depends on $\sigma \in \Gamma = G$ if G = U(1). For G = Z(2) or SU(2) there is only one choice $\sigma = -1$.

The result (9.8) for the nonabelian case is obtained by noting that Abelian configurations $U(b) \in H = U(1) \subset SU(2)$ bound the maximum in (9.7) from below by the result for the U(1) case. An upper bound of the same form is obtained from the inequality[43]

$$0 \geqslant \text{tr}\,(U(\partial\Xi) - 1) \geqslant |\Xi| \sum_{p\in\Xi} \text{tr}\,(U(\dot{p}) - 1) \tag{9.9}$$

This is valid for any rectangular surface of $|\Xi| = 2^{N_1} \times 2^{N_2}$ plaquettes, and arbitrary U.

The interpretation of results (9.8) is obvious. Since the continuum action involves the square of the magnetic field, a factor of $1/|\Xi|$ in the action can be gained by spreading a given magnetic flux over an area $|\Xi|$, for continuous gauge group G in four dimensions. In a Z(2) theory, the flux cannot be spread, since the flux through any plaquette must take values 0 or 1.

Result (9.8) is of very limited use. It cannot be used to obtain the asymptotic behavior as $|\Xi| \to \infty$ for any finite ß: As $ß \to \infty$, the correlation length ξ becomes infinite, if not before, and a vortex container of given thickness d in units of lattice spacing becomes very thin in physical units. In contrast, $|\Xi| \to \infty$ for fixed ß would mean $d/\xi \to \infty$.

Numerical Calculations and Effective Z(2) Coupling Constants

No computer calculation of vortex free energies has yet been performed. Such computations would however appear feasible[38] and desirable, for gauge group SU(2). The dependence on vortex thickness would be most interesting - and also the actual numerical values, for the following somewhat more speculative reasons.

Imagine that a lattice of block cells of $d \times \ldots \times d = d^{\nu}$ elementary hypercubes is superimposed on the original lattice. A vortex container of thickness d may be composed of such blocks. It looks like a thin vortex in the new lattice. If P is a plaquette in the new lattice, perpendicular to the vortex container, then $U(\partial P) \to -U(\partial P)$ under a singular gauge transformation (8.3). There are matching conditions between the configurations on the boundaries of neighboring block cells. They imply, for instance, that the soul of a vortex (to be defined in section 10) has to enter a neighboring block cell where it leaves the old one. These matching conditions affect the entropy in the block cells, and therefore in a vortex container composed from them. We may hope that this entropy remains approximately the same if we replace the matching conditions by periodic boundary conditions (p.b.c.) for each cell. This suggests to regard

$$-\frac{1}{2} \ln Z(\text{block cell, (p.b.c.)}_{\sigma})/Z(\text{block cell, p.b.c.}) = ß_{eff}(d) \tag{9.10}$$

as effective coupling constant of a Z(2) theory on the block lattice. $(\text{p.b.c.})_{\sigma}$ are boundary conditions obtained from periodic ones by a singular gauge transformation $U \to U_{\sigma}$, $\sigma = -1$, cp. (8.3).

It is tempting to speculate that confinement will prevail if

$$\beta_{eff}(d) < \beta_c \qquad (9.11)$$

for sufficiently large side length d of the block cells, $\beta_c \approx$ the critical value of the coupling constant for the standard Z(2) model: $\beta_c = 0.44$ in 4 dimensions, and $\beta_c = 0.76$ in 3 dimensions, see Balian et al.[15]

The effective Z(2) coupling constant could in principle be computed by the MonteCarlo method[38], since the ratio of partition functions on the left hand side is expectation value of a positive observable according to Eq. (8.8').

At high temperatures, $\beta_{eff}(d)$ behaves as

$$\beta_{eff}(d) \sim d^{\nu-2}\, e^{-\alpha d^2} \quad \text{for small } \beta\,, \qquad (9.12)$$

with α from (9.4a).

For fixed block size d, the bound (9.11) will of course be violated for sufficiently large β where β_{eff} will behave like

$$\beta_{eff}(d) \propto \beta \cdot d^{\nu-4} \quad \text{for fixed } d, \text{ and } \beta \to \infty \qquad (9.13)$$

in ν dimensions. This obtains in the same way as (9.8b).

For finite β, the dependence on block size d will be changed from the (zeroth order perturbation theory) behavior (9.13) by renormalization effects. It may be useful to distinguish* between not necessarily nonperturbative renormalization effects which will hopefully bring $\beta_{eff}(d)$ below β_c in 4 dimensions** for large enough d,

---

* The standard Z(2) model in four dimensions appears to have a first order phase transition at $\beta = \beta_c$. Therefore, even at temperatures above but close to β_c^{-1}, there should exist a metastable phase without vortex condensation and confinement. Making the distinction in the text amounts to talking about a similar hypothetical metastable phase in the SU(2) theory. It is supposedly unstable against condensation of vortices of sufficiently large thickness d (depending on β) at any β. Nevertheless it may be useful to imagine its existence.

** The factor $d^{\nu-4}$ in (9.13) suggests that in $\nu > 4$ dimensions it will be otherwise.

and the vortex condensation mechanism which is triggered by this. The latter is a nonperturbative effect, but is familiar from the standard Z(2) gauge theory model. It would lead to quark confinement, and would also lower $\beta_{eff}(d)$ still further for d larger than the critical value $d = d_c(\beta)$ at which (9.11) begins to be satisfied.

Currently, the behavior of the 4 - dimensional standard SU(2) model around $\beta = 2$ appears to be of considerable interest[38,49] Is the bound (9.11) fulfilled there, for a reasonable sized block? This question could be studied by computer calculation. Sufficiency of (9.11) could also be studied by computations for Higgs models[33].

Abelian Theories

If the gauge group G is a continuous Abelian group, its center $\Gamma = G$ is the whole group, and the elements $\sigma\in\Gamma$ which parametrize singular gauge transformations (8.3) may take values arbitrarily close to 1. As $\sigma \to 1$, the vortex free energy tends to zero. As a consequence, at any nonzero temperature, thin vortices with arbitrarily large extension but σ sufficiently close to 1 will survive. Thick vortices with σ away from 1 can be composed from thin vortices with σ close to 1.

Consider a vortex container Λ' of cross section $|\Xi|$, and regard it as composed from $|\Xi|$ ("thin") vortex containers Λ_i of cross section 1. The composition law for the partition functions $Z(\cdot, U_\sigma)$ leads to the following composition law for the corresponding normalized probability distributions (8.4) on Γ. ($*$ is the convolution product on Γ.)

$$p_{\Lambda',U}(\sigma) = \int d\nu(U) \mathop{*}_{i} p_{\Lambda_i,U}(\sigma) \tag{9.14}$$

$d\nu$ is a normalized measure. Integration is over variables U(b) attached to the links b in the interior of Λ'; the result depends on the remaining variables U(b) with b in $\partial\Lambda'$. Explicitly,

$$\begin{aligned} d\nu(U) &= \bar{Z}(\Lambda', U)^{-1}\left\{\prod_i \bar{Z}(\Lambda_i, U)\right\}\prod_b dU(b) \quad , \\ \bar{Z}(\cdot,U) &= \int d\sigma\, Z(\cdot, U_\sigma) \end{aligned} \tag{9.15}$$

One can derive estimates[29] for thin vortex containers Λ_i. It suffices to consider $G = U(1) = \Gamma$ with elements $\sigma = \exp i\varphi$, $\varphi = 0, \ldots 2\pi$, and Haar measure $d\sigma = d\varphi/2\pi$. We assume that the Lagrangean is of the form

$$\mathcal{L}(e^{i\vartheta}) = \beta \cos\vartheta \tag{9.16}$$

Characters are of the form

$$\omega_q(e^{i\varphi}) = e^{iq\varphi} \quad , \quad q = 0, \pm 1, \pm 2, \ldots \tag{9.17}$$

The partition function of a thin vortex container is

$$Z(\Lambda_i, U_\gamma) = \exp \Sigma\, \mathcal{L}(U(\dot{p})\gamma) \tag{9.18}$$

Summation is over all plaquettes that are not on the boundary of Λ_i. Each of them contains a link of T_i, and is supposed to have orientation inherited from that link. There are $|T_i|$ such plaquettes. There are no integrations involved, all variables U(b) in (9.18) are fixed by the boundary conditions U on $\partial\Lambda_i$.

Inserting the form (9.16) of the action, one obtains for $\gamma = e^{i\varphi}$

$$Z(\Lambda_i, U_\gamma) = \exp \beta\, \Sigma \cos(\vartheta_p + \varphi) = \exp \beta |T_i| \varkappa(U) \cos(\bar{\vartheta} + \varphi)$$

with $0 \leq \varkappa(U) \leq 1$. Therefore

$$\hat{p}_{\Lambda_i,U}(\omega_q) \quad = \quad I_q(\beta|T_i|\varkappa(U))/I_o(\beta|T_i|\varkappa(U)) \quad \leq \quad \exp - \frac{q^2}{2\beta}|T_i|^{-1} \tag{9.19}$$

for large $|T_i|$.

From these bounds one may draw two conclusions for pure Yang Mills theories with continuous Abelian gauge group U(1).

(1) They possess no Higgs phase in any number of dimensions (in the sense of definition (8.11), with Λ_i a box).

(2) In three dimensions, static quarks of any charge $q \neq 0$ are confined by a potential

$$V(L) \geq \frac{q^2}{16\beta} (\ln L + \text{const.}) \tag{9.20}$$

Property (1) obtains by inserting (9.18) and definition (8.4) of $p_{..}(\cdot)$ into composition law (9.14). The large $|T_i|$-behavior of the $\star$-product can be estimated by the saddle point method. This produces a bound on $p_{\Lambda',U}(\cdot)$ because the measure ν is normalized. Finally, one compares with the definition (8.11) of a Higgs phase.

To learn as much as possible from our derivation of property (2), the reader should study Fisher's domain wall argument[27] and compare with it.

Property (2) is obtained by inserting bound (9.19) into our

basic inequality (8.6). Let C be a rectangular path which encloses an area of $L \cdot T$ plaquettes, $T \gg L$. Divide the lattice into T-1 slices perpendicular to the long legs of C, each of them one lattice spacing thick. Into each slice pack the maximum number L of (quadratic) thin vortex containers that can be wound around one penetration point of C. They have extension (length) $4(2j+1)$, $j = 0 \ldots L-1$. Therefore, inequality (8.6) gives

$$\langle \omega_1(C) \rangle \leq \exp - T \left[\frac{g^2}{16\beta} \sum_{j=0}^{L-1} (j + \frac{1}{2})^{-1} \right] \tag{9.21}$$

For large L, the sum behaves like ln L + const.. Comparing with (8.1) gives the bound (9.20).

The Coulomb law (9.20) as a lower bound on V(L) was derived before by Glimm and Jaffe by another method.[39] The derivation given here serves to illustrate the use of our basic inequality (8.6). It may be viewed as a generalization of the Dobrushin Shlosman argument[28] for two dimensional ferromagnets.

10. TOPOLOGICALLY DETERMINED Z(2) VARIABLES IN SU(2) THEORIES

Preliminary Considerations: Abelian Higgs Model

To fix ideas, let us first consider a Higgs model with U(1) gauge group on three dimensional continuous space time. There will be a charged complex field $\varphi(x)$ which carries charge 1 (i.e. transforms according to the faithful fundamental representation of U(1)), and the space time components $A_\mu(x)$ of the vector potential take real values.

Consider a smooth field configuration with a (Nielsen Olesen[40]) vortex of winding number $k > 0$ along the x^3- axis. Near the x^3-axis, the fields will behave like

$$\begin{aligned} \varphi(x) &= a(x^1 + ix^2)^k + \ldots \\ A_\mu(x) &= A_\mu(0) + \ldots \end{aligned} \tag{10.1}$$

if they are invariant under translations in the x^3- direction.

Let us make the transition to the unitary gauge. It is defined by the requirement that $\varphi(x) \geq 0$. The unitary gauge is a special case of what the author calls a "local gauge"[32]. It is convenient to use cylinder coordinates

$$x^1 = \rho \cos\vartheta \quad , \quad x^2 = \rho \sin\vartheta$$

The gauge transformation to the unitary gauge involves $S(x) = e^{-ik\vartheta} \in U(1)$. Scalar fields $\hat{\varphi}(x)$ and vector potential $\hat{A}_\mu(x)$ in the unitary gauge are given by

$$\begin{aligned} \hat{\varphi}(x) &= a\,\rho^k + \ldots \\ \hat{A}_3(x) &= A_3(0) + \ldots \\ \hat{A}_1(x) &= -\frac{k}{\rho}\sin\vartheta + A_1(0) + \ldots \\ \hat{A}_2(x) &= \frac{k}{\rho}\cos\vartheta + A_2(0) + \ldots \end{aligned} \tag{10.2}$$

We see that the vector potential $\hat{A}_\mu(x)$ in the local gauge is singular on the x^3- axis although we start from a perfectly smooth vector potential. This is an instance of what the author calls a "local gauge singularity"[32]. Consider now a path C of infinitesimal length which winds once around the x^3- axis. Then the contour integral

$$\frac{1}{2\pi} \oint_C \hat{A}_\mu \, dx^\mu = k \quad . \tag{10.3}$$

It counts vorticity (winding number k) which need not be zero even for a perfectly smooth field configuration φ, A_μ. The "soul of the vortex" is here located at the zero of the scalar field.

The local gauge used here is a complete gauge; it leaves no freedom of gauge transformations whatsoever. Had we chosen a charged field of charge two then any local gauge would still leave the freedom of Z(2) gauge transformations (rotations by π). On a lattice, these could depend on space and time.

In the following we will introduce variables $\bar{W}(b)$ for the SU(2) theory. They will be exponentials of the vector potential in a certain <u>local gauge</u>. This gauge (and any other conceivable local gauge in a pure SU(2) Yang Mills theory) leaves the freedom of Z(2) gauge transformations. This is why we will be left with Z(2) gauge variables $\sigma(b)$ in addition to the local gauge invariants $\bar{W}(b)$.

The New Variables for SU(2) Theories

Given the matrix U(b) in G = SU(2), let

$$\bar{U}(b) = U(b)\Gamma \in G/\Gamma \quad . \tag{10.4}$$

$\bar{U}(b)$ may be regarded as a real 3×3 matrix $(\bar{U}(b)^{ac})$ in SO(3). Let p be the plaquette with corner points x, $x+e_i$, $x+e_i+e_j$, $x+e_j$ (e_μ =

unit vector in μ-direction.), and ijk = 123 or cyclic permutation. We define 3×3 matrices

$$B(x) = (B^a{}_k(x)) \ , \quad B^a{}_k(x) = \frac{1}{2}\,\varepsilon^{abc}\,\bar{U}(\dot{p})^{bc} \qquad (10.5)$$

If $U(b) = \exp A^a{}_\mu(x)\tau^a/2$ with small $A^{\cdot}$., then $B^a{}_k$ are the components of the magnetic field associated with the vector potential A.

The configurations U which are such that for a fixed x on the lattice one has det $B(x) = 0$ form a set of (Gibbs) measure zero. This remains true in the infinite volume limit. This follows from the Markov property by a standard argument that has been described in Glimms lectures at this school[4].

Every nonsingular real 3×3 matrix admits a unique decomposition $B = OP_+$, where O is orthogonal and P_+ is a positive matrix. Therefore we may decompose

$$B(x) = \bar{S}(x)P(x) \quad \text{with} \quad \bar{S}(x) \in SO(3) \qquad (10.6)$$

and $P(x)$ is either positive or negative definite, depending on sign det $B(x)$. $\bar{S}(x)$ is uniquely defined if det $B(x) \neq 0$.

We define

$$\bar{W}(b) = \bar{S}(x)^{-1}\bar{U}(b)\bar{S}(y) \in SO(3) \quad \text{for} \quad b = (x,y) \ . \qquad (10.7)$$

It is easily verified that $\bar{W}(b)$ is <u>gauge invariant</u>. It is also <u>local</u> in the sense that it depends only on gauge fields U(b') attached to links within a neighborhood of one lattice spacing of b. One lattice spacing becomes infinitesimally small in the continuum limit.

Now we turn to the definition of the Z(2) variables σ(b). The group $G = SU(2)$ is a twofold covering of $G/\Gamma = SO(3)$. To every matrix $\bar{W}$ resp. $\bar{S}$ of SO(3) there correspond two matrices ±W resp. ±S in SU(2). We select the one with positive trace. Thus , W(b) and S(b) will be defined by the requirements that

$$W(b)\Gamma = \bar{W}(b) \quad , \ \mathrm{tr}\ W(b) \geqslant 0 \ , \qquad (10.8a)$$

and

$$S(x)\Gamma = \bar{S}(x) \quad , \ \mathrm{tr}\ S(x) \geqslant 0 \ . \qquad (10.8b)$$

The variables σ(b) can now be defined by the formula

$$W(b)\sigma(b) = S(x)^{-1}U(b)S(y) \qquad \text{for} \quad b = (x,y). \qquad (10.9)$$

Taking Γ-cosets of both sides of this equation we see by comparison with (10.7) that

$$\sigma(b) = \pm 1 \quad .$$

We leave it to the reader to verify that an SU(2) gauge transformation V(·) of variables U(·) induces a Z(2) gauge transformation of variables $\sigma(b)$, viz.

$$\sigma(b) \longrightarrow \upsilon(x)\sigma(b)\upsilon(y)^{-1} \qquad \text{for} \quad b = (x,y) \ , \tag{10.10}$$

with $\upsilon(\cdot) = \pm 1$. It follows that $\sigma(\dot{p})$ is gauge invariant. It is also local in the same sense as $\bar{W}(b)$. For any closed path C, in particular for the boundary $C = \dot{p}$ of a plaquette, $\sigma(C)$ is defined as in Eq. (3.1).

If C is boundary of a surface Ξ consisting of plaquettes p then

$$\sigma(C) = \prod_{p\in\Xi} \sigma(\dot{p}) \qquad (C = \partial\Xi) \tag{10.11}$$

The local gauge invariants $\bar{W}(b)$ and $\sigma(\dot{p})$ determine a configuration U up to the action of a gauge transformation.

If b are spacelike links in the t=0 hyperplane Σ, $\bar{W}(b)$ can act as multiplication operators on wave functions Ψ in the Hilbert space (4.1) of physical states. They commute (at t=0). These operators are thus gauge invariant local SO(3)-valued field operators whose action does not lead out of the physical Hilbert space. In contrast, Z(2) gauge fields $\sigma(b)$ can never share these properties, only their field strengths $\sigma(\dot{p})$ do.

The reader may find it interesting to compare Eq.(10.9) with work of Glimm and Jaffe[39] on vortices in Abelian theories.

Wilson Loop and Action in Terms of the New Variables

Let C be a closed loop consisting of links $b_1..b_n$ which is boundary of a surface Ξ. For instance, Ξ may be a rectangle of L·T plaquettes. It follows from Eqs. (10.9) and (10.11) that

$$\text{tr } U(C) = \sigma(C) \text{ tr } W(C) = \Big(\prod_{p\in\Xi} \sigma(\dot{p})\Big)\text{tr } W(C) \ . \tag{10.12}$$

The last formula expresses the Wilson loop observable tr U(C) as a sum of products of local gauge invariants. This is so because both $\sigma(\dot{p})$ and $W_{\alpha\beta}(b)$ are local gauge invariants, and

$$\text{tr } W(C) = \sum_{(\alpha_1\ldots\alpha_n)} W_{\alpha_1\alpha_n}(b_n)\ldots W_{\alpha_3\alpha_2}(b_2)W_{\alpha_2\alpha_1}(b_1) \ . \tag{10.13}$$

More generally, let χ_l be the character of the 2l+1 dimensional irreducible representation of SU(2). If l is integer, it is at the same time a character of SO(3). From Eqs.(10.9) and (10.11) one finds that

$$\chi_l(U(C)) = \begin{cases} \chi_l(W(C)) \prod_{p\in\Xi} \sigma(\dot{p}) & \text{if } l = \frac{1}{2}, \frac{3}{2}, \ldots \quad (10.14a) \\ \chi_l(W(C)) & \text{if } l = 0, 1, 2, \ldots \quad (10.14b) \end{cases}$$

There is reason to believe that the fluctuations of the Z(2) variables $\sigma(\dot{p})$ are crucial in producing an area law decay of the Wilson loop observable $\langle(10.12)\rangle$, whereas the factor $\chi(W(C))$ (which is a sum of products of local gauge invariants that are localized near the path) is relatively unimportant. It is known[2] that $\langle \chi_l(U(C)) \rangle$ will not have an area law decay if l is integer (see the introduction). This is in agreement with the absence of the factor $\Pi\, \sigma(\dot{p})$ in (10.14b).

Next, we rewrite the action. Specializing Eq. (10.12) to a single plaquette , we obtain

$$\operatorname{tr} U(\dot{p}) = \operatorname{tr} W(\dot{p})\, \sigma(\dot{p}) \qquad (10.15)$$

Therefore

$$L(U) \equiv L(\bar{W}, \sigma) = \sum_p K_p(W) \sigma(\dot{p}) + \text{const.} \qquad (10.16)$$

The fluctuating Z(2) coupling constants K_p are given by

$$K_p(W) = \frac{\beta}{2} \operatorname{tr} W(\dot{p}) \;. \qquad (10.17)$$

Fluctuations of Z(2) Variables and Topological Interpretation

At high temperatures β^{-1}, fluctuations of the Z(2) variables $\sigma(\cdot)$ will confine static quarks according to the result of section 5. (The variables W(b) defined by Eqs.(10.7), (10.8) are special representatives of the classes of configurations that were considered in section 5.)

At low temperatures, we have instead the limiting behavior

$$\sigma(\dot{p}) \longrightarrow \operatorname{sign} K_p(W) \qquad \text{as} \quad \beta \longrightarrow \infty \qquad (10.18)$$

This follows from Eq.(10.15) and (3.8). It means that the probability that the right hand side is different from the left hand side tends to zero as $\beta \rightarrow \infty$, for any fixed plaquette p.

We see that the Z(2) field strengths freeze to values that are determined by the SO(3) variables $\bar{W}(b)$. (W is obtained from $\bar{W}$ as explained before (10.8)). The Z(2) coupling constants $K_p(W)$ are not positive definite (neither in the standard model nor in the modified model) but their sign is determined by topological properties of the SO(3) gauge field. This will be explained below. As a result, the Z(2) variables $\sigma(\dot{p})$ will continue to fluctuate at large ß, as a consequence of the fluctuations of the topological properties of the SO(3) gauge field. Such fluctuations can in principle confine static quarks because of the factor $\Pi\, \sigma(\dot{p})$ in (10.12).

We turn now to the topological interpretation of our Z(2) variables. Consider continuous space time R^4 and imagine that a lattice of arbitrarily small lattice spacing is superimposed on it. Let $A_\mu(x) \in su(2)$ be a smooth vector potential. Parallel transporters along paths C are then defined by

$$U(C) = T \exp \int_C A_\mu(x)dx^\mu \quad ; \; T = \text{path ordering} . \qquad (10.19)$$

This holds in particular for links b of the lattice.

The magnetic field is

$$\partial_i A_j(x) - \partial_j A_i(x) + [A_i(x), A_j(x)] = i\, \Sigma\, B^a{}_k(x)\tau^a/2 \, . \qquad (10.20)$$

This defines the 3×3 matrix $B(x) = (B^a{}_k(x))$.

Let $\mathcal{S}$ be the set of points x where det B(x) = 0. For x not in $\mathcal{S}$ we may define $\bar{S}(x)$ in SO(3) by the unique decomposition (10.6). A local gauge is defined by the requirement $\bar{S}(x) = 1$. The vector potential $\hat{A}_\mu$ in the local gauge is given in terms of A_μ by

$$\hat{A}_\mu(x) = \bar{S}(x)^{-1}A_\mu(x)\bar{S}(x) + \bar{S}(x)^{-1}\partial_\mu\bar{S}(x) \qquad (10.21)$$

This formula makes sense since the Lie algebra of SU(2) and of SO(3) is the same. $\hat{A}_\mu$ may be singular on $\mathcal{S}$ even if A_μ is smooth.

Consider now a closed path C and the homotopy class $[\bar{S}]_C$ of the map

$$\bar{S}: C \rightarrow SO(3) \qquad (10.22)$$

$[\bar{S}]_C = \pm 1$ since the first homotopy group $\pi_1(SO(3)) = Z(2)$. Moreover, $[\bar{S}]_C = -1$ is only possible when C winds an odd number of times around the set $\mathcal{S}$ of local gauge singularities. Let Ξ be a surface whose boundary is C. If $[\bar{S}]_C = -1$ we say that the soul of a vortex passes through Ξ. Souls of vortices are counted modulo 2 (for SU(2)).

Let us write

$$W(C) = T \exp \int_C \hat{A}_\mu(x)dx^\mu \tag{10.23}$$

For paths C of small length*

$$U(C) \approx 1 \quad \text{and} \quad W(C) \approx [\bar{S}]_C = \pm 1 \ . \tag{10.24}$$

Let us now restrict attention to paths which consist of links of the superimposed lattice. In the limit of small lattice spacing, $U(b) \to 1$, and the magnetic field defined by Eq.(10.20) agrees with (10.5). We assume that the lattice spacing is sufficiently small on a physical scale so that (10.24) holds, yet C stays sufficiently many lattice spacings away from $\mathcal{S}$ so that $\bar{S}(x)$ varies little along a link in C. Then W(C) as defined in (10.23) also agrees with the quantity W(C) that was introduced previously.

If we insert Eqs.(10.24) into (10.12) we obtain

$$\prod_{p\in\Xi} \sigma(\dot{p}) = [\bar{S}]_{C=\partial\Xi} = \pm 1 \ . \tag{10.25}$$

We see that the Z(2) variables $\sigma(\dot{p})$ count the number of souls of vortices that pass through the surface Ξ. According to Eq. (10.25) they have the meaning of a topological density (similar to the Pontrjagin density that is used in instanton physics). Because of (10.18) this interpretation carries over to the sign of the coupling constants K_p.

The reader can see now that it is absolutely crucial that the coupling constants $K_p(W)$ of the Z(2) theory with variables $\sigma(b)$ as defined here are not positive definite. This is also the reason why the dynamics of an SU(2) theory (in contrast with its kinematics, cp. section 8 and 't Hoofts lectures[23]) is not symmetric under a Kramers Wannier duality transformation when it is interpreted as a Z(2) theory in the manner of this section. If one applies a Kramers Wannier duality transformation to a Z(2) gauge theory with coupling constants that are not positive, one obtains something that is no longer statistical mechanics (the action is not real). Therefore the existence of a confining high temperature phase does not imply existence of a nonconfining low temperature Higgs phase.

To avoid misunderstanding, let us point out that the modified SU(2) model may also be interpreted as a Z(2) gauge theory in another way, with positive coupling constants and different Z(2) variables

---

* In a lattice theory with large ß, $U(C) \approx 1$ follows for such C from chessboard estimates via (3.8) and (9.9)

$\gamma(\dot{p})$ that freeze to 1 as $\beta \to \infty$. A Kramers Wannier duality transformation can be applied to this Z(2) theory and establishes the existence of two distinct phases as was shown in section 6. But the relation (10.12) for the Wilson loop involves the variables $\sigma(\dot{p})$ and not $\gamma(\dot{p})$ (= $\sigma(\dot{p})$sign $K_p(W)$). Therefore nothing can be learned from this duality transformation about the behavior of the Wilson loop observable in the low temperature phase. We know of no reason to believe that it is nonconfining in the modified model[12,36].

Example of a Field Configuration with a Vortex Soul

We shall exhibit an analog of the field configuration (10.1) for an SU(2) theory in four dimensions. It will have a vortex soul located in the x^3x^4 plane. In the neighborhood of this plane let

$$B(x) = \mu \begin{pmatrix} x^1 & x^2 & 0 \\ -x^2 & x^1 & 0 \\ 0 & 0 & \lambda/\mu \end{pmatrix} + \dots \qquad (10.26)$$

This can be obtained from a vector potential $A_\mu = i\ \Sigma\ A^a{}_\mu \tau^a/2$ of the form

$$\begin{aligned} A^1{}_3 &= \mu x^1 x^2 + \dots \quad , \quad A^2{}_3 = \tfrac{1}{2}\,\mu\left[(x^1)^2 - (x^2)^2\right] + \dots \\ A^3{}_2 &= \lambda x^1 + \dots \quad , \quad A^3{}_1 = -\lambda x^2 + \dots, \quad \text{others} = 0\ . \end{aligned} \qquad (10.27)$$

In cylinder coordinates $x^1 = \rho\cos\vartheta$, $x^2 = \rho\sin\vartheta$, this gives

$$\bar{S}(x) \approx \begin{pmatrix} \cos\vartheta & \sin\vartheta & 0 \\ -\sin\vartheta & \cos\vartheta & 0 \\ 0 & 0 & 1 \end{pmatrix} \qquad (10.28)$$

If C is a path that winds once around the vortex soul at $x^1 = x^2 = 0$ then the map $\bar{S}$: C $\to$ SO(3) belongs to the nontrivial homotopy class (since the image $\bar{S}(C)$ is projection to SO(3) of a path in SU(2) that links antipodes).

Note also that for $(x^1, x^2) \neq (0,0)$, $B(x)$ belongs to the class of elements of GL(3,R) with one real and two complex conjugate eigenvalues.

In the Abelian Higgs model with a potential $V(\varphi(x))$ of the usual double well shape, presence of a vortex soul costs energy proportional to length, since $V(\varphi(x)=0) > \min V$. No similar concentration of energy is associated with a vortex soul in a pure SU(2) gauge field theory. This is in accordance with the fact that the energy of a quantum of magnetic flux can be lowered by spreading the flux, as was discussed in section 9 (Eqs. (9.8b)).

Some solutions of the Yang Mills field equations with nonvanishing vorticity were found by Vinciarelli[45].

11. RELATION BETWEEN THEORIES WITH GAUGE GROUPS SU(2) AND SO(3)

In a pure Yang Mills theory on continuous space time, the dynamical variables are vector potentials $A_\mu(x)$. They take values in the Lie algebra $\mathfrak{g}$ which is the same for gauge groups G = SU(2) and SO(3). There is therefore no visible difference between pure Yang Mills theories in the continuum with either gauge group. (If one wants to add matter fields, they should transform according to a representation of the gauge group, though).

In lattice gauge theories there is a difference to begin with. The continuum limit is supposed to be approached when $\beta \to \infty$. One should therefore expect that it will be possible to express expectation values of the SU(2) theory in terms of expectation values of an SO(3) theory, at least for large values of β. This is indeed the case. In the present section we will describe formulae which achieve this.

We will work on a finite lattice Λ throughout. The infinite volume limit $\Lambda \to \infty$ is supposed to be taken only at the end. We will consider only the modified SU(2) model. Results can be generalized to the standard model, but some of the formulae below would have a somewhat more complicated form there.

SU(2) is a twofold covering of SO(3), viz. SO(3) $\cong$ SU(2)/Z(2) = G/Γ. In the last section we have "split" the variables U(b) of the SU(2) theory into gauge invariants $\bar{W}(b) \in$ SO(3), and Z(2) gauge fields $\sigma(b) = \pm 1$. Together they determine U up to a gauge transformation. $\bar{W}(b)$ depends only on the cosets $\bar{U}(b) = U(b)\Gamma \in$ SO(3). If we had started from a theory with gauge group SO(3), the variables $\sigma(b)$ would not be needed, and the gauge invariants $\bar{W}(b)$ would determine the configuration $\bar{U}$ up to a gauge transformation. In the SU(2) theory, the Z(2) variables freeze to values determined by the SO(3) variables $\bar{W}(b)$ when $\beta \to \infty$, according to (10.18). This suggests to treat the fluctuations of $\gamma(\dot{p}) = \sigma(\dot{p})\text{sign } K_p(W)$ at large but finite β as perturbations. The complete SO(3) action $\bar{L}$ is obtained by integrating out these fluctuations.

Consider first observables $F = F(\bar{U})$ of the SU(2) theory which depend on the configuration U only through cosets $\bar{U}(b) = U(b)\Gamma \in$ SO(3). They can be regarded as observables in a SO(3) theory with variables $\bar{U}(b)$, and it follows from our definition of the action $\bar{L}$ of the SO(3) theory that

$$\langle F \rangle_{SU(2)} = \langle F \rangle_{SO(3)} \quad \text{for} \quad F = F(\bar{U}) \tag{11.1}$$

For observables F of the SU(2) theory which are not SO(3) observables (= functions of an SO(3) configuration $\bar{U}$), the situation is not so simple. We limit our attention to the Wilson loop tr U(C). In this case the answer is suggested by Eqs. (10.12) and (10.18). One finds that

$$\langle \text{tr } U(C) \rangle_{SU(2)} = \langle z_C(W) \text{ tr } W(C) \prod_{p \in \Xi} \text{sign } K_p(W) \rangle_{SO(3)} \tag{11.2}$$

This is true for any surface Ξ whose boundary is C. W is determined by $\bar{W}$, and therefore by the configuration $\bar{U}$ of the SO(3) theory by Eqs. (10. 8a) and (10.7), (10.6), viz.

$$W = W(\bar{U})$$

According to the discussion in the last section, the topologically determined factor Π sign $K_p(W)$ will be most important in (11.2). The origin of the "wave function renormalization" factor $z_C(W)$ will be explained below. It depends essentially only on variables W(b') attached to links b' close to the path C if ß is large. It satisfies

$$0 \leq z_C(W) \leq 1 \; .$$

The path measure of the SO(3) theory is of the form

$$d\mu(\bar{U}) = \frac{1}{Z} \prod d\bar{U}(b) \; e^{\bar{L}(\bar{U})} \cdot (\text{step functions}) \tag{11.4}$$

$d\bar{U}(b)$ is normalized Haar measure on SO(3). The step functions enforce the constraints (3.5). They are the same as in (3.6). Making a substitution $U(b) \rightarrow \pm U(b)$ will convince the reader that constraints (3.5) are in fact constraints on the cosets $\bar{U}(b)$.

The measure (11.4) is obtained from the path measure (3.6) by integration out the Z(2) variables $\sigma(b)$. Thus

$$\bar{L}(\bar{U}) = \ln Z_{W(\bar{U})}/Z \tag{11.5}$$

with Z_W from Eq. (5.6). Due to the constraint (3.5), we may make a variable transformation $\sigma(b) \rightarrow \gamma(b) = \pm 1$ in (5.6) which is such that $\sigma(\dot{p}) = \gamma(\dot{p})$sign $K_p(W)$. As a result,

$$Z_{W(\bar{U})} = \int \prod_b d\gamma(b) \; \exp \sum_p |K_p(W)| \gamma(\dot{p}) \tag{11.6}$$

Expression (11.6) is partition function of a Z(2) theory. We want to bring ln Z_W into a more useful form. We use low temperature expansions in order to exploit the fact that large thin vortices are rare and small ones dilute at low temperatures.

A configuration γ is determined up to a gauge transformation by its vortices, i.e. by the coclosed set T of plaquettes p where $\gamma(\dot{p}) = -1$ (see section 7). Two sets P_1 and P_2 of plaquettes will be called disjoint if no plaquette in P_1 coincides with or touches a plaquette in P_2 along a link. A coclosed set T of plaquettes will be called a contour if it cannot be split into two disjoint coclosed subsets P_1 and P_2. Any T as above can be decomposed uniquely into disjoint contours $P_1 \ldots P_k$.

For every contour P we define

$$A(P) = \prod_{p\varepsilon P} \exp -2|K_p(W)| \tag{11.7}$$

Let X_Λ be the set of all contours in Λ. We convert the sum over γ in (11.6) into a sum over vortices. This gives

$$Z_{W(\bar{U})} = 2^{-|\Lambda|} \varsigma(X_\Lambda) \exp \sum_p |K_p(W)| \tag{11.8}$$

with

$$\varsigma(X) = 1 + \sum_{D\subset X}{}' \prod_{p\varepsilon D} A(P) \tag{11.9}$$

The prime ' indicates that summation is only over nonempty subsets D of X which consist of dijoint contours $P_1 \ldots P_n$.

We take the logarithm of (11.8) and convert the result into a linked cluster expansion by use of a Möbius inversion[44]. Let X, Y,... be subsets of X consisting of n(X), n(Y),... contours respectively. We define

$$\vartheta_{YX} = \begin{cases} (-1)^{n(Y)-n(X)} & \text{if } X \subseteq Y \\ 0 & \text{otherwise} \end{cases} \tag{11.10}$$

It satisfies

$$\sum_{\substack{Y \\ X\subseteq Y\subseteq X'}} \vartheta_{YX} = \begin{cases} 1 & \text{if } X = X' \\ 0 & \text{otherwise} \end{cases} \tag{11.11}$$

We use this to define

$$\zeta_Y(W) = \sum_{X\subseteq Y} \vartheta_{YX} \ln \varsigma(X)$$

Explicitly

$$\zeta_Y(W) = \sum_{X \subseteq Y} \vartheta_{YX} \ln\,(1 + \sum_{D \subset X}{}' \prod_{P \in D} A(P))$$

$$= \sum_{X \subseteq Y} \vartheta_{YX} \ln\,(1 + \sum_{D \subset X}{}' \exp\left[-2 \sum_{P \in D} \sum_{p \in P} |K_p(W)|\right]) \quad . \qquad (11.12)$$

Σ' is again summation over nonempty subsets D which consist of disjoint contours P.

It follow from (11.5), (11.8) and (11.11) that

$$\bar{L}(\bar{U}) = \sum_p |K_p(W)| + \sum_{Y \subseteq X_\Lambda} \zeta_Y(W) + \text{const.} \qquad (11.13)$$

with const. $= -\ln(2^{|\Lambda|}Z)$

The sum over Y in (11.13) can be restricted to linked clusters Y of distinct contours in Λ, because $\zeta_Y = 0$ if Y is not linked. (This follows from Eq.(11.11) with X empty, and factorization properties of ς's.) "Linked" means that Y is not union of any two subsets Y_1 and Y_2 which are such that all contours in Y_1 are disjoint from contours in Y_2.

Eq.(11.13) together with (11.12) is our final result for the action $\bar{L}$ of the SO(3) theory.

Suppose that Y consists of contours $P_1 \ldots P_k$. By expanding the logarithm in (11.12) into a Taylor series one obtains

$$\zeta_Y(W) = \sum_{n_1 \geq 1} \ldots \sum_{n_k \geq 1} a(P_1^{n_1} \ldots P_k^{n_k})\, A(P_1)^{n_1} \ldots A(P_k)^{n_k}. \qquad (11.14)$$

These are essentially the low temperature expansions of Marra and Miracle-Solé[25]. a(Q) are certain combinatorial factors. They depend on the linked cluster Q of not necessarily distinct contours. (Contour P_j may occur $n_j \geq 1$ times.)

Series (11.14)need not (and will not) always converge, even on a finite lattice and for large ß. For any finite ß there is a set of configurations $\bar{U}$ of small but finite path measure which has the property that $|K_p(W(\bar{U}))|$ for $p \in P_j \in Y$ are not sufficiently large to make activities $A(P_j)$ small enough to ensure convergence of series (11.14).

To escape from this difficulty, we stick to Eqs. (11.12), (11.13) as our definition of the SO(3) action. We work on a finite lattice Λ. All expectation values are to be computed on this finite lattice, and the infinite volume limit is supposed to be taken only at the end. On a finite lattice, Eqs. (11.12) and (11.13) involve finite sums, so there is no convergence problem.

Expansions (11.14) are useful for estimates of the correction terms ζ_Y in the action, however. They, and the result of substituting them into (11.13), converge even on an infinite lattice if the ξ-norm $\|A\|_\xi$ of the activities A (11.7) is less than 1 for some $\xi > 1$. The ξ-norm is defined as in section 2.[11]

Starting from the bounds (3.9) and suitable estimates[26] of the combinatorial factors a(Q), one can deduce estimates of the probability that series (11.14) converges and sums to a result of modulus less than some small number. These estimates are uniform in the lattice size. They show that, generally speaking, the correction terms ζ_Y in $\bar{L}$ are small for large ß, and decrease exponentially with ß times the size of the cluster Y. In this sense, the action $\bar{L}(\bar{U})$ is local at large ß.

To conclude that these "small" correction terms are actually unimportant, one would need information on the stability properties of the low temperature phase of the SU(2) theory. They have not been established. Therefore we must be content with the statement that our results are consistent with the idea of universality.

It remains to establish formula (11.2). We adapt the treatment[25,26] of the low temperature phase of the standard Z(2) model. We start from Eq.(5.3) and make a variable transformation $\sigma \to \gamma$ as before. This gives

$$\begin{aligned} \langle \operatorname{tr} U(C) \rangle_W &= \operatorname{tr} W(C) \prod_{p\in\Xi} \operatorname{sign} K_p(W) \cdot \\ &\quad \cdot \left\{ Z_W^{-1} \int \prod_b d\gamma(b) \left[\prod_{p\in\Xi} \gamma(\dot{p}) \right] \exp \sum_p |K_p(W)|\, \gamma(\dot{p}) \right\} \\ &\equiv z_C(W) \operatorname{tr} W(C) \prod_{p\in\Xi} \operatorname{sign} K_p(W) \end{aligned} \tag{11.15}$$

The result (11.2) follows from this because of the formula

$$\langle \operatorname{tr} U(C) \rangle = \frac{1}{Z} \int \prod d\bar{U}(b)\, Z_{W(\bar{U})} \langle \operatorname{tr} U(C) \rangle_{W(\bar{U})}$$

To see this, substitute Eq.(11.5) for Z_W/Z into it. An analogous formula (6.5) was used before in section 6.

To analyze $z_C(W)$, we expand the integral in $\{\}$ in (11.15) in contours and proceed as before. We write

$$z_C(W) = \exp - \left[\ln Z_W - \ln Z_W^- \right] . \tag{11.16}$$

The linked cluster expansion for $\ln Z_W = \bar{L}$ + const. was found before in Eq. (11.13). To obtain $\ln Z_W^-$ one substitutes

$$A^-(P) = (-1)^{n(P,C)} A(P) \quad , \; n(P,C) = |P \cap \Xi| \pmod 2 \qquad (11.17)$$

for A(P) in the correction terms ζ_Y in this expansion. n(P,C) = (number of plaquettes in $P \cap \Xi$) is odd only if the contour P winds an odd number of times around C. Therefore, in the difference $\ln Z_W - \ln Z_W^-$, all those terms in the linked cluster expansion cancel out that come from clusters Y which do not contain a contour that winds around C. In conclusion

$$z_C(W) = \exp - \sum_Y \left[\zeta_Y(W) - \zeta_Y^-(W)\right] \quad . \qquad (11.18)$$

Summation is over linked clusters Y of distinct contours, with the property that at least one of these contours winds around C. ζ_Y^- is obtained from expression (11.12) for ζ_Y by substituting $A^-(P)$ for A(P).

$z_C(W)$ is positive by the first Griffith's inequality and ≤ 1, by its definition (11.15).

Because ζ_Y and ζ_Y^- decrease exponentially with the size of the cluster Y (times ß) in the sense explained earlier, $z_C(W)$ depends essentially only on variables W(b') attached to links b' near the path C.

In conclusion, the factor $z_C(W)$ may be interpreted as a renormalization of the factor tr W(C) in (11.2). Otherwise, formula (11.2) looks entirely natural in view of the general discussion[32] of locality properties in gauge field theories.

REFERENCES

1. A.M. Polyakov, Phase transitions and quark confinement, internal ICTP-report IC/78/4, Trieste (Feb. 1978).
   G. 't Hooft, reference 16 below.
   G. Mack, references 6, 8 below.
2. J. Kogut and L. Susskind, Phys. Rev. D11:395 (1975).
3. K. Wilson, Phys. Rev. D10:2445 (1974).
   K. Osterwalder, E. Seiler, Ann. Phys. (N.Y.) 110:440 (1978), and references given there.
   M. Lüscher, Commun. Math. Phys. 54:283 (1977).
4. J. Glimm, lectures presented at this school.
   J. Glimm and A. Jaffe, Nucl. Phys. B149:49 (1979).
5. E.S. Fradkin and S.H. Shenker, Phys. Rev. D19:3682 (1979).

6. G. Mack, Quark and colour confinement through dynamical Higgs mechanism, DESY-report DESY 77/58, Hamburg (Aug. 1977).
7. J. Finger, D. Horn and J.E. Mandula, Quark condensation in QCD, preprint, Cambridge/Princeton (1979).
8. G. Mack, Phys. Letters B78:263 (1978).
9. G.F. De Angelis, D. De Falco and F. Guerra, Acta Physica Austriaca Suppl. XIX:205 (1978), and references in ref. 5 above.
10. D. Ruelle, "Statistical mechanics: rigorous results", Benjamin, New York (1969).
11. Ch. Gruber and A. Kunz, Commun. Math. Phys.22:133 (1971).
12. G. Mack and V.B. Petkova, Comparison of lattice gauge theories with gauge groups Z_2 and SU(2), DESY-report DESY 78/68 Hamburg, (Nov. 1978), to appear in Ann. Phys. (N.Y.)
13. J. Fröhlich, R. Israel, E.H. Lieb and B. Simon, Commun. Math. Phys.62:1 (1978).
B. Simon, New rigorous existence theorems for phase transitions in model systems, preprint, Princeton (1977).
14. E. Wegner, J. Math. Phys. 12: 2259 (1971).
15. B. Balian, J.M. Drouffe and C. Itzykson, Phys. Rev. D10:3376 (1974); D11:2098 (1975).
16. G. 't Hooft, Nucl. Phys. B138:1 (1978).
17. J.M. Drouffe, to appear.
18. J. Fröhlich, Phys. Letters 83B:195 (1979).
19. C.P. Korthals Altes, Inequalities for order and disorder parameters in SU(N) gauge theories, preprint, Utrecht (May 1979).
20. G. Mack, Commun. Math. Phys. 65:91 (1979).
21. R.B. Griffiths, J. Math. Phys. 8:978 (1967).
D.G. Kelly and S. Sherman, J. Math. Phys.9:466 (1968).
J. Ginibre, Commun. Math. Phys.16: 310 (1970).
22. A. Messager, S. Miracle Solé and C. Pfister, Commun. Math. Phys. 58:19 (1978).
23. G. 't Hooft, lectures presented at this school.
24. R. Peierls, Proc. Camb. phil. Soc. 32:477 (1936).
D. Ruelle, reference 10.
25. R. Marra and S. Miracle Solé, Commun. Math. Phys.67:233 (1979).
26. M. Göpfert, Peierls arguments for Z(2) lattice gauge theory, Diplomarbeit and DESY-report, Hamburg (in preparation).
27. M. Fisher, J. Appl. Phys. 38:981 (1967).
28. R.L. Dobrushin and S.B. Shlosman, Commun. Math. Phys.42:31 (1975).
29. G. Mack and V.B. Petkova, Sufficient condition for confinement of static quarks by a vortex condensation mechanism, DESY-report DESY 78/69 Hamburg (Nov. 1978), to appear in Ann. Phys. (N.Y.).
30. G. 't Hooft, Nucl. Phys. B153:141 (1979).
31. H.C. Tze and Z.F. Ezawa, Phys. Rev. D14:1006 (1976).
F. Englert, Cargèse lectures 1977, in: Hadron structure and lepton-hadron interactions, M. Lévy et al., eds., Plenum Press, New York (1979).
F. Englert and P. Windey, Les Houches lectures 1978.
Z.F. Ezawa, Quantum field theory of vortices and instantons, talk presented at DESY (June 1977); Phys. Rev. D18:2091 (1978).

32. G. Mack, Physical principles, geometrical aspects and locality properties in gauge field theories, Max Planck Institut preprint München (1979).
33. G. Münster, On the characterization of the Higgs phase in lattice gauge theories, DESY-report DESY 79/72 Hamburg (Nov. 1979).
34. G. Gallavotti, F. Guerra and S. Miracle Solé, in: Lecture Notes in Physics 80, Springer, Heidelberg (1978).
35. G. Münster, Ph.D. thesis and DESY-report, Hamburg (in preparation).
36. L.G. Yaffe, Confinement in SU(N) lattice gauge theories, preprint, Princeton University (Nov. 1979).
37. T. Yoneya, Nucl. Phys. B144:195 (1978).
38. K. Wilson, lectures presented at this school.
M. Creutz, Phys. Rev. Letters 43:553 (1979).
K. Binder, Monte Carlo investigations of phase transitions and critical phenomena, in: Phase transitions and critical phenomena, C. Domb and M. Green, eds., Academic Press, New York (1976), vol. 5B.
39. J. Glimm and A. Jaffe, Commun. Math. Phys. 56:195 (1977); Phys. Letters 66B:67 (1977)
A. Jaffe, Lattice instantons: What they are and why they are important, in: Lecture Notes in Physics 80, Springer, Heidelberg (1978).
40. H. B. Nielsen and P. Olesen, Nucl. Phys. B61:45 (1973)
41. W. Feller, "An introduction to probability and its applications" Wiley, New York (1961), vols. I,II.
B. Simon, "Functional integration and quantum physics", Academic Press, New York (1979).
M.C. Reed, Functional analysis and probability theory, in: Lecture Notes in Physics 25, Springer, Heidelberg (1973).
42. A. Ukawa, P. Windey and A.H. Guth, Dual variables for lattice gauge theories and the phase structure of Z(N) systems, preprint Princeton/Cornell (1979).
43. G. Mack and V.B. Petkova, Z_2 monopoles in the standard SU(2) lattice gauge theory model, DESY-report DESY 79/22 Hamburg (Apr. 1979), unpublished.
44. G.S. Rushbrooke, G.A. Baker and P.J. Wood , in: Phase transitions and critical phenomena, C. Domb and M. Green, eds., Academic Press, New York (1974), vol. 3.
C. Domb, ibid., p. 77; G.S. Rushbrooke, J.Math.Phys. 5:1106 (1964).
45. P. Vinciarelli, Phys. Letters B78:485 (1978).
46. J.R. Fontaine and Ch. Gruber, Commun. Math. Phys. 70:243 (1979), and references given there.
47. C. Callan and D. Gross, Phys. Rev. D16:2526 (1977).
48. J. Goldstone and R. Jackiw, Phys. Letters B74:81 (1978).
49. C. Callan, lectures presented at this school.

GEOMETRY OF THE SPACE OF GAUGE ORBITS AND THE YANG-MILLS DYNAMICAL SYSTEM*

P. K. Mitter

Laboratoire de Physique Théorique et Hautes Energies**
Université Pierre et Marie Curie (Paris VI)
Tour 16, 1er étage, 4 Place Jussieu
75230 Paris, Cedex 05, France

and

Department of Physics, New York University
4 Washington Place, New York, New York 10003

* Lectures at the 1979 Cargèse Summer Institute on "Recent developments in Gauge Theories"

** Laboratoire associé au C.N.R.S. Permanent address.

0. INTRODUCTION

In these notes we shall be mainly concerned with some global aspects of classical continuum Yang-Mills theory which may be of relevance to non-perturbative considerations in the quantum theory of the continuum Yang-Mills field. In doing so we make some assumptions about which one should be clear from the outset. In classical field theory it is necessary to introduce boundary conditions in order to solve the dynamics. In order to construct a (Euclidean) quantum theory of interacting fields it is customary to put in at the outset some cutoffs: a space (space-time) volume cutoff, as well as an ultraviolet cutoff. The latter is to be removed after renormalization, and then the burden is to take the infinite volume limit. We shall adhere to this philosophy [1]. A volume cutoff is introduced in these notes by compactifying space (space-time) in some way, right from the start in the classical theory. This is because in the functional integral approach to the quantum theory one integrates over classical field configurations. Ultraviolet regularization is not mentioned because we deal mostly with the classical theory, but it should be kept in mind. The other major hypothesis is with respect to regularity of classical field configurations.

In these notes all classical fields are suitably embedded in the space of continuous (but not necessarily differentiable) functions. We assume that in doing so we are not too far from a realistic set-up at least for the envisioned regularized theory. Given the type of assumptions explained above the global considerations of these notes follow inexorably.

Our primary concern was to have a set-up in which one may rigorously define a regularized continuum quantum theory of the Yang-Mills field, via functional integral methods. This has not yet been achieved, partly because of the complexity of the space of gauge orbits which, as these notes emphasize, is the true configuration space of the Yang-Mills theory. It is from this space that dynamical variables are to be chosen (in the classical theory), and to be integrated (in the quantum theory). In contrast the lattice gauge theory [2] is globally defined [3], although its relationship to the continuum quantum theory is not sufficiently well understood. One major point is that the configuration space of the lattice gauge theory is not in any simple sense an approximation to the true configuration space of the continuum theory.

The content of these notes is divided in two parts. In part I we give an account of recent progress in the differential geometry of the space of gauge orbits (the space of connections modulo the group of gauge transformations with some restrictions) using some natural function space topologies. The main conclusion is that the space of gauge orbits is an infinite dimensional C^∞ Hilbert manifold on which there is a natural Riemannian metric. The manifold is topologically non-trivial, and needs a non-trivial atlas for its description. A natural differential structure is supplied. The account is based on [4,5] together with [6], where necessary details have been added. We do not discuss the finite dimensional self-dual critical submanifolds (instantons and generalizations), for which see [9,16,17].

In part II* we develop an invariant formulation of the Dirac constraint analysis, adapted to the Yang-Mills theory, in order to cast the latter in the form of a non-singular dynamical system. The manifold of gauge orbits emerges as the <u>true</u> configuration space, and the (non-singular) Lagrangian is a function on its tangent bundle. The Riemannian metric of part I sits in the Kinetic term of this Lagrangian. Thus classical Yang-Mills dynamics is that of a "particle" in an infinite dimensional Riemannian manifold. Formal path integral quantization takes account of the (formal) invariant "volume element" which in local coordinates is nothing but de Witt-Faddeev-Popov quantization [7], as shown recently [8]. Finally we offer some conclusions and comments.

I. DIFFERENTIAL GEOMETRY OF THE GAUGE ORBIT SPACE

1.1 Preliminaries

In what follows M_n represents a compact, connected, oriented

*Based on joint work with M.S. Narasimhan and T.R. Ramadas.

smooth Riemannian manifold of dimension n (without boundary), $n = 4$ in the covariant approach to field theory, and $n = 3$ in the canonical (Hamiltonian) approach. Compactness amounts to a volume cutoff. Thus working in a box with periodic boundary conditions on fields is tantamount to working on a torus T^n. Another popular choice is S^n etc.

G stands for a compact, connected Lie group e.g $SU(N)$ or $SU(N)/Z_N$ etc. $\underline{P}(M_n,G)$ is a principal G-bundle on M_n. $P \xrightarrow{\pi} P/G = M_n$. $\underline{P}$ is labelled by integers, corresponding to isomorphism classes (e.g. Chern classes for $G = SU(N)$). Henceforth $\underline{P}$ is taken to be in a fixed class.

A stands for a connection in $\underline{P}(M_n,G)$. $\{U_\alpha, \psi_\alpha\}$ is a bundle atlas for $\underline{P}$ i.e. $\{U_\alpha\}$ system of neighborhoods covering M_n and ψ_α trivializing maps: $\pi^{-1}(U_\alpha) \to U_\alpha \times G$. $\{\sigma_\alpha\}$ is a system of local sections: $U_\alpha \to P$, $\pi \circ \sigma_\alpha =$ identity. Then A can be identified with the system $\{A^\alpha\}$, $A^\alpha = \sigma_\alpha^* A$ is a $L(G)$ valued 1-form on U_α, ($L(G) =$ Lie algebra of G) with transition law on $U_\alpha \cap U_\beta$:

$$A^\beta = ad_{\psi_{\alpha\beta}^{-1}} A^\alpha + \psi_{\alpha\beta}^{-1} d\psi_{\alpha\beta} \tag{1.1}$$

and

$$\psi_{\alpha\beta} : \quad U_\alpha \cap U_\beta \to G$$

transition functions of the bundle. This gives a global characterization of the connection A, realized as a system of local $L(G)$ valued 1-forms on M_n. In what follows, $\mathcal{A}$ = space of all C^∞ connections. [Note: Later we shall pass out of space of smooth connections.]

Gauge transformations. A gauge transformation of $\underline{P}(M_n,G)$ is a bundle automorphism which induces the identity transformation on the base space M_n. i.e.

$$f: P \to P \;, \quad u \mapsto f(u) \;; \quad f(ua) = f(u)a \,, \; a \in G$$
$$M_n \to M_n \quad (\textit{identity map}) \tag{1.2}$$

$\mathcal{G}$ = group of all C^∞ gauge transformations. If A is a connection in $\underline{P}$, then the pulled back connection f^*A is the gauge transform of A. Thus $\mathcal{G}$ acts as a transformation group on $\mathcal{A}$. There is another way to say all this. Since a gauge transformation f preserves fibres, we can realize each such f via

$$f(u) = u\,g(u) \tag{1.3}$$

where $g: \underline{P} \to G$ satisfying

$$g(ua) = a^{-1} g(u) a \,, \quad a \in G \tag{1.4}$$

We now define:

$$g_\alpha : U_\alpha \longrightarrow G, \quad g_\alpha = g \circ \sigma_\alpha \tag{1.5}$$

and for $x \in U_\alpha \cap U_\beta$: $g_\beta = ad_{\psi_{\alpha\beta}^{-1}}\, g_\alpha$ (1.6)

since $\sigma_\beta = \sigma_\alpha \psi_{\alpha\beta}$ and $g(ua) = ad_{a^{-1}}\, g(u)$

We can then identify $\mathcal{G}$ with the system $\{g_\alpha\}$ with pointwise group operations. This is the same as saying:

$$\mathcal{G} = \Gamma(E_G) \tag{1.7}$$

the space of C^∞ sections of the associated bundle of groups $E_G = \underline{P} \times_G G$ where G has right action on P, and adjoint action on G, the second factor. Finally realizing $\mathcal{G}$ as a transformation group on $\mathcal{A}$:

$$\begin{gathered} \mathcal{A} \times \mathcal{G} \longrightarrow \mathcal{A} \\ (A_\alpha, g_\alpha) \longmapsto A_\alpha \cdot g_\alpha = ad_{g_\alpha^{-1}} A_\alpha + g_\alpha^{-1} dg_\alpha \end{gathered} \tag{1.8}$$

over each U_α. Finally we have the space of gauge orbits as the quotient space $\mathcal{M} = \mathcal{A} / \mathcal{G}$.

In order to make further progress we have to introduce suitable function space topologies. We always start with spaces of C^∞ sections of various bundles and complete these spaces in some Sobolev norm k, thus obtaining Sobolev spaces of sections. If $U \subset \mathbb{R}^n$, an open set, and f: $U \to \mathbb{R}^m$ a C^∞ function, then recall

$$\|f\|_K = \left\{ \int_U d^n x \sum_{p=0}^{k} |D^p f|^2 \right\}^{1/2} \tag{1.9}$$

where $|u|^2 = \langle u, u \rangle$, scalar product on $\mathbb{R}^m$

For a Riemanian manifold M_n, with atlas (U_α, ϕ_α), and f: $M \to \mathbb{R}^m$, C^∞ function then $f \circ \phi_\alpha^{-1} : \phi_\alpha(U) \longrightarrow \mathbb{R}^m$, where $\phi_\alpha(U)$ is an open set in $\mathbb{R}^n$. Then

$$\|f\|_K^2 = \sum_\alpha \|(\rho_\alpha \circ \phi_\alpha^{-1}) \cdot (f \circ \phi_\alpha^{-1})\|_K^2 \tag{1.10}$$

where $\{\rho_\alpha\}$ is a partition of unity subordinate to the atlas and where the integral on $\phi_\alpha(U)$ is computed with measure $\sqrt{g}\, d^n x$ Sobolev norms of sections of vector buncles are similarly defined using also the bundle atlas, and the (invariant) inner product on the fibres. In what follows we shall always hold $k > n/2$ $n = \dim M_n$, for technical reasons. This will imply, in particular, that our function spaces are embedded in the space of continuous (but not necessarily differentiable) functions (Sobolev embedding).

1.2 The group of gauge transformation $\mathcal{G}$ as a Hilbert Lie group

For the purpose of realizing $\mathcal{G}$ as a (Hilbert) Lie group, it is convenient to embed the original structure group G as a closed subspace of the vector space of m x m complex matrices, $\mathbb{M}(m,\mathbb{C})$ for some m. Then we can introduce the associated bundle of matrices $E_{\mathbb{M}} = P \times_G \mathbb{M}(m,\mathbb{C})$ where G has adjoint action on $\mathbb{M}(m,\mathbb{C})$. Then E_G is a sub-bundle of $E_{\mathbb{M}}$ and

$$\mathcal{G} = \Gamma(E_G) \subset \Gamma(E_{\mathbb{M}})$$

$\Gamma(E_{\mathbb{M}})$ has the structure of a vector space, which we realize as a Hilbert space $\Gamma_k(E_{\mathbb{M}})$, the Sobolev space of sections in class k. For $k > n/2$ we have (i) the continuous Sobolev embedding $\Gamma_k(E_{\mathbb{M}}) \subset C^0(E_{\mathbb{M}})$, the space of continuous sections (ii) (Pointwise) multiplication is well defined and continuous in $\Gamma_k(E_{\mathbb{M}})$, in other words we have the structure of a Banach algebra. Henceforth $k > n/2$, and held fixed.

We give $\mathcal{G}$ the induced topology. Then $\mathcal{G}_k$ is a closed subspace of $\Gamma_k(E_{\mathbb{M}})$. Using property (ii) above one verifies $\mathcal{G}_k$ is a topological group (i.e. group operations are continuous in the above topology).

A C^∞ atlas, compatible with group operations, is introduced in $\mathcal{G}_k$ as follows. Let ad $\underline{P} = \underline{P} \times_G L(G)$, where G has adjoint representation on L(G) its Lie algebra, be the associated adjoint bundle, and $\Gamma_k(\mathrm{ad}\,P)$, the Sobolev space of sections in class k. The exponential map from L(G) onto a sufficiently small neighborhood of the identity in G is now "lifted" as follows: Let $V_k(0) \subset \Gamma_k(\mathrm{ad}\,P)$ be a neighborhood of the origin, $N_k(e) \subset \mathcal{G}_k$, a neighborhood of the identity. Then

$$\exp: \quad V_k(0) \longrightarrow N_k(e), \quad \xi \longmapsto \exp \xi$$

defined pointwise, is a local diffeomorphism, for sufficiently small neighborhood. This is verified by using property (ii) above.

Thus $\mathcal{G}_k$ is a local Lie group, with $\Gamma_k(\mathrm{ad}\,P)$ its Lie algebra. Finally, using the continuity of group operations, the chart provided by the exponential map is transported everywhere by translation. This immediately gives a C^∞ atlas, compatible with group structure. Thus $\mathcal{G}_k$ is indeed a (Hilbert) Lie group, modelled on the Hilbert space $\Gamma_k(\mathrm{ad}\,P)$ which is canonically identified as the Lie algebra of $\mathcal{G}_k$. The details of this fairly standard construction, outlined above, can be found in [6].

1.3 $\mathcal{G}_{K+1}$ action on $\mathcal{A}_K$

If A_1, $A_2 \in \mathcal{A}$ the space of connections introduced in 1·1 then $A_1 - A_2 = \tau \in \Gamma(\mathrm{ad}\,P \times \Lambda^1)$, the space of 1-forms on M_n with values in ad P. (Locally, Lie algebra valued 1-forms). The vector space $\Gamma(\mathrm{ad}\,P \times \Lambda^1)$ is realized as a Hilbert space $\Gamma_K(\mathrm{ad}\,P \times \Lambda^1)$, Sobolev space of sections in class k, and this topologizes $\mathcal{A}$. $\mathcal{A}$ with the above topology is called $\mathcal{A}_K$. Now the exterior covariant derivative d_A is a continuous map $\Gamma_{K+1}(\mathrm{ad}\,P) \longrightarrow \Gamma_K(\mathrm{ad}\,P \times \Lambda^1)$. Thus $\mathcal{G}_{K+1}$ has continuous action on $\mathcal{A}_K$. Using local charts, the continuity of d_A, and the Banach algebra property for Sobolev spaces H_K $(K > n/2)$, one verifies straightforwardly that the $\mathcal{G}_{K+1}$ action on $\mathcal{A}_K$ is C^∞.

It is important, for what follows, that the group action be free (i.e. no fixed point). For this purpose we restrict ourselves to the subgroup $\bar{\mathcal{G}}_{K+1} \equiv \mathcal{G}_{K+1}/Z$, where Z is the center, and to the subspace $\bar{\mathcal{A}}_K$ of irreducible connections (If $A \in \bar{\mathcal{A}}_K$) then the restricted holonomy group of A i.e. the group generated by parallel transport around homotopically trivial closed curves in M_n coincides with G the structure group of $\underline{P}(M,G)$). $\bar{\mathcal{A}}_K$ is open and dense in $\mathcal{A}_K$. Irreducibility, and absence of center in $\bar{\mathcal{G}}_{K+1}$, implies that the $\bar{\mathcal{G}}_{K+1}$ action on $\bar{\mathcal{A}}_K$ is free.

Let
$$\Delta_A = d_A^* d_A : \Gamma_{K+1}(\mathrm{ad}\,P) \longrightarrow \Gamma_{K-1}(\mathrm{ad}\,P) \tag{1.11}$$
the covariant Laplacian. Then $\ker \Delta_A = 0$, for $A \in \bar{\mathcal{A}}_K$ (This follows from the positivity in the L^2 norm on $\Gamma_K(\mathrm{ad}\,\underline{P} \times \Lambda^1)$ and irreducibility). Thus the Green's operator $G_A = \Delta_A^{-1}$ is uniquely defined. We have that:

$$G_A : \Gamma_{K-1}(\mathrm{ad}\,P) \longrightarrow \Gamma_{K+1}(\mathrm{ad}\,P) \tag{1.12}$$
is continuous, as follows from the work in [5]

Another fact is that:
$$T_A(\bar{\mathcal{A}}_K) = V_A \oplus H_A \tag{1.13}$$
is the direct sum of closed subspaces. V_A may be identified with $d_A \Gamma_{K+1}(\mathrm{ad}\,P)$, the tangent space to orbit through A and H_A with $\ker d_A^*$. Note that V_A is orthogonal to H_A in the L^2 scalar product on $\Gamma_K(\mathrm{ad}\,P \times \Lambda^1)$.

1.4 $\bar{\mathcal{A}}_K \longrightarrow \bar{\mathcal{A}}_K/\bar{\mathcal{G}}_{K+1} = \mathcal{M}_K$ as a C^∞ locally trivial fibration with $\mathcal{M}_K$ a C^∞ Hilbert manifold

We have seen that $\bar{\mathcal{G}}_{K+1}$ has C^∞ free action on $\bar{\mathcal{A}}_K$. We are let to consider the quotienting:

$$\bar{\mathcal{A}}_K \xrightarrow[\pi]{} \bar{\mathcal{A}}_K/\bar{\mathcal{G}}_{K+1} = \mathcal{M}_K \tag{1.14}$$

where π is the canonical projection and $\mathcal{M}_K$ is the space of gauge orbits. In order to prove that the above is a principal smooth $\bar{\mathcal{G}}_{K+1}$ bundle we have to prove C^∞ local triviality. Actually in [4], Singer had announced that $\bar{\mathcal{A}} \longrightarrow \bar{\mathcal{A}}/\bar{\mathcal{G}} = \mathcal{M}$ is a principal $\bar{\mathcal{G}}$ bundle using C^∞ topologies. In [5], Narasimhan and Ramadas proved that $\bar{\mathcal{G}}_{K+1}$ has proper action on $\bar{\mathcal{A}}_K$, and a theorem of N. Bourbaki (for which a proof is not readily available) then implies that we have a principal fibration. In [6], however, we have supplied an explicit proof of C^∞ local trivality. This proof supplies, at the same time, a C^∞ differential structure on the orbit space $\mathcal{M}_K$. We explain the main steps involved in this.

Giving $\mathcal{M}_K$ the quotient topology we realize it as a topological space. Next we give $\mathcal{M}_K$ a system of coordinate neighborhoods covering $\mathcal{M}_K$ (this will later turn out to be a C^∞ atlas) by giving a system of effective local "background" gauge conditions very much as in [9], where self dual submanifolds of $\mathcal{M}_K$ (necessarily finite dimensional) were studied.

Given any $A \in \bar{\mathcal{A}}_K$, we take a neighborhood of A as a ball of radius δ centred at A:

$$N_K^\delta(A) = \{A' \in \bar{\mathcal{A}}_K \mid \|A' - A\|_K = \|\tau\|_K < \delta\} \tag{1.15}$$

We also define:

$$\mathcal{H}_K(A) = \{A' \in \mathcal{A}_K \mid d_A^*(A' - A) = d_A^*\tau = 0\} \tag{1.16}$$

$\mathcal{H}_K(A)$ is a closed subspace of $\mathcal{A}_K$, isomorphic to H_A of (1.13) and so identified as a Hilbert space. Next we define:

$$\mathcal{S}_K^\delta(A) = N_K^\delta(A) \cap \mathcal{H}_K(A) \tag{1.17}$$

as an open set in $\mathcal{H}_K(A)$.

The main fact about $\mathcal{S}_K^\delta(A)$ is that no two distinct points are gauge related for sufficiently small δ, i.e. if $A_1 \neq A_2$, and $A_1 = A_2 \circ g \Longrightarrow g = \text{identity}$.
This is proved exactly as in [9]. See also [10]. First one shows that $\|A_1 - A_2\|_K < \delta$, implies that such a $g \in \bar{\mathcal{G}}_{K+1}$ is necessarily small, i.e. in the neighborhood of the identity (one

uses the fact that a $L(G)$ valued function is controlled by its derivative, which follows from $G_A: \Gamma_{K-1} \longrightarrow \Gamma_{K+1}$ is continuous). For g in a neighborhood of the identity, using the exponential map $g = \exp \xi$, $A_1 = A_2 \bullet g$ becomes a non-linear elliptic equation for ξ, whose unique solution is $\xi = 0$. (One uses standard things: the above continuity of G_A, continuity of d_A: $\Gamma_{K+1}(ad\,P) \longrightarrow \Gamma_K(ad\,P \times \Lambda^1)$ and the Banach algebra property for the Γ_K, $k > \frac{n}{2}$)

We restrict the canonical projection π of

$$\pi: \quad \bar{A}_K \longrightarrow \bar{A}_K / \bar{\mathcal{G}}_{K+1} = \mathcal{M}_K \tag{1.18}$$

to $\mathcal{J}_K^S(A)$,

$$\pi / \mathcal{J}_K^S(A): \quad \mathcal{J}_K^S(A) \longrightarrow \eta_A \subset \mathcal{M}_K \tag{1.19}$$

Then η_A is an open set in $\mathcal{M}_K$ and, because points in $\mathcal{J}_K^S(A)$ are gauge unrelated, $\pi / \mathcal{J}_K(A)$ is actually a homeomorphism. Let σ_A be its inverse, i.e.

$$\sigma_A: \quad \eta_A \longrightarrow \mathcal{J}_K^S(A) \subset \bar{A}_K, \quad \pi \circ \sigma_A \big|_{\eta_A} = \text{identity}. \tag{1.20}$$

σ_A is a local (gauge) section; it provides a local chart for η_A in $\mathcal{H}_K(A)$, since $\mathcal{J}_K^S(A)$ is an open set in the Hilbert space $\mathcal{H}_K(A)$ (the model space for η_A).
Finally,

$$\{ \eta_A, \sigma_A, \mathcal{H}_K(A) \}_{A \in \bar{A}_K} \tag{1.21}$$

will play the role of a smooth atlas on the space of orbits $\mathcal{M}_K$, provided we can prove that the transition maps (coordinate changes) over overlaps are C^∞ i.e.

$$\sigma_{A_2} \circ \sigma_{A_1}^{-1}: \quad \sigma_{A_1}(\eta_{A_1} \cap \eta_{A_2}) \longrightarrow \sigma_{A_2}(\eta_{A_1} \cap \eta_{A_2}) \tag{1.22}$$

is a C^∞ map between open sets in $\mathcal{H}_K(A_1)$ and $\mathcal{H}_K(A_2)$, respectively. That this is indeed the case follows very simply once one has shown C^∞ local triviality. Note that the model spaces $\mathcal{H}_K(A)$ are isomorphic Hilbert spaces (for different A) but there is no canonical isomorphism between them.

Next we turn to C^∞ local triviality. We have to display diffeomorphisms:

$$\Phi_A: \quad \eta_A \times \bar{\mathcal{G}}_{K+1} \longrightarrow \pi^{-1}(\eta_A) \subset \bar{A}_K \tag{1.23}$$

i.e. $\pi^{-1}(\eta_A)$ looks like a product.

We define Φ_A by:

$$(m', g) \mapsto \Phi_A(m', g) = \sigma_A(m') \circ g \tag{1.24}$$

$\sigma_A(m') \in \mathcal{S}_K^g(A)$ and $\sigma_A(m') \circ g$ is its gauge transform. We also define:

$$\chi_A : \pi^{-1}(\eta_A) \longrightarrow \eta_A \times \bar{\mathcal{G}}_{K+1} \tag{1.25}$$
$$A' \longmapsto \chi_A(A') = (\pi A', g_A(A'))$$

Here

$$g_A : \pi^{-1}(\eta_A) \longrightarrow \bar{\mathcal{G}}_{K+1} \tag{1.26}$$

is defined (uniquely) by

$$A' \circ g_A(A')^{-1} = \sigma_A(m') \tag{1.27}$$

where $m' = \pi A'$. Thus m' is the orbit through A', the local section σ_A intersects the orbit at $\sigma_A(m')$, and $g_A(A')$ is the gauge transformation which takes $\sigma_A(m')$ to A'. Uniqueness follows from free group action. It is easy to verify that χ_A is the inverse of Φ_A and vice-versa. Thus Φ_A is indeed an isomorphism. Moreover in local charts Φ_A is C^∞, since the $\bar{\mathcal{G}}_{K+1}$ action on $\bar{\mathcal{A}}_K$ is C^∞. In [6] it is shown that $(\Phi_A)_*$, the linear differential of Φ_A, leads to an isomorphism of the appropriate tangent spaces at any point. By the inverse function theorem it follows that χ_A is also C^∞, in local charts. Thus Φ_A is a diffeomorphism. This proves C^∞ local triviality, and $\bar{\mathcal{A}}_K \longrightarrow \bar{\mathcal{A}}_K/\bar{\mathcal{G}}_{K+1} = \mathcal{M}_K$ is a C^∞ locally trivial $\bar{\mathcal{G}}_{K+1}$ bundle.

Now let $m' \in \eta_{A_1} \cap \eta_{A_2}$. Then $\tau_i = \sigma_{A_i}(m') - A_i$, $i = 1,2$ are local coordinates for m' in H_{A_i}. The transition function for the bundle is

$$\Psi_{A_1 A_2}(m') = g_{A_1}(A') g_{A_2}(A')^{-1} = g_{A_2}(\sigma_{A_1}(m'))^{-1} \tag{1.28}$$

We have the change of coordinates formula in $\mathcal{M}_K$:

$$\sigma_{A_1}(\eta_{A_1} \cap \eta_{A_2}) \longrightarrow \sigma_{A_2}(\eta_{A_1} \cap \eta_{A_2}) \tag{1.29}$$

as a map between open sets in $\mathcal{H}_K(A_1)$ and $\mathcal{H}_K(A_2)$

$$\sigma_{A_2}(m') = \sigma_{A_1}(m') \circ g_{A_2}(\sigma_{A_1}(m'))^{-1} \qquad (1.30)$$

Since the $\bar{\mathcal{G}}_{K+1}$ action is C^∞, and g_{A_2} is C^∞ (in local charts) (follows from C^∞ nature of χ_A) it follows that (1.29) is C^∞. Thus we have the main

<u>Theorem 1</u> $\bar{\mathcal{A}}_K \longrightarrow \bar{\mathcal{A}}_K/\bar{\mathcal{G}}_{K+1} = \mathcal{M}_K$ is a C^∞ locally trivial $\bar{\mathcal{G}}_{K+1}$ bundle. Moreover $\mathcal{M}_K$ is a C^∞ Hilbert manifold.

We also have:

<u>Theorem 2</u>. Let $M_n = S^n$ on $T^n = S^1 \times S^1 \times \cdots \times S^1$ (n-factors), $n = 3$ or 4, and $G = SU(N)$ or $SU(N)/Z_N$. Then the bundle

$$\bar{\mathcal{A}}_K \longrightarrow \bar{\mathcal{A}}_K / \bar{\mathcal{G}}_{K+1} = \mathcal{M}_K$$

is non-trivial.

This implies that the bundle does not admit a continuous global section, and also that $\mathcal{M}_K$ cannot be described by a single coordinate chart in $\bar{\mathcal{A}}_K$. Theorem 2 was actually proved in [4,5] for the case $M_n = S^n$ but it is not hard to see that it remains true for the case T^n. By [4], $\bar{\mathcal{A}}_K$ is homotopically trivial, so that $\mathcal{M}_K$ inherits (by a shift of one unit) the homotopy of $\bar{\mathcal{G}}_{K+1}$

$$\pi_{n+1}(\mathcal{M}_K) \simeq \pi_n(\bar{\mathcal{G}}_{K+1})$$

The computation of the homotopy groups of $\bar{\mathcal{G}}_{K+1}$, for the case $M_n = S^n$, is given in [4]. This explains the "Gribov ambiguity".

It is also worth remarking that when $N \to \infty$, an infinite set of homotopy groups of $\mathcal{M}_K$ are non-vanishing. The dimension of self-dual critical submanifolds (in any finite Chern class) [9] becomes negative when $N \to \infty$. Thus the manifold of gauge orbits remains highly non-contractible, although self-dual critical manifolds disappear. (We are on S^4).

1.5 Riemannian metric on $\mathcal{M}_K$, the manifold of gauge orbits

We have seen that $\bar{\mathcal{A}}_K \longrightarrow \bar{\mathcal{A}}_K/\bar{\mathcal{G}}_{K+1} = \mathcal{M}_K$ is a principal fibre bundle. Thus we may introduce a connection. A natural connection [5], see also [4], is supplied by the continuous linear map $\hat{\omega}$:

$$\hat{\omega}_{A'} : T_{A'}(\bar{A}_K) \longrightarrow L(\bar{\mathcal{G}}_{K+1}) = \Gamma_{K+1}(\mathrm{ad}\,P)$$
$$\tau \mapsto \hat{\omega}_{A'}(\tau) = G_{A'}\, d^*_{A'}\, \tau \qquad (1.31)$$

$\hat{\omega}$ satisfies:

(i) $R^*_g\, \hat{\omega} = \mathrm{ad}_{g^{-1}}\, \hat{\omega}\,,\ g \in \bar{\mathcal{G}}_{K+1}$

(ii) If $v_{A'} \in V_{A'} \subset T_{A'}(\bar{A})$, where $V_{A'}$ is the tangentspace to the fibre (gauge orbit) through A', then $v_{A'}$ has the form

$$v_{A'} = d_{A'}\, \xi$$

where

$$\xi \in L(\bar{\mathcal{G}}_{K+1}) = \Gamma_{K+1}(\mathrm{ad}\,P)$$

Then

$$\hat{\omega}_{A'}(v_{A'}) = \xi$$

Becasue of (i) and (ii) $\hat{\omega}$ is indeed a connection and we can identify:

$$H_{A'} = \ker \hat{\omega}_{A'} = \{\tau \in \Gamma_K(\mathrm{ad}\,P \times A') \mid d^*_{A'}\,\tau = 0\} \qquad (1.32)$$

the horizontal tangentspace at A'. We have $(R_g)_*\, H_{A'} = H_{A' \circ g}$ Moreover,

$$T_{A'}(\bar{A}_K) = V_{A'} \oplus H_{A'} \qquad (1.33)$$

where $V_{A'}$, $H_{A'}$ are mutually orthagonal (in the L^2 scalar product on $\Gamma_K(\mathrm{ad}\,P \times A')$). The invariant L^2 scalar product on $T_{A'}(\bar{A}_K) = \Gamma_K(\mathrm{ad}\,P \times A')$ induces a scalar product on

$$H_{A'} \subset T_{A'}(\bar{A}_K)$$

denoted $(\ ,\ )$. Now

$$\pi_* : T_{A'}(\bar{A}_K) \longrightarrow T_{m'}(\mathcal{M}_K),\ m' = \pi A' \qquad (1.34)$$

Note that $V_{A'} = \ker \pi_*$. Restriction to $H_{A'}$ gives:

$$\pi_* : H_{A'} \longrightarrow T_{m'}(\mathcal{M}_K) \qquad (1.35)$$

is an isomorphism. If $X_1, X_2 \in T_{m'}(\mathcal{M}_K)$, then we can define the scalar product:

$$g_{m'}(X_1, X_2) = (X_1^h, X_2^h)_{A'} \tag{1.36}$$

where X_i^h are the (unique) horizontal lifts to any point A' of the fibre (orbit) on m', and $(\,,\,)_{A'}$ is the induced invariant L^2 scalar product on $H_{A'}$. The R.H.S. is actually independent of which point A' of the orbit where we choose to lift (since $(R_g)_* H_{A'} = H_{A'\circ g}$ and the induced scalar product on H_A is gauge invariant). This defines a Riemannian structure g on $\mathcal{M}_K$.

We can express the Riemannian structure, g in terms of the local charts introduced in Sec. 1.4. Let $m' \varepsilon \eta_A$. If $X_i \varepsilon T_{m'}(\mathcal{M}_K)$ then in local chart σ_A :

$$X_i^h = P_{A'} \tau_i \tag{1.37}$$

Here $\tau_i \;\varepsilon\; T_{A'}(\mathcal{S}_A)$ which can be identified with H_A. $P_{A'}$ is the horizontal projector onto H_A. Thus:

$$g_{m'}(X_1, X_2) = (\tau_1, \tilde{g}(A')\tau_2), \; A' \varepsilon \mathcal{S}_A \tag{1.38}$$

where

$$\tilde{g}(A') = P_{A'} = 1 - d_{A'} G_{A'} d_{A'}^* \tag{1.39}$$

and

$$\tau_i \;\varepsilon\; H_A = \{\tau \;\varepsilon\; \Gamma_K(\mathrm{ad}P \times \Lambda') \,|\, d_A^* \tau = 0\} \tag{1.40}$$

Claim. ker $\tilde{g}(A') = 0$ (in H_A) (1.41)

Suppose $\tilde{g}(A')\tau_2 = 0$, where $\tau_2 \varepsilon H_A$, i.e. $d_A^* \tau_2 = 0$

Using

$$A' = A + \tau \;\varepsilon\; \mathcal{S}_A$$
$$d_{A'}^* = d_A^* - *[\tau, * \cdot\,] \tag{1.42}$$

we have

$$0 = \tau_2 - d_{A'} G_{A'} * [\tau, *\tau_2] \tag{1.43}$$

Using the continuity of $d_A : \Gamma_K \longrightarrow \Gamma_{K-1}$, and of $G_A : \Gamma_{K-1} \rightarrow \Gamma_{K+1}$ we have for sufficiently small τ (guaranteed since $A + \tau \;\varepsilon\; \mathcal{S}_A$) the unique solution $\tau_2 = 0$.

Thus g(A) is invertible. Moreover it is not hard to show that

$$(\tau_1, \tilde{g}(A')\tau_1) > 0 \tag{1.44}$$

We conclude that the Riemannian structure g on $\mathcal{M}_k$ is actually a (weak) Riemannian metric. [We say weak because the norm $\|\tau\|_g^2 = (\tau_1, \tilde{g}\,\tau_1)$ is weaker than the Sobolev k-norm which gives the Hilbert space structure of $T_{m'}(\mathcal{M}_k)$]. Note that if we attempted to extend the coordinate system σ_A, the metric becomes singular when $\mathcal{J}_k(A)$ is tangential to an orbit, and then the coordinate system is not legitimate. The above Riemannian metric has been exploited in [8] to give a geometrical interpretation of the Faddeev-Popov determinant arising in formal path integral quantization of gauge theories. This will become clearer in the succeeding sections.

II. THE YANG-MILLS DYNAMICAL SYSTEM* AND QUANTIZATION

2.0 In the second part of these notes we will explain how the manifold of gauge orbits emerges as the true (or physical configuration space) of the Yang-Mills theory, viewed as a (classical) dynamical system. In the process the role of the Riemannian metric, introduced earlier, will get clarified, as well as its relation to the problem of gauge field quantization. The invariant form of the Dirac constraint analysis developed in Section 2.2 was motivated by the appendix of Faddeev's classical paper.

2.1 A. Dynamical Systems (Rapid Review of Some Essential Concepts)

We shall use the invariant (i.e. geometrical) formulation of classical mechanics. For the finite dimensional case, see Arnold [11], and for the infinite dimensional set up relevant for us see Marsden [12]

The configuration space will be a smooth manifold $\mathcal{C}$ modelled on a Banach (or Hilbert space) E, assumed to be reflexive $((E^*)^* = E)$. E^* is the topological dual (the space of continuous linear functionals on E) of E.

The velocity space is the total space of the tangent bundle $\pi_T : T\mathcal{C} \to \mathcal{C}$. Here $T\mathcal{C} = \bigcup_{q \in \mathcal{C}} T_q\mathcal{C}$

and π_T the canonical projection. $T\mathcal{C}$ is a mooth manifold modelled on E x E. A point in $T\mathcal{C}$ is represented by $(q, \dot{q})$ where $q \in \mathcal{C}$ and $\dot{q} \in T_q\mathcal{C}$. In local chart

$$u = (q, \dot{q}) \longrightarrow (\underline{x}, \underline{\dot{x}}) \in E \times E$$

*This section is based on joint work with M.S. Narasimhan and T.R. Ramadas, who are, however, not to be held responsible for any inaccuracies in my presentation.

The phase space is the total space of the cotangent bundle $\tilde{\pi} : T^*\mathcal{C} \longrightarrow \mathcal{C}$ where $T^*\mathcal{C} = \bigcup_{q\in\mathcal{C}} T^*_q\mathcal{C}$ and $\tilde{\pi}$ the canonical projection. $T^*\mathcal{C}$ is a smooth manifold, modelled on $E \times E^*$, and will be denoted P (the phase space). A point $u \in T^*\mathcal{C}$ will be represented $u = (q, \omega_q)$ where $q \in \mathcal{C}$, and $\omega_q \in T^*_q\mathcal{C}$, in other words a 1-form at q. In a local chart: $u = (q, \omega_q) \longrightarrow (\underline{x}, \alpha) \in E \times E^*$,

The phase space P, comes with a canonical 1-form θ defined on it as follows: Let $(q, \omega_q) \in T^*\mathcal{C}$ and $v \in T_{(q,\omega_q)}(T^*\mathcal{C})$

Then

$$\theta_{(q,\omega_q)}(v) = \omega_q(\tilde{\pi}_* v) \tag{2.1}$$

In local charts: $(q, \omega_q) \longrightarrow (\underline{x}, \alpha) \in E \times E^*$

$$v \longrightarrow (\dot{\underline{x}}, \dot{\alpha}) \in E \times E^*$$

$$\theta_{(\underline{x},\alpha)}(\dot{\underline{x}}, \dot{\alpha}) = \alpha(\dot{\underline{x}}) \tag{2.2}$$

We now define the symplectic (closed) 2-form Ω on P by:

$$\Omega = d\theta \tag{2.3}$$

Expression in local charts:

Let $u = (q, \omega_q) \longrightarrow (\underline{x}, \alpha) \in E \times E^*$, a point in P

$X_i \in T_u(P) \longrightarrow (\dot{\underline{x}}_i, \dot{\alpha}_i) \in E \times E^*$, assume $[X_1, X_2] = 0$

$$\Omega_u(X_1, X_2) = X_1\theta_u(X_2) - X_2\theta_u(X_1) - \cancel{\theta_u([X_1, X_2])}$$

From

$$X_1\theta_u(X_2) = \frac{d}{dt}\Big|_{t=0} \theta_{(\underline{x}+t\dot{\underline{x}}_1, \alpha+t\dot{\alpha}_1)}(\dot{x}_2, \dot{\alpha}_2)$$

$$= \frac{d}{dt}\left(\alpha(\dot{\underline{x}}_2) + t\dot{\alpha}_1(\dot{x}_2)\right) = \dot{\alpha}_1(\dot{x}_2)$$

(use (2.2)

Thus, in local charts:*

$$\Omega_{(\underline{x},\alpha)}\left((\dot{\underline{x}}_1,\dot{\alpha}_1),(\dot{\underline{x}}_2,\dot{\alpha}_2)\right)=\dot{\alpha}_1(\dot{\underline{x}}_2)-\dot{\alpha}_2(\dot{\underline{x}}_1) \tag{2.4}$$

The symplectic 2-form Ω induces, for each $u \in P$, a mapping:

$$T_u(P) \longrightarrow T_u^*(P), \quad X \longmapsto i_X \Omega_u \tag{2.5}$$

defined by:

$$i_X \Omega_u(y) = \Omega_u(X,y), \quad \forall\, y \in T_u(P)$$

It is easy to check (use (2.4)) that the mapping is injective i.e. its kernel is trivial. Moreover it is onto (E is reflexive), see [12], thus it is an isomorphism. Ω is said to be non-degenerate.

Hamiltonian vector field. We use this isomorphism to set up the following map: Let $\mathcal{F}$ be the space of C^∞ functions on P and Γ(TP) the space of (continuous) vector fields on P. Then we have the map:

$$\mathcal{F} \rightarrow \Gamma(TP), \quad f \longmapsto X_f$$

defined by (2.6)

$$i_{X_f}\Omega = -df$$

Note that X_f exists and is uniquely defined since (2.5) is an isomorphism. If H is a privileged function, the Hamiltonian, X_H is the corresponding Hamiltonian vector field, the infinitesimal generator of motion (Hamiltonian flow) in the phase space. The pair (P,Ω) is a symplectic manifold and (P,Ω,H) a Hamiltonian dynamical system.

Expression in local chart.** If f is a C^1 function on P,

*In finite dimensions, $q=(q^1,\dots\dots,q^n)$, $\omega_q=\sum_{i=1}^{n} p_i\,dq^i$, so take $(q^1\cdots q^n;p_1\cdots p_n)$ as coordinates in phase space. Then $\theta=\sum_i p_i\,dq^i$ interpreted as a 1-form on P, and $\Omega = d\theta = \sum_i dp_i \wedge dq^i$

**In finite dimensions $X_f=\sum\left(\frac{\partial f}{\partial p_i}\frac{\partial}{\partial q^i}-\frac{\partial f}{\partial q^i}\frac{\partial}{\partial p_i}\right)$.

then in local chart $f(\underline{x},\alpha)$ function on an open set in E x E*.
Let $X_f \rightarrow (\dot{x}_f, \dot{\alpha}_f) \in E \times E^*$ in local chart.
Then

$$\dot{\alpha}(\dot{x}_f) = -\frac{d}{dt}\Big|_{t=0} f(\underline{x}, \alpha + t\dot{\alpha}), \forall \dot{\alpha} \in E^*$$
$$\dot{\alpha}_f(\dot{x}) = \frac{d}{dt}\Big|_{t=0} f(\underline{x} + t\dot{x}, \alpha), \forall \dot{x} \in E \tag{2.7}$$

Finally the Poisson bracket of two C^1 functions f, g is given by*

$$\{f, g\} = \Omega\,(X_f, X_g) \tag{2.8}$$

B. Legendre Transformation

The Lagrangian L is a function on the velocity space, or the tangent bundle $T\mathcal{C}$ of the configuration space $\mathcal{C}$. We write L(q,$\dot{q}$) where $(q,\dot{q}) \in T\mathcal{C}$. Using L, we get a bundle map

$$FL : T\mathcal{C} \rightarrow T^*\mathcal{C} = \mathcal{P}$$

(inducing the identity map $\mathcal{C} \rightarrow \mathcal{C}$ on the base space) defined by:

$$FL_q : T_q\mathcal{C} \rightarrow T_q^*\mathcal{C}, \; \dot{q} \mapsto FL_q\dot{q} = \pi$$
$$\pi(v) = FL_q\dot{q}\,(v) = \frac{d}{dt}\Big|_{t=0} L(q, \dot{q} + tv) \tag{2.9}$$

π is the "canonical momentum".

The Lagrangian L is said to be non-singular when FL is an isomorphism, otherwise it is said to be singular. When L is singular, FL is not onto and/or has non-trivial kernel. In general $FL(T\mathcal{C}) \subsetneq T^*\mathcal{C} = \mathcal{P}$, i.e. a sub-bundle of $T^*\mathcal{C}$, and this gives rise to constraints in phase space. In particular, Ω when restricted to $FL(T\mathcal{C})$ may become degenerate and then further analysis is necessary in order to define a dynamical system. (This will be the case for the Yang-Mills theory).

C. Riemannian Metric on $\mathcal{C}$

*In finite dimensions, $\{f,g\} = \sum_i \left(\frac{\partial f}{\partial p_i}\frac{\partial g}{\partial q^i} - \frac{\partial f}{\partial q^i}\frac{\partial g}{\partial p_i}\right)$

We will assume the existence of a <u>(weak)</u> <u>Riemannian metric</u> on $\mathcal{C}$. This means: (i) there exists a symmetric bilinear form $(\, , \,)_q$ on $T_q\mathcal{C} \times T_q\mathcal{C}$ depending continuously on $q \in \mathcal{C}$. Write

$$(\dot{q}_1, \dot{q}_2)_q = (\dot{q}_1, g(q)\dot{q}_2) \tag{2.10}$$

(ii) Using $(\, , \,)_q$ we define for each $q \in \mathcal{C}$, a map

$$\mathcal{R}_q : T_q\mathcal{C} \to T_q^*\mathcal{C}, \quad \dot{q} \mapsto \mathcal{R}_q\dot{q} = \pi$$

where

$$R_q\dot{q}(v) = (\dot{q}, v)_q = (g(q)\dot{q}, v) \tag{2.11}$$

$\mathcal{R}_q$ is assumed to be injective, i.e. trivial kernel. Thus $\mathcal{R}$ (specified by each $\mathcal{R}_q$, identity map on base) maps $T\mathcal{C}$ onto a sub-bundle of $T^*\mathcal{C}$ (if $\mathcal{R}_q$ is also onto then $\mathcal{R} : T\mathcal{C} \to T^*\mathcal{C}$ is an isomorphism, and we speak of a (strong) Riemannian metric).

<u>Remark</u>. Let i be the inclusion map $\mathcal{R}(T\mathcal{C}) \subset T^*\mathcal{C} = \underline{P}$, and Ω the canonical symplectic 2-form on P. Then $i^*\Omega$ is the symplectic 2-form on $\mathcal{R}(T\mathcal{C})$. It is easy to check that $i^*\Omega$ is (weakly) non-degenerate (use $\mathcal{R}_q$ is injective) i.e. the map (2.5), using $i^*\Omega$ instead of Ω and $\mathcal{R}(T\mathcal{C})$ instead of $\underline{P}$ is injective.

Finally note that since $\mathcal{R}$ gives an isomorphism between $T\mathcal{C}$ and $\mathcal{R}(T\mathcal{C}) \subset T^*\mathcal{C}$, a point in $\mathcal{R}(T\mathcal{C})$ is given by (q, π) where $\pi = R_q\dot{q}$ and may be identified with $g(q)\dot{q}$ (see (2.11)).

2.2 <u>The Yang-Mills Dynamical System</u>

In the context of Part I, Section 1.1, we choose $M_n = M_3$, a (space-like) 3-dimensional section of 4-dimensional space-time $M_3 \times \mathbb{R}$ with g the induced Riemannian metric.* (We shall freely use the definitions and results of Part I). In particular now:

$\bar{A}_k$ = space of irreducible connections of $\underline{P}(M_3,G)$, of Sobolev class k. $\underline{P}$ is labelled by integers (corresponding to characteristic classes).

= "space of vector potentials".

$A^0_{k+1} = \Gamma_{k+1}(\mathrm{ad}\,\underline{P})$ = space of "scalar potentials".

* $ds^2 = dt^2 - g_{ij}dx^i dx^j = g^{(4)}_{\mu\nu}\, dx^\mu dx^\nu$.

Our configuration space $\mathcal{C}$, to begin with is $\mathcal{C}=\bar{\mathcal{A}}_K \times \mathcal{A}^0_{K+1}$ which is modelled on

$$E = \Gamma_K (ad P \times \Lambda^1) \times \Gamma_{K+1}(ad P)$$

Note that E is reflexive, a Hilbert space. The velocity space $T\mathcal{C} = T(\bar{\mathcal{A}}_K \times \mathcal{A}^0_{K+1})$ . . . A point in $T(\bar{\mathcal{A}}_K \times \mathcal{A}^0_{K+1})$ is represented by $(A, A_0; \dot{A}, \dot{A}_0)$ The Lagrangian, $L : T(\bar{\mathcal{A}}_K \times \mathcal{A}^0_{K+1}) \to \mathbb{R}$, is given by:

$$L(A, A_0; \dot{A}, \dot{A}_0) = \frac{1}{2}\|\dot{A} - d_A A_0\|^2 - \frac{1}{2}\|F(A)\|^2 = -\frac{1}{4}\int d^3x \sqrt{g}\, Tr\, F_{\mu\nu}F^{\mu\nu} \tag{2.12}$$

where $-\frac{1}{4} Tr\, F_{\mu\nu} F^{\mu\nu}$ is the usual relativistically invariant Lagrange density. The time t is held fixed and considered a parameter. Moreover:

$$\|\omega\|^2 = (\omega, \omega)$$

$$(\omega, \eta) = \int_{M_3} Tr\,(\omega \wedge * \eta)\ , \quad \omega, \eta : p\text{-forms}$$

in local charts

$$\text{e.g} \quad = \int d^3x \sqrt{g} \sum_i Tr\,(\omega^i(x)\, \eta^i(x)) : 1\text{-forms}$$

$$d_A A_0 = dA + [A, A_0], \quad F(A) = dA + \frac{1}{2}[A, A]$$

is the curvature 2-form for a connection A of $\underline{P}(M_3, G)$. The above Riemannian metric on the space of L(G) valued forms (globally sections of ad $P \times \Lambda^1$, ad P) supplies us with a weak Riemannian metric $\mathcal{R}$ on the configuration space $\bar{\mathcal{A}}_K \times \mathcal{A}^0_{K+1}$.
Then, as in section 2.1C, the map:

$$\mathcal{R} : T(\bar{\mathcal{A}}_K \times \mathcal{A}^0_{K+1}) \to T^*(\bar{\mathcal{A}}_K \times \mathcal{A}^0_{K+1}) = \underline{P} \tag{2.13}$$

is injective. We denote the image by $\underline{P}^K$ (dense in $\underline{P}$). (The fibres of the image $\underline{P}^K$ are in Sobolev class k, whereas those of $\underline{P}$ in Sobolev class (-k)). The canonical symplectic 2-form Ω, restricted to $\underline{P}^K$, is weakly non-degenerate. Points in $\underline{P}^K$ may be represented as $(A, A_0; \pi, \pi_0)$, where

$$\pi(v) = (\pi, v)$$

$$\pi_0(v_0) = (\pi_0, v_0) \tag{2.14}$$

$v, v_0 \in T_A(\bar{\mathcal{A}}_K), T_A(\mathcal{A}^0_{K+1})$ isomorphic to $\Gamma_K(ad P \times \Lambda^1), \Gamma_{K+1}(ad P)$ respectively.

We now make a Legendre transform FL , using the Lagrangian L, as in Section (2.1)B. As the image we get:

$$\begin{aligned} \pi &= \dot{A} - d_A A_0 \\ \pi_0 &= 0 \end{aligned} \tag{2.15}$$

Clearly FL is singular. The image of FL is the $\pi_0 = 0$ subspace of $\underline{P}^K \subset \underline{P}$. We shall call this subspace $\underline{P}_0^K$. The corresponding Hamiltonian on the $\underline{P}_0^{(K)}$ subspace is:

$$\begin{aligned} H_0 &= \tfrac{1}{2}(\pi, \pi) + \tfrac{1}{2}\|F(A)\|^2 + (\pi, d_A A_0) \\ &= \tfrac{1}{2}\|\pi\|^2 + \tfrac{1}{2}\|F(A)\|^2 + (d_A^* \pi, A_0) \end{aligned} \tag{2.16}$$

As is well known, $\pi_0 = 0$ is called the primary constraint (following Dirac). Take an arbitrary extension of H_0 to the whole phase space $\mathcal{P} = T^*(\mathcal{A}_K \times \mathcal{A}^0_{K+1})$ This will be of the form

$$H_1 = H_0 + \pi_0 \hat{H}$$

The Hamiltonian vector field X_{H_1} on $\mathcal{P}$ can be uniquely calculated using the non-degenerate symplectic 2-form Ω (Section 2.1A). Its restriction to $\underline{P}^K_0$ is:

$$\begin{aligned} X_{H_1} &= X_{H_0} + \hat{H}\frac{\delta}{\delta A_0} \\ X_{H_0} &= \pi\frac{\delta}{\delta A} - d_A^* F(A)\frac{\delta}{\delta \pi} - d_A^* \pi \frac{\delta}{\delta \pi_0} \end{aligned} \tag{2.17}$$

(Here the derivatives are directional (Gateaux) derivatives. e.g.

$$d_A^* F(A)\frac{\delta}{\delta \pi} f(A, A_0; \pi, \pi_0) = \frac{d}{dt}\Big|_{t=0} f(A, A_0; \pi + t d_A^* F(A), \pi_0)$$

Also X_{H_1} is only densely defined (defined on Sobolev space of class (k+2)) .

Irrespective of the extension chosen,

$$X_{H_1} \pi_0 = 0 \implies d_A^* \pi = 0 \tag{2.18}$$

(the secondary constraint). This is just the condition that X_{H_1} is tangential to $\underline{P}_0^{(K)}$, i.e. $\pi_0 = 0$ constraint is preserved in time. It is easy to check:

$$X_{H_1}(d_A^* \pi) = 0$$

i.e. there are no further constraints. The $d_A^* \pi = 0$ subspace of $\underline{P}_o^{(K)}$ will be called $\underline{P}^{(K)}_{Gauss}$, ($d_A^* \pi = 0$ is the non-abelian Gauss Law, π is the "Electric" field interpreted as L(G) valued 1-form). On $\underline{P}^{(K)}_{Gauss}$, the Hamiltonian is:

$$H_{Gauss} = \frac{1}{2} \| \pi \|^2 + \frac{1}{2} \| F(A) \|^2 \tag{2.19}$$

Naively one might have thought that $\underline{P}^{(K)}_{gauss}$ was the physically appropriate phase space, and H_{Gauss}, the Hamiltonian. <u>But this is false</u>. This is because the canonical symplectic 2-form Ω which was non-degenerate on $\mathcal{P} = T^*(\mathcal{A}_K \times \mathcal{A}^0_{K+1})$, and weakly non-degenerate on $\underline{P}^K$, becomes <u>degenerate</u> when restricted to $\underline{P}^{(K)}_{Gauss}$. Hence

$$\left(\mathcal{P}^K_{Gauss}, \Omega |_{\mathcal{P}^K_{Gauss}}, H_{Gauss} \right)$$

is <u>not</u> a Hamiltonian dynamical system (not even in the weak sense where we allow for a weak non-degenerate symplectic structure, and densely defined Hamiltonian vector fields). From H_{Gauss} we cannot uniquely obtain $X_{H_{Gauss}}$, the infinitesimal generator of motion on $\underline{P}^{(K)}_{Gauss}$.

We now verify that $\Omega |_{\mathcal{P}^K_{Gauss}}$ is degenerate and determine the null vectors. $\underline{P}^{(K)}_{gauss}$ is determined by the constraints:

$$\pi_0 = d_A^* \pi = 0 \tag{2.20}$$

This implies:

$$(\pi_0, \xi) = 0 \quad , \quad \text{for any } \xi$$
$$(\pi, d_A \eta) = 0 \quad , \quad \text{for any } \eta \tag{2.21}$$

Here: $\xi, \eta \in \Gamma_{K+1}(\mathrm{ad}\,\mathcal{P})$

We consider the families of functions $(\pi_0, \xi), (\pi, d_A \eta)$. These functions are defined on $\underline{P}^K \subset \underline{P}$ (dense in $\underline{P}$). Using the (weak) non-degenerate symplectic structure Ω (on $\underline{P}^K$), the corresponding Hamiltonian vector fields can be uniquely defined:

$$X_{(\pi_0, \xi)} = \xi \frac{\delta}{\delta A_0} \tag{2.22}$$

$$X_{(\pi, d_A \eta)} = d_A \eta \frac{\delta}{\delta A} + [\pi, \eta] \frac{\delta}{\delta \pi} \tag{2.23}$$

Furthermore it is easy to verify the Poisson Bracket relations:

$$\{(\pi_0, \xi), (\pi_0, \xi')\} = 0$$
$$\{(\pi_0, \xi), (\pi, d_A\eta)\} = 0 \tag{2.24}$$
$$\{(\pi, d_A\eta), (\pi, d_A\eta')\} = (\pi, d_A[\eta, \eta'])$$

The last Poisson bracket vanishes on $\underline{P}^K_{gauss}$. (Thus the above families of functions give a system of first class constraints in the sense of Dirac.) We also have:

$$[X_{(\pi_0,\xi)}, X_{(\pi_0,\xi')}] = [X_{(\pi_0,\xi)}, X_{(\pi,d_A\eta)}] = 0$$
$$[X_{(\pi,d_A\eta)}, X_{(\pi,d_A\eta')}] = X_{(\pi,d_A[\eta,\eta'])} \tag{2.25}$$

Now it is easy to see that $\Omega|_{P^K_{gauss}}$ is degenerate. For:

(a) $X_{(\pi_0,\xi)}$, $X_{(\pi,d_A\eta)}$ are tangential to $\underline{P}^K_{gauss}$.

This follows from (2.24). e.g.

$$X_{(\pi,d_A\eta)}(\pi, d_A\eta') = \{(\pi, d_A\eta), (\pi, d_A\eta')\} = 0 \tag{2.26}$$

on $\underline{P}^K_{gauss}$.

(b) For any Y tangential to $\underline{P}^K_{gauss}$,

$$\Omega(X_{(\pi,d_A\eta)}, Y) = Y(\pi, d_A\eta) = 0$$
$$\Omega(X_{(\pi_0,\xi)}, Y) = Y(\pi_0, \xi) = 0 \tag{2.27}$$

on $\underline{P}^K_{gauss}$.

Thus $X_{(\pi,d_A\eta)}$, $X_{(\pi_0,\xi)}$ are null vectors of $\Omega|_{P^K_{gauss}}$, which is degenerate. Somehow we have "to get rid of" the null directions to define dynamics.

Because of (2.25), the null vector fields form an involutive system, whose integral submanifolds are leaves of a foliation . To quotient out the null vectors we have to project to the space of leaves.

In our case the situation is relatively simple and falls into the framework of Part I. To get rid of the null vectors (2.22) we simply put $A_0 = 0$ in $\underline{P}^K_{gauss}$. We call the resulting subspace $\hat{\underline{P}}^K_{gauss}$, on which the Hamiltonian is still H_{gauss} of (2.19). We are left with the null vectors $X_{(\pi, d_A\eta)}$ given by (2.23).

Now the group of gauge transformations $\bar{\mathcal{G}}_{K+1}$ act freely on $\bar{\mathcal{A}}_K$ (as in Part I), and hence on $\hat{P}^K_{gauss} \subset T^*(\bar{\mathcal{A}}_K)$. (Canonical lift) We notice that (2.23) is nothing but the fundamental vector field on $\hat{P}^K_{gauss}$ corresponding to the action of the one parameter group $g_t = \exp t\eta \in \bar{\mathcal{G}}_{K+1}$. Thus to get rid of the null vectors (2.23) we have to pass to the space of gauge orbits in $\hat{P}^K_{gauss}$, i.e. to $\hat{P}^K_{gauss}/\bar{\mathcal{G}}_{K+1} \equiv P^K_{true}$. This is the space on which the Yang-Mills Hamiltonian dynamics can be (uniquely) defined, and hence we call it the <u>true phase space</u>. (It has a naturally defined*(weak) non-degenerate symplectic structure $\tilde{\Omega}$). H_{gauss} of (2.19) being $\bar{\mathcal{G}}_{K+1}$ invariant is a function on $\underline{P}_{true}$, the Hamiltonian H_{true}. $(\underline{P}_{true}, \tilde{\Omega}, H_{true})$ <u>is the Yang-Mills (Hamiltonian) dynamical system</u>.

Corresponding to the Hamiltonian system on $\underline{P}_{true}$, there is a Lagrangian system on $T\mathcal{M}_K$, the tangent bundle of $\mathcal{M}_K$, the manifold of gauge orbits which was studied in Part I. The situation may be described as follows.

In Part I we showed $\bar{\mathcal{A}}_K \longrightarrow \bar{\mathcal{A}}_K/\bar{\mathcal{G}}_{K+1} = \mathcal{M}_K$ was a C^∞ principal fibration and $\mathcal{M}_K$ was a C^∞ Hilbert manifold with a weak Riemannian metric g (described in Section 1.5).

Then we have (easy to verify) the commutative diagram:

$$\begin{array}{ccc} \hat{P}^K_{gauss} & \longrightarrow & \bar{\mathcal{A}}_K \\ \downarrow \bar{\mathcal{G}}_{K+1} & & \downarrow \bar{\mathcal{G}}_{K+1} \\ P_{true} = T^*(\mathcal{M}_K) & \underset{p}{\longrightarrow} & \mathcal{M}_K \end{array} \qquad (2.28)$$

By $T^*\mathcal{M}_K$ we mean the sub-bundle of the cotangent bundle with fibres restricted to Sobolev class k. Here $\hat{P}^K_{gauss} = p^*\bar{\mathcal{A}}_K$ the induced bundle on $T^*\mathcal{M}_K$, and $T^*\mathcal{M}_K = P_{true}$.
(Thus $\hat{P}^K_{Gauss} \longrightarrow \hat{P}^K_{gauss}/\bar{\mathcal{G}}_{K+1}$

---
*See the remark preceding equation 2.32.

is also a non-trivial C^∞ principal fibration).

We now make a Legendre transformation:

$$FH_{gauss} : \hat{P}^K_{gauss} \longrightarrow T_{gauss}\bar{\mathcal{A}}_K \tag{2.29}$$

using H_{gauss} of (2.19). $T_{gauss}\bar{\mathcal{A}}_K \subset T\bar{\mathcal{A}}_K$ is a sub-bundle whose points are:

$$(A, \dot{A}), \; d_A^* \dot{A} = 0 \tag{2.30}$$

The Lagrangian on $T_{gauss}\bar{\mathcal{A}}_K$ is:

$$L = \frac{1}{2}(\dot{A}, \dot{A}) - \frac{1}{2}\|F(A)\|^2 \tag{2.31}$$

L being $\bar{\mathcal{G}}_{K+1}$ invariant descends as a function L_{true} on $T\mathcal{M}_K$. Notice that L as a function on $T\mathcal{M}_K$ has its kinetic term defined precisely with the (weak) Riemannian metric of Section 1.5, Part I. $\mathcal{M}_K$, the manifold of gauge orbits, is the true configuration space. (It is easy to verify that the Lagrangian L_{true} on $T\mathcal{M}_K$ is nothing but the Legendre transform of H_{true} on $\underline{P}_{true}$).

Remark. In part I Section 1.5, we gave an explicit connection for the bundle $\bar{\mathcal{A}}_K \to \bar{\mathcal{A}}_K/\bar{\mathcal{G}}_{K+1} = \mathcal{M}_K$. Using the bundle map of (2.28) we pull back this connection to a connection in the bundle

$$\hat{P}^K_{gauss} \longrightarrow \hat{P}^K_{gauss}/\bar{\mathcal{G}}_{K+1} = \underline{P}_{true}$$

If X,Y are tangent vectors (at a point) to P_{true} and $\hat{X}$, $\hat{Y}$ their horizontal lift to $\hat{P}^K_{gauss}$ (at any point), then the (non-degenerate) symplectic 2-form $\tilde{\Omega}$ on P_{true} is defined by:

$$\tilde{\Omega}(X, Y) = \Omega(\hat{X}, \hat{Y}) \tag{2.32}$$

$\tilde{\Omega}$, so defined, coincides with the canonical symplectic 2-form on $T^*\mathcal{M}_K$.

Observables: These are globally defined functions on the true configuration space (more generally on the true phase space, but that will not be relevant for the quantum theory). If we seek a description in terms of connections, then observables are necessarily gauge invariant functions on the space of connections. This is because on the overlap of two coordinate neighborhoods on the true configuration space (i.e. the orbit space $\mathcal{M}$), the value of the function in the two coordinate systems must agree, and as we have seen in Part I, Section 1.4, the coordinate change is a gauge transformation.

2.3 Quantization

By virtue of what we have shown in Section 2.2, quantization of Yang-Mills dynamics is that of a "particle" on a Riemannian manifold $\mathcal{M}$ (the true configuration space = space of gauge orbits) which was studied in Part I of these notes. The Lagrangian L_{true} on $T\mathcal{M}$ comes with its kinetic term defined with a (weak) Riemannian metric on $\mathcal{M}$, studied in Section 1.5, Part I. The standard formal prescription for path integral quantization (see e.g. [13,14]) on finite dimensional Riemannian manifold employs the invariant volume element. We take over this prescription formally in our infinite dimensional context (a non-formal meaning will have to be given later). Even in the finite dimensional case, great care must be paid in order to take the manifold structure into account. Here the situation is even more complicated.

The generating functional of Euclidean correlation functions of gauge invariant functions (these are the only functions which are globally defined on the orbit space as remarked at the end of Section 2.2) is given (formally) by: (f is the coupling constant)

$$Z\{\underline{J}\} = \int_{\Omega(\mathcal{M})} \mathcal{D}\mu_g \exp -\left[\int_{-\infty}^{\infty} dt \left\{ \frac{1}{f^2} L_{true}(A;\dot{A}) - i \sum_{j=1}^{\ell} (J^j, G^j(A)) \right\}\right] \tag{2.33}$$

where

$$L_{true}(A;\dot{A}) = \frac{1}{2}(\dot{A},\dot{A}) + \frac{1}{2}\|F(A)\|^2 \tag{2.34}$$

considered as a function on $T\mathcal{M}$ (hence $d_A^*\dot{A} = 0$). $L_{true}(A;\dot{A})$ depends on t, as in (2.12), through $A(x,t)$, and the scalar products are as in (2.12). $G^j(A)$ are gauge invariant functions, J^j are external sources, $(J^j, G^j(A))$ are globally defined on $\mathcal{M}$. $\mathcal{D}\mu_g$ is a formal measure on path space associated to the Riemannian metric on $\mathcal{M}$. $\Omega(\mathcal{M})$ = space of (continuous) maps: $\mathbb{R} \to \mathcal{M}$, in other words the path space.

As a first step we shall restrict ourselves to a contribution to (2.33) coming from paths in a coordinate neighborhood $\eta_{\hat{A}}$ of $\mathcal{M}$, where we can use the chart $\sigma_{\hat{A}}$ (Section 1.4, Part I). A global definition, involving paths stretching over overlaps of coordinate neighborhoods is not trivial and will not be attempted here.

Over $\eta_{\hat{A}}$, we have, by definition:

$$\int_{\Omega(\eta_{\hat{A}})} \mathcal{D}\mu_g \exp - [\quad]$$

$$= \int_{\Omega(\mathcal{J}_{\hat{A}})} \mathcal{D}A \sqrt{\det \tilde{g}(A)} \exp - \int_{-\infty}^{\infty} dt \Big[\frac{1}{2f^2} \big((\dot{A}, \tilde{g}(A)\dot{A}) + \|F(A)\|^2 \big) - i \sum (J^j, G^j(A)) \Big] \tag{2.35}$$

where $\mathcal{J}_{\hat{A}} = \sigma_{\hat{A}}(\eta_{\hat{A}})$, and $\tilde{g}(A)$ is the local expression of the metric, given in Section (1.5), (Part I) where $\mathcal{J}_{\hat{A}}$ is also defined. The formal calculation of [8] gives:

$$\sqrt{\det \tilde{g}(A)} = \frac{\det(d^*_{\hat{A}} d_A)}{\sqrt{\det \Delta_A} \sqrt{\det \Delta_{\hat{A}}}} \tag{2.36}$$

where $\Delta_A = d^*_A d_A$ etc. (2.35) is then seen to be identical to:

$$\int_{\Omega(\mathcal{J}_{\hat{A}} \times A_0)} \mathcal{D}A\,\mathcal{D}A_0 \det(d^*_{\hat{A}} d_A) \exp - \int d^4x \sqrt{g}\, \mathrm{Tr} \Big[\frac{1}{4f^2} F_{\mu\nu}F^{\mu\nu} - i \sum J^j \cdot G^j(A) \Big] \tag{2.37}$$

as is explicitly seen by performing the A_0 Gaussian integration. (In defining $F_{\mu\nu}F^{\mu\nu}$ the "Euclidean" 4-metric $ds^2 = dt^2 + g_{ij}\, dx^i dx^j$ is employed).

Within a coordinate patch the expansion of $F_{\mu\nu}F^{\mu\nu}$ into a quadratic piece and higher order perturbation has an invariant meaning. Thus we may attempt to define (2.37), in the usual way, e.g. via perturbation series. The reader will observe that we have recovered de Witt-Faddeev-Popov quantization, except that the range of integration is limited because of the very definition of $\mathcal{J}_{\hat{A}}$, which employs local (in function space) gauge fixing. How does this effect perturbation theory?

In order to do perturbation theory we choose $(\hat{A}, \hat{A}_0)$ as an extremal of the classical action, and make an asymptotic expansion in f. If we extend integration range to obtain complete Gaussian integrals it is easy to see that we make an apriori error of the type $\exp - 1/f^2$. This error is unimportant in perturbation theory only when we expand around the "trivial vacuum" where no $\exp - 1/f^2$ effects are around.

Remark. What we have alluded to above is due to the fact that the manifold of gauge orbits is topologically non-trivial and cannot be described by a single chart (Part I, Section 1.4, Theorem 2). This is true both in the "canonical" picture which we have adopted in Part II of these notes, as well as in the fully covariant Euclidean formalism. Let us stick for the moment to the canonical picture, and, for simplicity, take $M_3 = S^3$ and $G = SU(N)$. Then $P(M_3,G)$ is a trivial bundle, and $\mathcal{G} = \Omega_3(SU(N)/Z_N)$ is the group of gauge transformations (space of maps $S^3 \longrightarrow SU(N)/Z_N$). Now $\pi_0(\mathcal{G}) \simeq \pi_3(SU(N)/Z_N) \simeq Z$ (In other words, $\mathcal{G}$ is not connected). Correspondingly $\pi_1(\mathcal{M}) \simeq Z$, i.e. the orbit space is not simply connected. In this picture, "instanton paths" in the orbit space are those closed paths which are not homotopic to a constant. Lifting these paths horizontally to the space of connections (a connection in the bundle of connections was provided in Part I, Section 1.5) one arrives at gauge related connections, where the gauge transformation belongs to a nontrivial component of $\mathcal{G}$ (a so called "singular gauge transformation"). Suppose we ignore such paths, i.e. we work in the trivial component of $\mathcal{G}$. The trivial component of $\mathcal{G}$ is itself topologically complicated (as was emphasized in [15]), as measured by higher nonvanishing homotopy groups. Quite a few even homotopy group of $\mathcal{G}$ (odd homotopy groups of $\mathcal{M}$) are non-vanishing, and as $N \to \infty$, an infinite set of non-vanishing groups are computed (isomorphic to Z). This means the trivial component of $\mathcal{M}$ cannot be described by a single chart, and so on.

Conclusion

In these notes we have tried to give a coherent account of classical Yang-Mills theory formulated globally as a dynamical system. In doing so we were led naturally to describe the geometry, and to a limited extent the topology, of the space of gauge orbits which turned to be the true configuration space of Yang-Mills dynamics. That the Yang-Mills Lagrangian is singular, that this leads to constraints in phase space, etc. is well known to physicists for a long time. The importance of the invariant (read: geometrical) approach that we have adopted is that it focuses attention on the true issues. Of course the motivation for all this is to understand one day non-perturbative aspects of the quantum

Yang-Mills field. The invariant formulation of the classical dynamical system already makes clearer the formal quantization of the theory, and at the same time its limitation. There is already a volume cutoff. The next step would be to introduce a space-like (gauge invariant) ultraviolet regularization and try to understand better the regularized theory as a quantum theory. The proper arena is the space of gauge orbits.

Acknowledgements

I wish to thank M.S. Narasimhan and T.R. Ramadas for fruitful and instructive conversations and the physicists and mathematicians of the Tata Institute of Fundamental Research for their generous hospitality during the summer of 1979. I thank my colleagues at New York University for their kind hospitality and G.F. Dell'Antonio for stimulating conversations which should not go unacknowledged.

References

1. A. Wightman in "Renormalization Theory", G. Velo and A. S. Wightman (editors), D. Reidel (Boston) 1976."
2. Lectures of J. Fröhlich, J. Glimm, C. Itzykson, G. Mack, K. Wilson (these proceedings) and references therein.
3. K.G. Wilson, K. Osterwalder in "Recent developments in Quantum field theory and Statistical Mechanics", Cargèse 76, M. Lévy, P.K. Mitter (editors), Plenum Press (London/New York).
4. I.M. Singer: Commun. Math. Phys. 60 (1978), 7.
5. M.S. Narasimhan and T.R. Ramadas: Commun. Math. Phys. 67 (1979), 121.
6. P.K. Mitter and C.M. Viallet: "On the bundle of connections and the gauge orbit manifold in Yang-Mills theory", Paris, LPTHE 79/09 (preprint).
7. L.D. Faddeev: Theor. Math. Phys. Vol 1, No. 1, pp. 3-18, 1969.
8. O. Babelon and C.M. Viallet: Phys. Lett.
9. M.F. Atiyah, N. Hitchin and I.M. Singer: Proc. Roy. Soc. London A362 (1978), 425.
10. M. Daniel and C.M. Viallet: Phys. Lett. 76B (1978), 458.
11. V.I. Arnold: Méthodes mathématiques de la méchanique classique Édition MIR (Moscou) 1974.
12. J. Marsden: Applications of Global analysis in Mathematical Physics. Publish or Perish Inc. (1974) Boston.
13. B.S. de Witt: Rev. Mod. Phys. 29, 377 (1957).
14. K.S. Cheng, J. Math. Phys. 13, 1723 (1972).
15. M.F Atiyah and J.D.S. Jones: Comm. Math. Phys. 61, 97 (1978).

16. M.F. Atiyah, V.G. Drinfeld, N.J. Hitchin, Yu. I. Manin, Phys. Lett. 65A (1978), 185.
17. J. Madore, J.L. Richard, R. Stora: F = *F , a review. Marseille preprint 79/P1077 (1979).

ON THE STRUCTURE OF THE PHASES IN LATTICE GAUGE THEORIES

Giorgio Parisi

INFN
Laboratori Nazionali
00044 - Frascati (Italy)

In recent years a lot of work has been concentrated on the study of non-abelian gauge theories on a lattice[1]. The introduction of a lattice is crucial do define the theory in a non-perturbative way: in lattice gauge theories it is possible to use strong coupling techniques such as the high temperature expansion[2], the numerical simulations based on the Montecarlo method[3,4] and the real space renormalization group[4,5]. Of course we have to pay a price for having all these advantages: the theory can be interpreted as the Euclidean version of a relativistic invariant local gauge field theory only in the limit in which the coherence lenght ξ goes to infinity, when it is measured in units of the lattice spacing. In the language of statistical mechanics the divergence of the correlation lenght corresponds to a second order phase transition.

If a model has only first order transitions, the absence of second order phase transitions implies that the coherence lenght never becomes infinite: the model under consideration does not have a continuum limit and the corresponding local field theory does not exist.

These remarks clearly show the importance of knowing the structures of the phases in lattice gauge theories; unfortunately at the present moment only few results are firmly estabilished. Before discussing them, let us define our notation. The gauge fields U are defined on the links of a D-dimensional hypercubic lattice and have a value in the gauge group G (not in the Lie algebra of the group). We will concentrate our attention on non-abelian Lie groups, abelian groups have their own characteristics and they behave in a rather peculiar way.

The partition function can be computed using the following hamiltonian:

$$H = -\sum_p T_r(U_p) \tag{1}$$

where the sum runs over all the plaquettes of the lattice (four links forming a square are called a plaquette) and U_p stand for the product of the four U around the plaquette p. The temperature T, at which the partition function is evaluated, is proportional to the bare coupling constant g^2.

In the high temperature phase, when $T \to \infty$ ($\beta = 1/T \to 0$), the theory is confined and the correlation lenght is proportional to β^4. At low temperature the perturbative expansion can be trusted only if the dimensions D are greater than 4; if $D < 4$ the standard perturbative expansion cannot be constructed because of strong infrared divergences; renormalization group arguments suggest that in dimensions $D = 4 + \varepsilon$, a second order transition is present at temperature $T_{II}(\varepsilon)$ proportional to ε: the value of the critical temperature and the corresponding critical exponents can be computed as a power expansion in ε. In this low temperature phase the theory is not confined: it is believed that for $D < 4$ the theory is always confined, for any value of the temperature, while if $D > 4$ the second order phase transition at $T_{II}(\varepsilon)$ separates the unconfined from the confined phase. This picture is the simplest one compatible with the firmly established results; the phase diagram in the temperature-dimensions plane would be the one shown in Fig. 1. There are only one confined and one unconfined phase, separated by a line of second order phase transitions. There are strong indications which suggest that the situation is more complex: they will be discussed later. For the moment let us stick to this naive picture.

In Fig. 1 the star denotes the upper critical dimension D_S: if $D > D_S$ the appropriate mean field theory holds at the phase transition point and the critical exponents have a trivial dependence on the dimensions. This diagram is constructed in analogy with the one known for the non linear model, shown in Fig. 2, where D_S is known to be equal to 4.

It is possible that, as it happens in the non linear σ-model, the critical exponents can be computed also in powers of $\varepsilon' = D_S - D$. In ref. 6 it has been suggested that $D_S = 8$; two ways are allowed in order to construct the ε' expansion: in the first approach one should introduce "prepotentials" in such a way that the field A_μ would become a composite field and that the gauge theory would look like the high derivative version of the non linear σ-model; in the second approach one should show that gauge theories are the equivalent of a generalized string model in which off-shell Green functions are computable and finite. It is very hard to implement these suggestions, however I believe that something can be done in this direction using the techniques introduced by Wallace and Zia[7] in the study of the

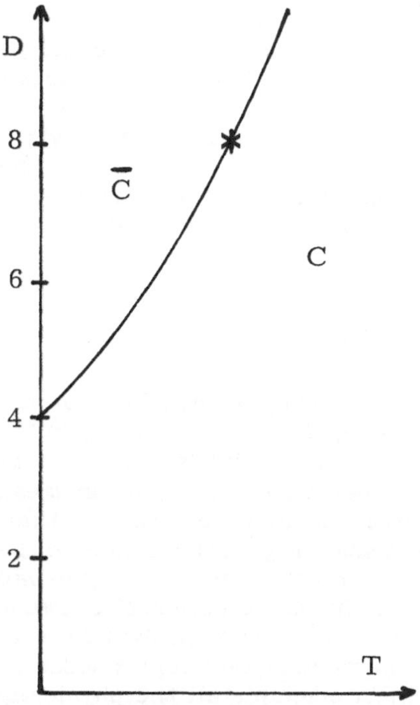

FIG. 1 The naive phase diagram for non abelian gauge theories in the temperature dimensions plane: there are two phases, the confined (C) and the unconfined one ($\bar{C}$), which are separated by a line of second order phase transitions; the star denotes the upper critical dimensions, as explained in the text.

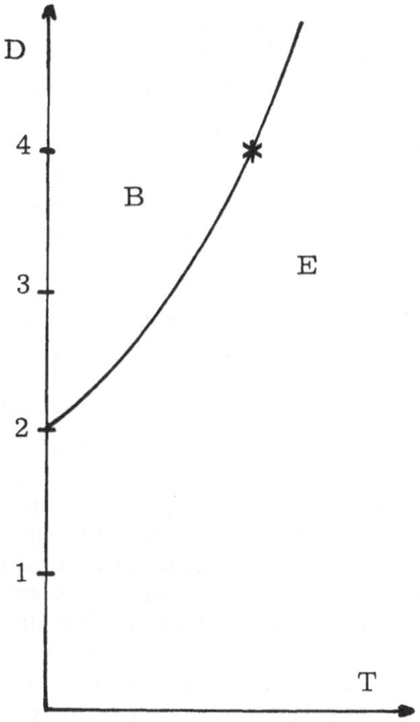

FIG. 2 The phase diagram for the non-linear O(N) invariant σ model ($N > 2$): there are two phases: in phase (B) the O(N) symmetry is spontaneously broken while in phase (E) the symmetry is exact.

"gas-liquid interface" in $1 + \varepsilon$ dimensions.

Let us forget these speculations and let us come back to the study of the consequence of the conventional picture: contrary to the naive intuition, the correcteness of Fig. 1 does not imply that the theory is confined in the continuum limit. In order to discuss more carefully this point, it is convenient to define the surface tension σ using the expectation values of large Wilson's loops:

$$\langle \Pi_{i \in C} U_i \rangle \simeq \exp(-\sigma S) \qquad (2)$$

where S is the minimal area enclosed by the large contour C; σ is different from zero only if the lattice theory is confined; by continuity it should go to zero at the deconfinement transition ($T_{II}(\varepsilon)$. Now, if $D > 4$, two different theories can be constructed in the continuum limit: the one is obtained at T_{II}^-, the other at T_{II}^+. The first theory will be not confined for obvious reasons; for the second theory the situation is more subtle: in the continuum limit we must measure everything in units of the coherence lenght ξ(i.e. the inverse of the mass gap). Now three possibilities are given:

$$(3) \qquad \lim_{T \to T_{II}^+} \xi^2 \sigma \equiv R = \begin{cases} 0 \\ 0(1) \\ \infty \end{cases} .$$

If our expectations are correct and the theory is confined in the continuum limit, R should be non zero and finite; if R = 0 the theory in the continuum limit is not confined (or at least the potential is not linear at large distances); if $R = \infty$, the quarks are hyperconfined, i.e. their mean separation would be zero and we would never be able to discover their existence in experiments like deep inelastic scattering. In this scheeme the quantity R (which can be expressed in terms of the ratio of the glueball mass with the coefficient of the linear term of the static potential between two quarks) can be computed as an expansion in powers of β, evaluated at $\beta_c = T_{II}^{-1}$. If D = 4, $\beta_c = \infty$ and appropriate numerical techniques must be used to sum the high temperature expansion. In principle all the physically interesting quantities can be extracted from the high temperature expansion: the success of this program crucially depends on the correctness of the phase diagram shown in Fig. 1, i.e. from the absence of any phase transition at finite temperature at D = 4. (We recall to the reader that here we consider only pure gauge theory: one hopes that the fermions can be succesfully studied in perturbation theory at a later stage).

A shadow of doubt on the whole picture was trown some years ago by the paper of Balian, Drouffe and Itzykson (BDI)[8]: in a mean field approximation they found that gauge theories undergo a first order transition. However it was not clear which was the range of validity of their mean field approximation and their prediction was not considered with the due weight. New results obtained this years, show that the conventional picture of Fig. 1, is too naive and that the true situation is more complex.

The first published Montecarlo simulations of gauge theories[3] have shown a clear signature for a first order transition for the Z_2 gauge group in dimensions D = 4,5 and for SU(2) in D = 5 (no Montecarlo simulations have been done in non integer dimensions). For SU(2) with D = 4 no first order transition is observed, but there is a peak in the specific heat which may be due to a second order phase transition[4].

Apart from these "experimental" results, new theoretical results where contained in the paper by Drouffe, Sourlas and myself in which gauge theories where studied in the limit $D \to \infty$[9]: indeed it is known that in this limit the lattice version of the non linear σ-model become soluble and the standard mean field theory holds[10].

When $D \to \infty$, the high temperature expansion can be exactly summed in a closed form, and the free energy becomes a function of the scaling variable $\tilde{\beta} = \beta/D^{1/4}$. In very high dimensions many diagrams are negligeable; the only diagrams which survive are those which at fixed power of β, have the maximal power of D; these diagrams are very simple: they corresponds to the surface of polybranched polymers of cubes. The enumeration of these diagrams is a rather simple combinatorical problem; however one must be quite careful in estabilishing the correct diagrammatical rules, in particular in computing the contribution of disconnected diagrams: we do not enter in the detailed derivation of these rules; which can be found in the original papers[8,9,11] and we present only the final results. The free energy we obtain after resumming the dominant diagrams has second order transition at $T_c \propto D^{1/4}$. This transition is characterized by the divergence of the specific heat, i.e. by the plaquette-plaquette susceptibility: at the transition point the plaquette-plaquette correlation lenght is going to infinity, but the surface tension σ remains finite ($R = \infty$ in eq.(3)). This transition is not a deconfinement transition and corresponds to the condensation into vacuum of the boxitons of ref.3, i.e. the glueballs in a more phenomenological language.

These results strongly suggest the possibility of having two different confined phases: the standard high temperature phase C_1 and a new low temperature phase C_2 in which boxiton are condensed. The possible existence of two qualitatively different phases is a rather general phenomenon which can be simply understood in geometrical terms.

Let us consider a geometrical picture in which we neglect all the group theoretical quantum numbers: gauge theories become theories of interacting surfaces, as can be readly seen from the high temperature expansion. When the surface tension decreases, the possibility of thin, long deformations of the surface increases: these hydra-like configurations are local deformations of the surface which are allowed also when the surface still keeps its local rigidity (i.e. $\sigma \neq 0$): as an explicit computation shows, these local deformations have an entropy much higher than global deformations, at least in high dimensions.

With an abuse of language one could say that the surface roughening transition happens at a temperature greater than the critical temperature at which the surface tension goes to zero.

Unfortunately we have been unable to find a good order parameter which characterizes this surphases roughening transition and we do not know if the transition between C_1 and C_2 is really a second order phase transition, or if the second order phase transition we have found is only the end of the metastable region and there is a first order transition from C_1 and C_2 at a higher temperature.

However this problem is purely academic in high dimensions: if one compare the high temperature expansion either with the low temperature expansion or with the rigorous lower bounds for the free energy of BDI, one finds a first order transition at $T = T_I \propto D^9$. This BDI transition is a deconfinement transition and there is no field theory associated to it in the continuum limit; the temperature of our roughening transition is much lower (it is proportional to $D^{1/4}$). The condensation of boxitons may happen only in the metastable phase for very high dimensions and therefore it is not relevant. The reasons for which the BDI first order transition can hardly be seen in the high temperature expansion, are explained in a small appendix on first order transitions.

Now which is the substitute of Fig. 1. We definitely do not know, we do not have enough informations. Waiting for careful analysis of longer high temperature expansions and for more accurate Montecarlo simulations I will try to put forward a new picture, which has very small chances of being correct, but it will be quite interesting to be able to disprove it. This picture is shown in Fig. 3 and 4 for the gauge groups Z_2 and SU(2) respectively.

In both Fig. 3 and 4 we indicate by full lines real phase transitions and by dashed lines virtual phase transitions in the metastable region. In Fig. 3 line (a) denotes the BDI first order transition between the confined and the unconfined phase, line (b) is the second order transition between the confined and the unconfined phase, which is supposed to be a virtual transition: the true transition (line (a)) happens at a lower temperature. Line (c) is the second (first) order roughening transition, which at high dimensions is in the metastable region, and becomes a real transition only for enough low dimensions. In Fig. 4 line (a) denotes a first order deconfinement phase transition for $D > 3$, and line(b) the second order roughening transition, which for dimensions greater than 3 is a virtual transition and comes to physical reality only for dimensions 3. The detailed form of Fig. 3 and 4 for D smaller than 4 and 3 respectively, is not very important and should not be taken too seriously.

It would be easy to find other scheemes which are as reasonable as the one presented here: for example we could suppress line (c) in Fig. 3 and declare than line (a) is a first order transition if $D > 4.5$, but a second order one if $D < 4.5$.

The diagram of the phases of gauge theories deserves further investigations and may be further surprises are waiting for us: ice, which is a relatively simple system, has a dozen of different phases.

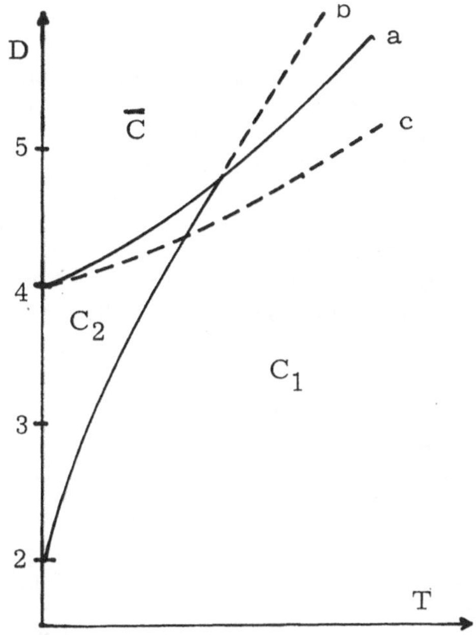

FIG. 3 The conjectured phase diagram for SU(2) gauge theories: there are two confined phases C_1 and C_2 and one unconfined phase $\bar{C}$: line (a) separates the confined from the unconfined phase and it is a first order transition, line (b) should separate the two confined phases C_1 and C_2, but it may physically realized only for low dimensions: for high dimensions it is located in the metastable region and it is masked by the first order transition line (a). Line (c) is the second order phase transition which can be seen using the renormalization group and the low temperature (low coupling) expansion; we suppose that this second order transition only indicates the end of the metastable region, the true transition (a) happens at a lower temperature.

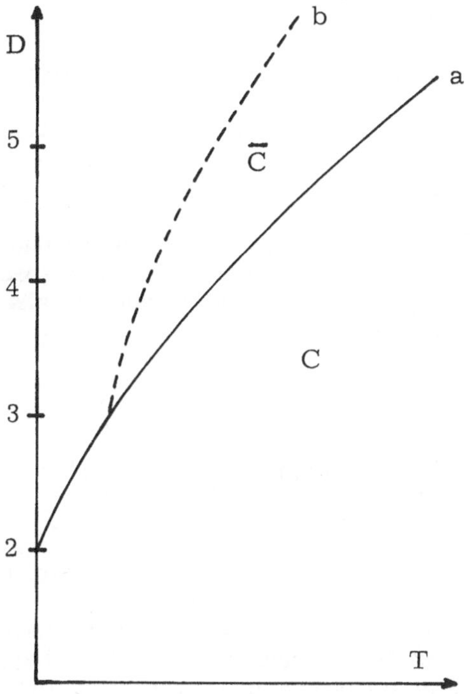

FIG. 4 The conjectured phase diagram for Z_2 gauge theories: line (a) is the only real phase transition (first order for $D > 3$) which separates the confined from the unconfined phase; line (b) is the second order phase transition which can be seen from the analysis of the high temperature expansion.

Appendix

In this appendix we recall some know (proved or conjectured) facts on first order phase transitions. A second order phase transition is characterized by the absence of latent heat, the correlation lenght goes to infinity and the free energy has a power-like singularity at the critical temperature.

Let us consider a first order phase transition from the phase A to the phase B. It is usual to define two free energies F_A and F_B, which describe the system in phase A and in phase B respectively; they satisfy the inequalities:

$$\text{(A.1)} \qquad \begin{aligned} F(T) &= F_A(T) < F_B(T) \qquad T > T_c \\ F(T) &= F_B(T) < F_A(T) \qquad T < T_c \end{aligned}$$

where F(T) is the true free energy of the system.

In writing eq. (A.1) we have implicitely assumed that both F_A and F_B are analytic function at T_c: that happens in the mean field approximation and at all orders in perturbation theory. In reality both functions are only C^∞ at the transition point; it is believed that their analytic continuation below (or above) the transition has an exponentially small imaginary part due to "instanton" effects[12].

In not very sofisticated analysis there is no difference between a C^∞ function and an analytic function: the singularity at T_c of the free energy is practically invisible in the high temperature expansion unless one consider incredible high orders. The only viable method to find first order transitions consistes in constructing the two free energies F_A and F_B and to compare them. If we only consider the properties of the free energy F_A it is practically impossible to find the first order transition; as a consequence, if we forget to construct the free energy F_B, we cannot suspect the existence of the first order transition, unless we discover that F_A violates rigorously established inequalities.

References

1. K.G. Wilson, Confinement of quarks, Phys. Rev. D10:2445 (1974); for a review see J.M. Drouffe and C. Itzykson, Lattice gauge fields, Phys. Repts. 38C:133 (1978).
2. J. Kogut and L. Susskind, Hamiltonian formulation of Wilson's lattice gauge, Phys. Rev. D11:395 (1975);
   T. Banks, R. Myerson and J. Kogut, Phase transitions in Abelian lattice gauge theories, Nucl. Phys. B129:493 (1977);
   J. Kogut, D.K. Sinclair and L. Susskind, A quantitative approach to low-energy quantum chromodynamics, Nucl. Phys. B114:199 (1976).
3. M. Creutz, L.Jacobs and C. Rebbi, Experiments with a gaugeinvariant Ising system, Phys. Rev. Letters 42:1390 (1979);
   M.Creutz, Confinement and the critical dimensionality of space-time, Phys. Rev. Letters 43:553 (1979).
4. K.G. Wilson, in the Proceedings of this School.
5. A.A. Migdal, Phase transitions in gauge and spin-lattice systems, Sov. Phys.-JEPT 42:743 (1976).
6. G. Parisi, Hausdorf dimensions and gauge theories, Phys. Letters 81B:357 (1979);
   Gauge theories and the dual model, Frascati Preprint LNF-79/43 (1979); to be published in Proceedings of the "Third Workshop on Current Problems in High Energy Particle Theory", Firenze (1979).
7. D. Wallace and R.K.P. Zia, Euclidean group as a dynamical symmetry of surface fluctuations: The planar interface and critical behavior, Phys. Rev. Letters 43:808 (1979).
8. R.Balian, J.M.Drouffe and C. Itzykson, Gauge fields on a lattice. I. General outlook, Phys. Rev. D10:3376 (1974);
   Gauge fields on a lattice. II. Gauge-invariant Ising model, Phys. Rev. D11:2098 (1975);
   Gauge fields on a lattice. III. Strong-coupling expansions and transition points, Phys. Rev. D11:2104 (1975).
9. J.M. Drouffe, G. Parisi and N.Sourlas, Strong phase in lattice gauge theories at large dimensions, Saclay Preprint DPh-T/104/79 (1979); to be published on Nuclear Physics.
10. F.Englert, Linled cluster expansions in the statistical theory of ferromagnetism, Phys.Rev. 129:567 (1963);
    M.E. Fisher and D.S. Gaunt, Ising model and self-avoiding walks on hypercubical lattices and "high-density" expansions, Phys. Rev. 133:A224 (1964).
11. J. M. Drouffe, Transitions and duality in gauge lattice systems, Phys. Rev. D18:1174 (1978).
12. G.S. Langer, Theory of the condensation point, Ann. Phys. (N.Y.) 41:108 (1967).

SUPERALGEBRAS AND CONFINEMENT IN CONDENSED MATTER PHYSICS

V. Poénaru

Université Paris-Sud, Mathématiques
91405 Orsay cedex - France

This is a companion of Toulouse's lecture at this conference and large part of the material described below is joint work of Toulouse and the author (see [1,2] and also [3]).

The general topic under discussion will be graded Lie algebras ("Superalgebras") coming from topology in connection with "confinement", in the context of condensed matter physics. "Confinement" is meant here rather metaphorically ; on the other hand, there are many analogies between field theory and condensed matter physics (ordered media), so the kind of mechanisms we will describe could turn out to be useful for field theory too.

I) ENTANGLEMENT AND CROSSING OF DEFECT LINES

I will describe a certain situation in which non-commutativity (in the sense of the non-vanishing of certain Lie brackets) leads to "slavery" or "confinement". In prentice, non-commutativity is by no means the only "topological confinement mechanism". In [4] (see also [3]) a completely different mechanism coming from the so-called Hopf invariant is described and actually shown to be physically realized in the realm of cholesteric liquid crystals. In contrast, for the non-commutative effects described below, theory is still largely ahead of the experimental work. Our description will be very sketchy. For more details, I refer to the original papers [1,2] .

From an abstract mathematical standpoint, an "ordered medium" is a d-dimensional smooth manifold ("the physical space"), and a "manifold of internal states" V , characteristic for the order in question. Outside a set of "defects" $\Sigma \subset V$, for every $p \in V - \Sigma$, an order parameter, $\Phi(p) \in V$, varying continuously with p , is defined.

Generally speaking, the physical system under consideration has two phases, a high energy phase (corresponding to the defects), of symmetry group G , and a low energy phase (corresponding to the points which are not defects) with a smaller symmetry group : $H \subset G$. The manifold of internal states is $V = G/H$.

According to a general framework of Kleman and Toulouse [15] , a defect "hypersurface" or "pellicule" , P^p, of dimension p is surrounded by a sphere $S^q \subset M - P$ where $p+q+1 = d$, and defects are classified according to the homotopy class $[\Phi(S^q)] \in \pi_q V$. In particular, if $[\Phi(S^q)] = 0$, the defect is topologically (and hence physically) unstable and one can "cut" the defect pellicule P .

Imagine now the following "gedanken process" : two defect pellicules, of dimensions p and q (with $p+q+1 = d$) come close to each other along a line L (contained in $M - \Sigma$) , merge at some point, and then continue, in such a way that they cross each other without staying hooked or entangled. Figure 1 describes schematically this crossing process.

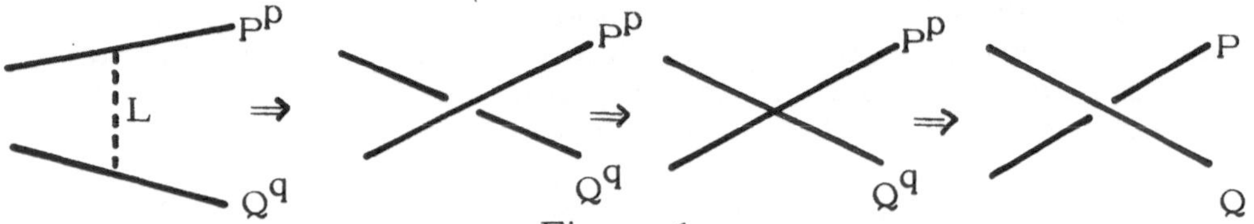

Figure 1

The question is when is this process physically possible without an exceedingly large input of energy (defects are in the high energy phase, hence costly). To make things precise we will set the following "rules of the game" :

We can move defects around, inflate them a bit or add defect <u>lines</u> (but not pellicules of dimension > 1) . Somehow these operations are supposed to correspond to small inputs of energy.

We can also de-inflate defects or cut them, whenever <u>this is topologically possible</u>. Of course these last two operations do not cost anything.

The answer to the question requires to talk about the so called "Whitehead products" from homotopy theory (see 2 for more details and bibliography).

For the sequence of homotopy groups : $\pi_1 V$, $\pi_2 V$, ... , there is a "bilinear" operation, called the "Whitehead product" :

$$\pi_n V \times \pi_m V \xrightarrow[{[\ ,\ ]}]{} \pi_{n+m-1} V$$

which has the following properties :

1) If $n = m = 1$, $[\ ,\ ]$ is the usual commutator :

$$[\alpha,\beta] = \alpha\beta\alpha^{-1}\beta^{-1}$$

2) If $n = 1$, then, up to a sign :

$$[\alpha,\beta] = \beta - \alpha . \beta$$

where . denotes the π_1-action on π_m .

3) For $n, m > 1$, the Whitehead product has the formal properties of a graded Lie product (graded anticommutativity and graded Jacobi identity).

A nice geometric way to define the Whitehead product is the following. Let $x \in S^n$, $y \in S^m$ and define $S^n \vee S^m = (S^n \times y) \cup (x \times S^m) \subset S^n \times S^m$. If one thickens slightly $S^n \vee S^m$ to get an (n+m)-dimensional submanifold $N^{n+m} \subset S^{n+m}$, one can notice that $\partial N^{n+m} = S^{n+m-1}$. So, by crushing N back to $S^n \vee S^m$, one gets a continuous map, defined once for all $S^{n+m-1} \xrightarrow[F]{} S^n \vee S^m$ (see figure 2) .

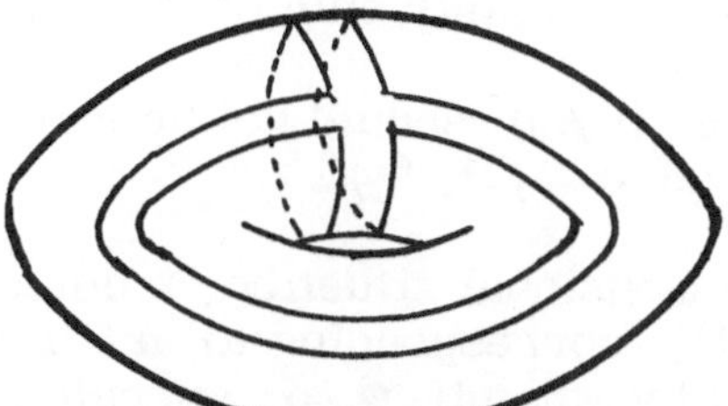

Figure 2

Now if $\alpha \in \pi_n V$, $\beta \in \pi_m V$ are represented by $a : (S^n, x) \to (V, *)$, $b : (S^m, y) \to (V, *)$, then $[\alpha,\beta]$ is just the homotopy class of the composite map :

$$S^{n+m-1} \xrightarrow{F} S^n \vee S^m \overset{a}{\underset{b}{\rightrightarrows}} V .$$

With this tool at our disposal, we can answer our original question.

Surround the pellicules P^p, Q^q by spheres S^q, S^p (figure 3) .

If π is the middle of L, we can push these spheres to π and define $\alpha \in \pi_q(M-\Sigma,\pi)$, $\beta \in \pi_p(M-\Sigma,\pi)$.

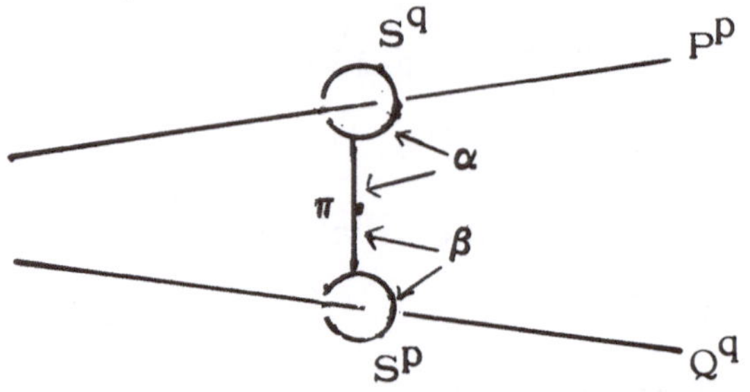

Figure 3

<u>The crossing of</u> P <u>and</u> Q <u>is possible if and only if the elements</u> $\Phi_*\alpha \in \pi_q V$ <u>and</u> $\Phi_*\beta \in \pi_p V$ <u>commute, in the sense their Whitehead product is zero</u> :

$$[\Phi_*\alpha, \Phi_*\beta] = 0 .$$

<u>Examples</u>. 1) For a liquid crystal of "biaxial nematic" or "cholesteric" type, one has $V = G/H$ with $G = SO(3)$ and $H = \{$the group of 180°-rotations around the three directions of a given frame$\}$ (see [4, 5]).

Notice that H is commutative. But $\pi_1 V$ is the lift of H to SU(2), denoted by Q. The group Q consists of 8 spinors :

$$1, \ \mathcal{J}, \ \pm e_1, \ \pm e_2, \ \pm e_3$$

and $e_1 e_2 = -e_2 e_1$, $e_i^2 = e_1 e_2 e_3$, $\mathcal{J} e_i = e_i \mathcal{J} = -e_i$, $\mathcal{J}^2 = 1$.
So Q is non-commutative. Its center consists only of 1 and $\mathcal{J}$.

So $\pm e_i$ and $\pm e_j$ defect lines $(i \neq j)$ should not be able to cross each other but stay entangled (see also [1, 3, 4]).

2) In a hypothetical 4-dimensional situation, a defect bag S^2 and a defect loop S^1 (figure 4), corresponding to $\alpha \in \pi_1(M-\Sigma)$ and $\beta \in \pi_2(M-\Sigma)$ will stay hooked together if $\Phi_*\alpha$ operates non trivially on $\Phi_*\beta$:

$$\Phi_*\alpha \cdot \Phi_*\beta \neq \Phi_*\beta .$$

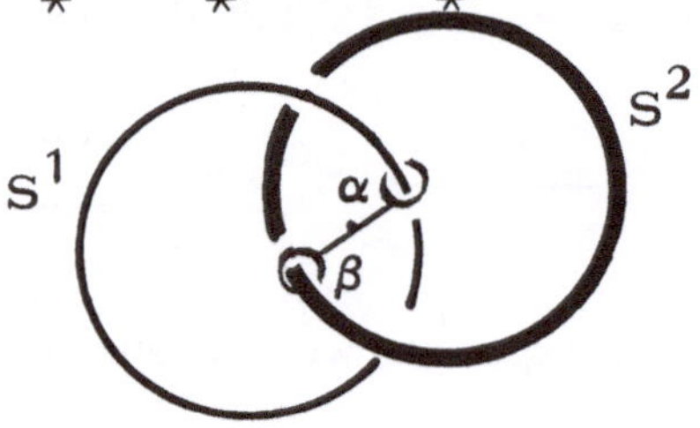

Figure 4

II) LINKS

Here I will describe some mathematics hopefully useful for the theory of polymers ([7], [8]) .

We will consider links, L, which means collections of disjoined simple closed curves in R^3 (or S^3) .

Figure 5 gives examples of 2-component links.

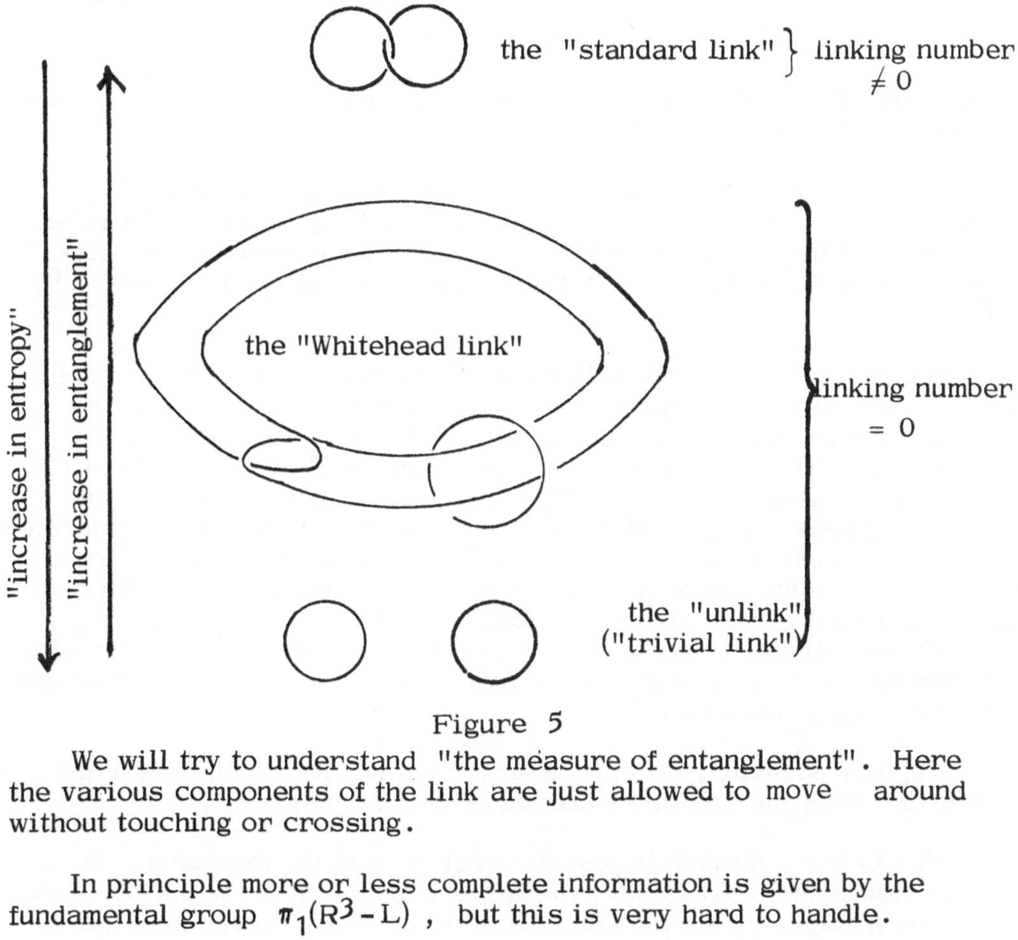

Figure 5

We will try to understand "the measure of entanglement". Here the various components of the link are just allowed to move around without touching or crossing.

In principle more or less complete information is given by the fundamental group $\pi_1(R^3 - L)$, but this is very hard to handle.

So one has to look for weaker but more manageable substitutes.

Now, for any discrete group Γ , one can attach a graded Lie algebra (see [6] for details). One starts by considering the sequence

$$\Gamma_0 = \Gamma, \ \Gamma_1 = [\Gamma,\Gamma], \ \Gamma_2 = [[\Gamma,\Gamma],\Gamma], \dots, \Gamma_k = [\Gamma_{k-1},\Gamma], \ \dots \ .$$

This is also called the "lower central series" of the group Γ . The following quotient groups, constructed from the lower central series are all abelian :

$$A_1(\Gamma) = \Gamma/[\Gamma,\Gamma] = \Gamma_0/\Gamma_1 , \quad A_2(\Gamma) = \Gamma_1/\Gamma_2 , \ldots$$
$$\ldots, \quad A_k(\Gamma) = \Gamma_{k-1}/\Gamma_k , \ldots .$$

By taking the commutators, one can define an operation :

$$A_i \times A_k \xrightarrow[{[\ ,\ ]}]{} A_{i+k} ,$$

which turns the graded group $A_1 + A_2 + \ldots$ into a graded Lie algebra, which we will denote by $L\Gamma$.

Dennis Sullivan has developed a "de Rham homotopy theory" , which in particular allows us to compute $L\Gamma \otimes R$, with $\Gamma = \pi_1(R^3-L)$ in terms of differential forms on R^3-L . Actually Sullivan computes a dual, equivalent object, a "graded differential algebra" (see 9, 10, 11).

The general idea is the following. Suppose, for simplicity's sake that L has only two components, like in figure 5. One chooses two closed 1-forms X_1 , X_2 , defined on R^3-L and generating the cohomology. The non-vanishing of the linking number is equivalent to the fact that the closed 2-form $X_1 \wedge X_2$ is not exact. If this is the case, the process stops and our differential graded algebra is generated by X_1 ,X_2 with $dX_1 = dX_2 = 0$. If $X_1 \wedge X_2$ is exact, one chooses a 2-form Z such that $dZ = X_1 \wedge X_2$. This will be another piece of the differential graded algebra ; notice now that $Z \wedge X_1$, $Z \wedge X_2$ are closed. Again, if there are not exact the process stops, otherwise one throws in more generators U_i (which are also 1-forms), such that $dU_i = Z \wedge X_i$, and so on (*) .

Vaguely speaking, this process (or rather its dual) is for the group Γ what the Taylor expansion is for a function f .

The linking number is just the first term in the expansion. It cannot distinguish, let is say, the Whitehead link from the unlink. But climbing a couple of more steps in this expansion, one can distinguish them.

(*) Notice the "philosophical" difference of this procedure with Gauss's formula for the linking number. Gauss computes an integral "on the link itself" while Sullivan's "invariants" are tested by integrals computed in the ambient space, outside the link.

Now, for the standard link our Lie algebra is trivial (everything commutes), while for the unlink it has the richest structure. So here, unlike for defects, non vanishing commutators are rather a mechanism of "freedom" and not of "slavery".

The richness of the Lie algebra is a sort of "entropy" property very related to what is called the "order of growth" of the group Γ (see [12, 13] for more details) about which I will say just a few words.

Let S be a finite system of generators for Γ, such that $S = S^{-1}$. For given n, consider N(n) = {the number of distinct elements of Γ obtained by writing all words of length n with letters from S}. The issue is how does N(n) behave asymptotically when $n \to \infty$. This is very related to the random walk on the graph of the group determined by S.

By choosing the standard two generators, one finds that for the standard link N(n) grows polynomially and for the unlink it grows exponentially. In general, the fewer relations there are in $\pi_1(R^3-L)$ hence the less entanglement there is, the <u>faster</u> N(n) grows.

Lack of entanglement means increase in entropy !

A projected future joint paper of F. Axel and the author will hopefully give some concrete applications of these kind of mathematical ideas to polymer physics.

III) REMARKS ON THE "DEFECTS" OF A GAUGE FIELD

So far only matter fields and their distorsions have here discussed. But gauge fields also should exhibit similar "defects".

We set the scene by considering a principal G-bundle over the physical space M, endowed with a connection τ. Here G is a Lie group (the gauge group), τ is the potential and the curvature F is the field.

There are at least two kind of "defects" we can consider. Firstly there are the "monopoles" (or "instantons"). Here the bundle is not defined over a "defect" region (the monopole itself, see [14, 15]). From a formal homotopical standpoint, we can treat this like a defect in an ordered medium (with H = 1), but using everywhere $\pi_{i-1}G$ instead of $\pi_i G$. In mathematical terms, the manifold of internal states V is the classifying space BG.

But there is also another kind of possible "defect" which is exhibited in the famous Bohm-Aharonov experiment [14].

Here the bundle is everywhere defined and in the unexcited region of the physical space the connection is flat $(F = 0)$. One says that the field is in pure gauge form.

The "Bohm-Aharonov defect" is the region of space where $F \neq 0$. That is where all the energy of the system lives.

Here one has to distinguish inbetween G itself, and the abstract, discrete group G_d which is G stripped of its topology. Notice that the identity map $G_d \to G$ is continuous.

This inclusion $G_d \to G$ (plus a shift in the dimension of homotopy groups) is somehow the analogue of the symmetry break $H \subset G$ which produces the defects of matter fields. But the complete discussion of this kind of defects will probably require more than just homotopy theory, since integrability conditions are involved too. This should not be to surprising since similar problems occur already for the defects of smectic liquid crystals (see for instance [16, 17]) .

BIBLIOGRAPHY

1. V. Poenaru - G. Toulouse, The crossing of defects in ordered media and the topology of 3-manifolds, J. de Physique 8 887:895 (1977).

2. V. Poénaru - G. Toulouse, Topological solitons and graded Lie algebras, J. of Math. Phys. 20 (1), 13:19 (1970).

3. V. Poénaru, Lectures on elementaty topology in connection with defects and textures, in the volume "Ill condensed matter, Les Houches 1978 Summer school", to appear.

4. Y. Bouligand, B. Derrida V. Poénaru, Y. Pomeau, G. Toulouse, Distortions with double topological character, the case of cholesterics, J. de Physique, 39, 863:867 (1978).

5. G.E. Volovik - V.P. Miveev, Study of singularities in ordered systems by homotopic topology methods, Zh. Eksp. Theor. Fiz. 72, 2256 (1977).

6. J.P. Serre, Lie algebras and Lie groups, Benjamin (1965)

7. P.G. de Gennes, Theory of polymers, book to appear.

8. S.F.Edwards, Statistical mechanics with topological constraints, Proc. Phys. Soc. 91, 513:519 (1967).

9. D. Sullivan, Differential form and the topology of manifolds, in the volume Manifolds - Tokyo 1973, 37:49 (1975).

10. D. Sullivan, Infinitesimal computations in topology, Publ. Math. IHES, 47, 269:331 (1978).

11. D. Sullivan, Lectures given at Orsay, 1973-74 (handwritten notes).

12. J. Milnor, A note in curvature and the fundamental group, J. of diff. Geometry, 2, 1:7 (1968).

13. Séminaire d'Orsay, Travaux de Thurston sur les surfaces, Astérisque 66-67, 1979.

14. T.T. Wu - C.N. Yang, Concept of non integrable phase factor and global formulation of gauge field, Phys. Rev. D., vol. 12, 3845:3857 (1975).

15. G. Toulouse and M. Kleman, Principles of a classification of defects in ordered media, J. Phys. Letters 37, p. 149 (1976).

16. V. Poénaru, Remarks on the punctual defects of 2-dimensional smectics, to appear.

17. D. Mermin, The topological theory of defects in ordered media, Rev. of Mod. Phys. 3, 591:648 (1979).

CUTOFF DEPENDENCE IN LATTICE ϕ^4_4 THEORY

K. Symanzik

Deutsches Elektronen-Synchrotron DESY

Hamburg

0. INTRODUCTION

This seminar is logically the continuation of the one [1] I gave here three years ago. The stimulus to the present work came from recent results [2] on lattice ϕ^4_4 theory, obtained by high-temperature expansions, which seem to require to study the Λ^{-2} corrections (Λ^{-1} = a = lattice constant) to the formulae of [1] which dealt in detail only with ln Λ terms. We shall see that the numerical ϕ^4_4 results offer a test of the merely technical (!) idea of "asymptotic freedom". This is the link between the present topic and nonabelian gauge theory.

The tool to analyse the $\Lambda^{-2}(\ln\Lambda)^{\cdot}$ corrections is, as for the $(\ln\Lambda)^{\cdot}$ terms in [1], the renormalization group. While we could work entirely in four dimensions (as we shall indicate at the end of sect. 2) it is simpler to work initially in 4 + ε dimensions, with ε generic, and to set ε = 0 only later. That a 4 +ε-dimensional lattice offers no difficulty in perturbation theory (and this suffices for our purpose) we shall show in sect. 1. This allows us to write in sect. 2 a simple effective Lagrangean for small-a dependence, and to analyse its consequences, using the renormalization group, in sect. 3. The results in terms of formulae that can directly be compared with numerical data are given in sect. 4. Final remarks are offered in sect. 5.

1. LATTICE ϕ^4 THEORY IN 4 + ε DIMENSIONS

The Euclidean Lagrangean of hypercubic lattice ϕ^4 theory in D = 4 + ε dimensions is

(1.1) $$L = a^D \sum_i \Big[-\tfrac{1}{2} \sum_\mu a^{-2} (\phi_i - \phi_{i+\hat\mu})^2 - \tfrac{1}{2} \Delta m_B^2 \phi_i^2 - - \tfrac{1}{4!} g_B \phi_i^4 - \tfrac{1}{2} m_{B0}^2 \phi_i^2 \Big]$$

where $i \in \mathbb{Z}^D$, $i+\hat\mu$ is the lattice point next to i in the positive μ-direction, and the μ-sum is from 1 to D.

$m_{B0}^2 = a^{-2} f(g_B a^{-\epsilon}, \epsilon)$ is the bare-mass-squared of the massless (critical) theory. For $\Delta m_B^2 > 0$, and we shall only consider this case, we are in the symmetric $(\langle\phi\rangle = 0)$ phase. $\langle\cdots\rangle$ is the "true" vacuum expectation value, i.e. the Gibbs-ensemble expectation in the thermodynamic limit.

We define the theory formally by perturbation expansion in g_B, i.e. by Feynman graphs with bare propagators

(1.2) $$\langle \phi_i \phi_j \rangle_0 =$$

$$= (2\pi)^{-D} \prod_\mu \Big(\int_{-\pi/a}^{\pi/a} dk_\mu \Big) \exp\Big[i a \sum_\mu (i-j)_\mu k_\mu \Big] \cdot$$

$$\cdot \Big[\Delta m_B^2 + \sum_\mu 4 a^{-2} \sin^2(\tfrac{1}{2} k_\mu a) \Big]^{-1} =$$

$$= \frac{1}{2} a^{2-D} \int_0^\infty dt \exp\Big[-Dt - \tfrac{1}{2} a^2 \Delta m_B^2 t \Big] \cdot$$

$$\cdot \prod_\mu I_{(i-j)_\mu}(t).$$

Noting that

$$a^D \sum_i \exp[i a i_\mu k_\mu] = (2\pi)^D \sum_l \delta^D (k_\mu - 2\pi l_\mu a^{-1})$$

where $l \in \mathbb{Z}^D$ one finds that summation over lattice points gives results completely analogous to continuum Feynman graphs: momentum conservation holds modulo $2\pi/a$, and all loop momenta are integrated over one Brillouin zone the precise location of which is immaterial due to the $2\pi/a$ periodicity of all factors. E.g., for a one-loop graph with external momenta $p_1 \ldots p_n$ (in order 1 ... n along the loop) one finds

$$(1.3)\quad I(p_1 \ldots p_n) = (2\pi)^D \sum_l \delta^D\Big(\sum_m p_m - 2\pi l a^{-1}\Big) \cdot$$

$$\cdot\, a^{-D} \prod_{m=1}^{n} \Big(\tfrac{1}{2} a^2 \int_0^\infty dt_m\Big) \exp\Big[-\big(D + \tfrac{1}{2}\Delta m_B^2 a^2\big) \sum_m t_m\Big] \cdot$$

$$\cdot \prod_\mu I_0\Big(\Big[\big(\sum_m t_m\big)^2 - 4 \sum_{l<m} t_l t_m \sin^2\big(\tfrac{1}{2} a \sum_{k=l+1}^{m} (p_k)_\mu\big)\Big]^{\frac{1}{2}}\Big)$$

where the t_m play the rôle of the usual Feynman parameters.

The use of momentum vectors with $4+\varepsilon$ components will, as in continuum dimensional regularization [3], not give rise to problems. E.g., we could think of external momenta with only the first four components different from zero. Moreover, the coefficients in the effective Lagrangean, to be described in the next section, can be expressed in closed form in terms of functions on the lattice at zero external momenta or derivatives there, see (2.3) below. – We do not know, however, whether the dimensional interpolation used here is the same as the one obtained from high-temperature expansions [4].

We shall always work with vertex functions (VFs), i.e. the full-propagator-amputated one-particle-irreducible connected parts of Green's functions. The Fourier transform of the VF to $g^{-l}\langle \phi_{i_1} \cdots \phi_{i_{2n}} \phi^2_{j_1} \cdots \phi^2_{j_l}\rangle$, with $(2\pi)^D \cdot$ momentum conserving delta function, as in (1.3), omitted we denote as $\Gamma_B(p_1 \cdots p_{2n}, q_1 \cdots q_l; g_B, \Delta m_B^2, \varepsilon, a)$, or $\Gamma_B((2n),(l))$ for short. We shall suppose these functions to have in perturbation theory the small-a-expansions

$$(1.4)\quad \Gamma_B((2n),(l);\varepsilon,a) = \sum_{j=0}^{\infty} \sum_{k=0}^{\infty} a^{2j-\varepsilon k} f_{jk}((2n),(l);\varepsilon).$$

The f_{jk} are finite for generic ε but have singularities at positive rational ε in such a fashion that the r.h.s. of (1.4) stays finite at all ε (if $\Delta m_B^2 = 0$, at all $\varepsilon \geq 0$ for nonexceptional momenta) due to the l.h.s. being finite. At $\varepsilon = 0$, (1.4) reduces to

$$(1.5)\quad \Gamma_B((2n),(l);0,a) = \sum_{j=0}^{\infty} \sum_{k=0}^{\infty} a^{2j} (\ln a)^k \bar{f}_{jk}((2n),(l)).$$

The argument for (1.4-5) is given in the next section.

2. EFFECTIVE LAGRANGEAN FOR LARGE-CUTOFF BEHAVIOUR

The expansion (1.4) arises from the following effective Lagrangean:

(2.1)
$$L_{eff} = -\tfrac{1}{2}\partial_\mu\phi_B\partial_\mu\phi_B - \tfrac{1}{2}\Delta m_B^2\phi_B^2 - \tfrac{1}{4!}g_B\phi_B^4 + \sum_{n=2}^{\infty}[(2n)!]^{-1}a^{2n-2}\sum_\mu \phi_B\partial_\mu^{2n}\phi_B + \sum_{r=0}^{\infty}\sum_{s=1}^{\infty}\sum_{t=0}^{\infty}\sum_{\nu=1}^{n_{rs}}(\text{"}D^{2r}\phi_B^{2s}\text{"})_\nu \Delta m_B^{2t} g_B^{s-1} \cdot f_{rst\nu}(g_B a^{-\varepsilon},\varepsilon)\, a^{-4+2r+2s+2t}.$$

The $(\text{"}D^{2r}\phi_B^{2s}\text{"})_\nu$ are all monomials of order 2r in derivatives and order 2s in ϕ_B, having the lattice symmetry and being linearly independent at zero momentum. (For r + s $\geqslant$ 3, $n_{rs} > 1$ in general.) Computations with L_{eff} are meant by use of continuum Feynman rules with bare propagator and vertex as obtained from the first three terms; all other vertices, including the two-vertices, are to be treated as insertion into graphs, with dimensional integration rules [3] to be used throughout.

All coefficients $f_{rst\nu}$ can be expressed by the regularized Γ_B, with $\Delta m_B^2 = 0$, at zero momenta and derivatives there: We define, with $\Box = \sum_\mu \partial_\mu^2$,

(2.2)
$$\bar{O}_1 = -\tfrac{1}{4!}g_B\phi_B^4 \qquad \bar{O}_2 = -\tfrac{1}{2}\Delta m_B^2\phi_B^2$$

$$\bar{O}_3 = -\tfrac{1}{2}\sum_\mu \partial_\mu\phi_B\partial_\mu\phi_B \qquad \bar{O}_4 = \tfrac{1}{2}\sum_\mu \phi_B\partial_\mu^4\phi_B$$

$$\bar{O}_5 = \frac{1}{2}\phi_B \Box^2 \phi_B \qquad \bar{O}_6 = \frac{1}{3!} g_B \phi_B^3 \Box \phi_B$$

$$\bar{O}_7 = \frac{1}{6!} g_B^2 \phi_B^6 \qquad \bar{O}_8 = \Delta m_B^2 \bar{O}_3$$

$$\bar{O}_9 = \Delta m_B^2 \bar{O}_1 \qquad \bar{O}_{10} = \Delta m_B^2 \bar{O}_2$$

Then the (total) coefficients in L_{eff} of these operators are

(2.3)

$$\bar{O}_1: \; -g_B^{-1}\Gamma_B(0000,;g_B,0,\varepsilon,a) \equiv \bar{Z}_1$$

$$\bar{O}_2: \; \Gamma_B(00,0;g_B,0,\varepsilon,a) \equiv \bar{Z}_2$$

$$\bar{O}_3: \; -[\partial/\partial p^2]\Gamma_B(p(-p),;g_B,0,\varepsilon,a)\big|_{p=0} \equiv \bar{Z}_3$$

$$\bar{O}_4: \; \frac{1}{24}D^{-1}(D-1)^{-1}\Big[(D+2)\sum_\mu(\partial/\partial p_\mu)^4 - 3\big(\sum_\mu(\partial/\partial p_\mu)^2\big)^2\Big]\cdot \Gamma_B(p(-p),;g_B,0,\varepsilon,a)\big|_{p=0} \equiv a^2\bar{Z}_4$$

$$\bar{O}_5: \; -\frac{1}{8}D^{-1}(D-1)^{-1}\Big[\sum_\mu(\partial/\partial p_\mu)^4 - \big(\sum_\mu(\partial/\partial p_\mu)^2\big)^2\Big]\cdot \Gamma_B(p(-p),;g_B,0,\varepsilon,a)\big|_{p=0} \equiv a^2\bar{Z}_5$$

$$\bar{O}_6: \; -\frac{1}{12}D^{-1}\sum_{i=1}^{3}\sum_\mu(\partial/\partial p_{i\mu})^2 \cdot \Gamma_B(p_1p_2p_3(-p_1-p_2-p_3),;g_B,0,\varepsilon,a)\big|_{p_i=0} \equiv a^2\bar{Z}_6$$

$$\bar{O}_7: \; g_B^{-2}\Gamma_B(000000,;g_B,0,\varepsilon,a) \equiv a^2\bar{Z}_7$$

$$\bar{O}_8: \; [\partial/\partial p^2]\,\Gamma_B(p(-p),0;g_B,0,\varepsilon,a)\big|_{p=0} \equiv a^2\bar{Z}_8$$

$$\bar{O}_9: \quad g_B^{-1}\, \Gamma_B(0000,0;\, g_B, 0, \varepsilon, a) \equiv a^2 \bar{Z}_9$$

$$\bar{O}_{10}: \quad \Gamma_B(00,00;\, g_B, 0, \varepsilon, a) \equiv a^2 \bar{Z}_{10}.$$

Hereby $\bar{Z}_1$, Z_2, Z_3 are defined by analytic continuation from $\varepsilon > 0$ being IR singular at (rational) $\varepsilon \leq 0$, and $\bar{Z}_4 \ldots \bar{Z}_{10}$ are defined by analytic continuation from $\varepsilon > 2$, being IR singular at (rational) $\varepsilon \leq 2$. The rôle of these singularities is seen by writing (2.1-3) as

(2.4) $$L_{eff} = \sum_{i=1}^{3} \bar{Z}_i \bar{O}_i + a^2 \sum_{i=4}^{10} \bar{Z}_i \bar{O}_i + a^4 \ldots =$$

$$= L_{eff(0)} + \sum_{K=1}^{\infty} a^{2K} L_{eff(K)}.$$

$L_{eff(0)}$ yields finite results to ℓ-loop order for (not only generic) $\varepsilon < 2/\ell$. $L_{eff(0)} + a^2 L_{eff(1)}$ does so for $\varepsilon < 4/\ell$, whereby the IR singularities of the $\bar{Z}_i$ (i = 4 ... 10) act as final subtractions for graphs computed from $L_{eff(0)}$ and no longer convergent for $\varepsilon \geq 2/\ell$ (and for graphs with lower-order $L_{eff(1)}$ insertions), etc. For $\varepsilon \sim 0$, $L_{eff(0)}$ is a dimensionally regularized Lagrangean with counter terms, however, not precisely of the 't Hooft [5] but of the Zinn-Justin form [6]: $\bar{Z}_i - 1$ (i = 1,2,3) are not pure pole terms in ε but have also regular parts, determined by the regularization (2.4) comes from and characterized by having in their perturbation theoretical expansion only powers of $a^{-\varepsilon}$ or, for $\varepsilon = 0$, $\ln a$ appearing. If we only consider $\varepsilon \gtrsim 0$, then $L_{eff(1)}$ is an altogether finite operator insertion, at zero momentum, of dimension six, and so on for higher insertions. Hereby the a^2 explicit in (2.4) can be chosen different from "normalization length" a that plays the rôle of μ^{-1} in 't Hooft's approach [5].

The reason [7] for validity of (2.3) is that at the momenta indicated, all non-Born contributions from L_{eff} with $\Delta m_B^2 = 0$ vanish for dimensional reasons as far as $\bar{Z}_1 \ldots \bar{Z}_7$ are concerned, and for $\bar{Z}_8 \ldots \bar{Z}_{10}$ one uses that for Γ_B, $\phi_B^2/2$ insertion at zero momentum is equivalent to differentiation w.r.t. $-\Delta m_B^2$.

In a fully perturbative construction, in (2.1)

$$f_{rstv}(g_B a^{-\varepsilon}, \varepsilon) = \sum_{\ell=1}^{\infty} f_{rstv\ell}(\varepsilon)(g_B a^{-\varepsilon})^{\ell}$$

with meromorphic $f_{r,s,t,v,\mathcal{L}}(\varepsilon)$ holds, where $\mathcal{L}$ is the number of loops in the VFs in (2.3). In this way, (1.4) is obtained and in the $\varepsilon \searrow 0$ limit (1.5). However, formulae (2.3) are meant "exactly", i.e. computing from (1.1) to arbitrary order in g_B. In this sense, L_{eff} is not mere perturbation theory.

The argument for validity of (2.1) (with only covariant operators needed) for Pauli-Villars regularization and $\Delta m_B^2 = 0$, using oversubtraction, is given elsewhere [7]. The extension to $\Delta m_B \neq 0$ is straight-forward, whereby one uses that differentiation with respect to the bare mass, all other bare parameters fixed, reduces UV divergence degree by two. (A related argument within merely dimensional regularization is used in [8]).

The present point is that (2.1) seems to hold also for lattice regularization at least to order a^2. Consider the one-loop graph contribution to the four-point VF. Using (1.2) we can write in obvious notation

$$I(a,p) = \prod_\mu \left(\int_{-\pi/a}^{\pi/a} dk_\mu \right) N(a,k)^{-1} N(a,k+p)^{-1}.$$

Then

$$(2.5) \quad [\partial/\partial a^2]\, I(a,p) =$$

$$= -\pi a^{-3} \sum_\mu \prod_{\nu \neq \mu}' \left(\int_{-\pi/a}^{\pi/a} dk_\nu \right) N(a,k)^{-1} N(a,k+p)^{-1} \Big|_{k_\mu = \frac{\pi}{a}} -$$

$$- \prod_\mu \left(\int_{-\pi/a}^{\pi/a} dk_\mu \right) \left[N(a,k)^{-1} \frac{\partial}{\partial a^2} N(a,k) \cdot N(a,k)^{-1} N(a,k+p)^{-1} + \ldots \right]$$

$\equiv$ Bound. term + Insert. term.

Due to $[\partial/\partial a^2] N(a,k) = -\frac{1}{12} \sum_\mu k_\mu^4 + O(a^2 k_\mu^6)$ the insertion term is a regularized form of the graphs with dimension-six insertions in the two lines. As $a \searrow 0$, this term would be quadratically divergent. Thus, with $T_{2n}(p)$ the Taylor operator around zero momentum, according to BPHZ [9] in the identity

Ins. term = $[1 - T_2(p)]$ Ins. term + $T_2(p)$ Ins. term

the first term on the r.h.s. has a limit as $a \searrow 0$, which is the square bracket applied to the unregularized insertion term. In order to split the square bracket up, we need to give an interpretation to the unregularized <u>unsubtracted</u> insertion term, which we do by the dimensional-integration rule [3]. It gives always a finite result for generic ε , and we denote its use by underlining. Thus,

$$(2.6)\ \text{Ins. term} = \underline{\text{Ins.term } (a=0)} + T_2(p)\left[\text{Ins.term} - \underline{\text{Ins.term}(a=0)}\right] + O(a^2).$$

The boundary term in (2.5) is a sum of $3+\varepsilon$ -dimensional graphs with mass $\Delta m_B^2 + 4a^{-2}$ and has a Taylor expansion in p_α with finite convergence radius. Thus,

$$(2.7)\ \text{Bound.term} = T_2(p)\ \text{Bound.term} + O(a^{2-\varepsilon})$$

Using (2.6-7) in (2.5) and integrating yields

$$(2.8)\ I(a,p) = a^2\ \underline{\text{Ins.term } (a=0)} + R(p) + T_2(p)\left[I(a,p) - a^2\ \underline{\text{Ins.term } (a=0)}\right] + O(a^{4-\varepsilon})$$

whereby $T_2(p)R(p) = 0$, with the interpretation: The first term on the r.h.s. is the contribution from the insertion of the a^2 part of the fourth term in (2.1). The third term is the sum of the f_{0201}, f_{0211} and f_{1201} contribution, plus part of the ordinary one-loop graph the remainder of which is in R(p). All integrations are to be done dimensionally. If $\Delta m_B^2 = 0$, the Taylor operator should be taken around some nonzero a-independent momentum. (2.8) as well as agreement of the constants therein with (2.3) can also be verified directly on the basis of (1.3). In a similar fashion, correctness of (2.1) for all one-loop graphs is easily shown.

For graphs with ≥ 2 loops, the BPHZ method requires to subtract all subdivergences before performing the final subtraction whereupon one proceeds as before. For the second-order self-energy part, I have verified that (2.1) does describe it to order a^2 correctly. Hereby the subdivergences are cancelled by the contributions from the $\bar{O}_{7,9}$ and $\bar{O}_6$ counter terms in (2.1) to the order stated. The total effect of the boundary terms is again absorbed by the constants in (2.1). On the basis of this, we here take (2.1) as valid at least to order a^2, the one we are interested in.

Working in four dimensions throughout would require to write, for reproducing (1.5), to use an effective Lagrangean with normal products in the sense of Zimmermann [10], which imply an elaborate subtraction prescription, in particular in the massless theory. The coefficients prop. a^2 in such effective Lagrangean would be proportional to the difference between VFs computed with regularization and computed from the a^0 terms alone, with ln a occurring in the finite renormalizations. This difference-taking is replaced in our case by the need of continuing analytically $\bar{Z}_4 \ldots \bar{Z}_{10}$ in (2.3) from $\varepsilon > 2$. (Effective Lagrangeans of the Zimmermann type, for large-mass rather than large-cutoff behaviour, have recently been used by C.K. Lee [11] and Kazama and Yao [12].)

As pointed out before [1], there is no local L_{eff} describing large-cutoff behaviour beyond the order $\Lambda^0(\ln\Lambda)$ for e.g. sharp cutoff, i.e. using propagators $\theta(\Lambda^2-p^2)\,[p^2+m^2]^{-1}$ in momentum space. The origin of this is that in this regularization, the cutoff-change effect is carried by boundary terms only. In contrast, in lattice regularization, the effect is mainly carried by insertion terms which are essentially local but depend on the choice of the Brillouin zone of the integration momenta; the main function of the boundary terms is to restore the Brillouin-zone independence, as in the transition from (2.6) to (2.8) above.

3. RENORMALIZATION GROUP PROPERTIES OF OPERATOR INSERTIONS

To obtain the consequences of (2.4), for convenience we temporarily use 't Hooft's Lagrangean [5]

(3.1) $$L_{'tH.} = \sum_{i=1}^{3} Z_i O_i$$

and define

(3.2) $$\hat{O}_i = \sum_j Z_{ij} O_j \quad (i,j = 4 \ldots 10).$$

The operators O_i are obtained from the $\bar{O}_i$ of (2.2) by the replacements

(3.3) $$\phi_B \to \phi,\ \Delta m_B^2 \to m^2,\ g_B \to g\mu^{-\varepsilon}$$

and we may identify μ with α^{-1}. The coefficients

(3.4) $$Z_{ij} = \delta_{ij} + \sum_{k=1}^{\infty} \varepsilon^{-k} f_{ijk}(g)$$

are so chosen, uniquely, as to give, as $\varepsilon \searrow 0$, finite operator insertions $\hat{O}_i$ of dimension six.

With

(3.5a) $$\beta(g,\varepsilon) = \varepsilon\left([\partial/\partial g]\ln(gZ_1Z_3^{-2})\right)^{-1} = \varepsilon g + b_0 g^2 + b_1 g^3 + \ldots$$

where

(3.5b) $$b_0 = 3\,(16\pi^2)^{-1}, \quad b_1 = -17\,(2^8\cdot 3\pi^4)^{-1},$$

(3.6) $$\gamma(g) = \beta(g,\varepsilon)\,[\partial/\partial g]\ln Z_3 = c_0\, g^2 + \cdots,$$

(3.7) $$\eta(g) = -\beta(g,\varepsilon)\,[\partial/\partial g]\ln(Z_2 Z_3^{-1}) = \tfrac{1}{3}\, b_0\, g + \cdots,$$

the VFs to (3.1) obey

(3.8a) $$Op_{2n}\,\Gamma((2n);\mu, m, g, \varepsilon) = 0$$

where

(3.8b) $$Op_{2n} = [\mu/\partial\mu] + \beta(g,\varepsilon)[\partial/\partial g] - 2n\gamma(g) + \eta(g)\,m^2[\partial/\partial m^2].$$

Then for the VFs with $\hat{0}_i$ inserted at zero momentum

(3.9) $$Op_{2n}\,\hat{\Gamma}_i((2n)) = \gamma_{ij}\,\hat{\Gamma}_j((2n))$$

holds, with an upper-right (4/567/89/10) block-triangular mixing matrix

(3.10) $$\gamma_{ij}(g) = b_0\, q_{ij}\, g + O(g^2)$$

with

(3.11) $$q_{..} = \left(\begin{array}{c|ccc|cc|c}
0 & 0 & 0 & -15 & 0 & 3 & 1/2 \\
\hline
 & 0 & 0 & -30 & 0 & 6 & 1 \\
 & 0 & 0 & -60 & -2/3 & 8 & -2/3 \\
 & 0 & 0 & -3 & 0 & 1/3 & 0 \\
\hline
 & & & & 1/3 & 2 & 2/3 \\
 & & & & 0 & -2/3 & -1/3 \\
\hline
 & & & & & & 1/3
\end{array}\right)$$

obtained from one-loop calculations. The $\gamma_{ij}(g)$ with i,j = 8, 9, 10 can be expressed, using the method of [13], in terms of β, γ, η and their derivatives alone.

From now on we consider $\varepsilon = 0$ only. We set

(3.12) $$\hat{\Gamma}_i = W_{ij}(g)\,\tilde{\Gamma}_j$$

and demand

$$Op_{2n}\, \tilde{\Gamma}_i((2n)) = 0 \tag{3.13}$$

and invertibility of the matrix W. Then we must solve

$$g[\partial/g]\, W_{ij}(g) = [q_{ik} + u_{ik}(g)]\, W_{kj}(g) \tag{3.14}$$

where $u_{..}(g) = O(g)$. The eigenvalues λ_α of q are its diagonal elements, and one finds that it can be diagonalized by a similarity transformation. Consequently, (3.14) can be solved by power series $g^{\lambda_\alpha}(X_0^\alpha + g X_1^\alpha + \ldots)$, with X_0^α the corresponding eigenvectors, except possibly lng appearing in

$$W_{..} = g^{-3}(X_0^{-3} + g X_1^{-3} + \ldots) + \ln g\,(Y_0 + g Y_1 + \ldots) \tag{3.15}$$

due to integer-spacing with the threefold eigenvalue zero. To determine whether the logarithm does appear or not requires to compute γ_{ik} and β to four loops. There is no $g^{1/3}\ln g$ in the solution to eigenvalue $-2/3$ as one finds using the known [13] higher terms in $\gamma_{ik}(g)$.

We find the following solution $(i = 4 \ldots 10)$

$$
\begin{aligned}
(3.16)\quad W_{i4} &= \{1, -,-,-,-,-,-\} \\
W_{i5} &= \{0,1,0,0,-,-,-\} \\
W_{i6} &= \{0,0,1,0,-,-,-\} \\
W_{i7} &= g^{-3}\{5,10,20,1,-,-,-\} + \ln g\,\{O(1),O(1),O(1),0,-,-,-\} \\
W_{i8} &= g^{1/3}\{0,0,-2,0,1,0,-\} \\
W_{i9} &= \tfrac{1}{7}\, g^{-2/3}\{-9,-18,-8,1,-14,7,-\} \\
W_{i10} &= \tfrac{1}{3}\, g^{1/3}\{0,0,-4,-\tfrac{1}{10},2,-1,3\}.
\end{aligned}
$$

Herein, in the wavy bracket only the constant terms are written and to be amended by terms with higher integer powers of g, except for the barred entries which are empty. For the inverse matrix we have

$$
\begin{aligned}
(3.17)\quad (W^{-1})_{4i} &= \{1,0,0,-5,0,2,1/2\} + g^3 \ln g .. \\
(W^{-1})_{5i} &= \{-,1,0,-10,0,4,1\} + g^3 \ln g .. \\
(W^{-1})_{6i} &= \{-,0,1,-20,2,8,-2\} + g^3 \ln g ..
\end{aligned}
$$

$$(W^{-1})_{7i} = \frac{1}{7}g^{3}\{-,0,0,7,0,-1,-\tfrac{1}{10}\}$$
$$(W^{-1})_{8i} = g^{-1/3}\{-,-,-,-,1,2,-\}$$
$$(W^{-1})_{9i} = g^{2/3}\{-,-,-,-,0,1,1/3\}$$
$$(W^{-1})_{10i} = g^{-1/3}\{-,-,-,-,-,-,1\}$$

where lng-terms appear if and only if they do in (3.16).

4. APPLICATION TO LATTICE VERTEX FUNCTIONS

We now can organize and resum the $a^2(\ln a)^K$ terms in (1.5) as it was done for the $(\ln a)^K$ terms in [1]. To this end we identify (2.4) with (3.1) amended by the appropriate linear combination of the $\hat{O}_i$. Returning for the moment to $\varepsilon > 0$, we set

$$a^{-1} = \mu \qquad (4.1)$$
$$g_B a^{-\varepsilon} = \bar{g}$$
$$\phi_B = c(g,\varepsilon)\phi$$
$$\Delta m_B^2 = d(g,\varepsilon) m^2$$

Defining in analogy to (3.5a)

$$\bar{\beta}(\bar{g},\varepsilon) = \varepsilon\left([\partial/\partial\bar{g}]\ln(\bar{g}\bar{Z}_1\bar{Z}_3^{-2})\right)^{-1} = \qquad (4.2)$$
$$= \varepsilon\bar{g} + \bar{b}_0(\varepsilon)\bar{g}^2 + \bar{b}_1(\varepsilon)\bar{g}^3 + \cdots =$$
$$= \varepsilon\bar{g} + \bar{\beta}_0(\bar{g}) + \varepsilon\bar{\beta}_1(\bar{g}) + \cdots$$

one finds that

$$d\bar{g}/dg = \bar{\beta}(\bar{g},\varepsilon)/\beta(g,\varepsilon) \qquad (4.3)$$

is uniquely solved by

$$\bar{g} = f_o(g) + \varepsilon f_1(g) + \varepsilon^2 f_2(g) + \cdots \qquad (4.4)$$

with $f_o(g) = g + O(g^2)$, whereby

(4.5) $f_0(g) = \bar{\rho}^{-1}(\rho(g))$

with the definition

(4.6)
$$\bar{\rho}(\bar{g}) = \int^{\bar{g}} d\bar{g}' \, \bar{\beta}_0(\bar{g}')^{-1},$$
$$\rho(g) = \int^{g} dg' \, \beta(g';0)^{-1}.$$

(The integration constant in (4.5) is determined via fitting (4.3) in order ε.) With $\bar{\gamma}$ and $\bar{\eta}$ defined in analogy to (3.6) and (3.7), one finds

(4.7a) $c(g,\varepsilon) =$
$$= \exp\left\{ 2\int_0^g dg' \, \beta(g';\varepsilon)^{-1} [\gamma(g') - \bar{\gamma}(\bar{g}';\varepsilon)] \right\}, =$$
$$= 1 + O(g^2) \quad \text{for } \varepsilon = 0,$$

(4.7b) $d(g,\varepsilon) =$
$$= \exp\left\{ -\int_0^g dg' \, \beta(g';\varepsilon)^{-1} [\eta(g') - \bar{\eta}(\bar{g}';\varepsilon)] \right\}, =$$
$$= 1 + O(g) \quad \text{for } \varepsilon = 0,$$

whereby (4.4) is to be used. One now sees that

(4.8) $L_{eff}(\text{lattice}) = L_{\text{'t Hooft}} + a^2 \sum_{i=4}^{10} \bar{c}_i(\bar{g},\varepsilon) \hat{O}_i + \ldots$

where the $\bar{c}_i(\bar{g},\varepsilon)$ are linear combinations of the $\bar{Z}_i(\bar{g},\varepsilon)$ ($i = 4\ldots 10$) of (2.3) with coefficients, obtained from (4.1), (4.4), (4.7), and (3.2), such that the $\bar{c}_i(\bar{g},0)$ are finite since the l.h.s. of (4.8) is finite at $\varepsilon = 0$. Explicitly, one finds from (2.3)

(4.9)
$$\bar{c}_i(\bar{g},0) = \begin{cases} \frac{1}{12} + O(\bar{g}^{-2}), & i = 4 \\ O(\bar{g}), & i = 6,7,9,10 \\ O(\bar{g}^{-2}), & i = 5,8. \end{cases}$$

(4.8) now gives at $\varepsilon = 0$

(4.10) $\Gamma_B((2n)_i;\, g_B, \Delta m_B^2, a) =$

$= c(g(g_B))^{-2n}\,\cdot$

$\cdot[\Gamma + a^2\sum_{i=4}^{10}\bar{c}_i(g_B)\hat{\Gamma}_i]((2n)_i;\, a^{-1}, g(g_B), d(g(g_B))^{-1}\Delta m_B^2) +$

$+ O(a^4(\ln a)^{\cdot})$

with the obvious insertions. The use herein of

(4.11) $\hat{\Gamma}_i(\ldots) = W_{ij}(g(g_B))\tilde{\Gamma}_j(\ldots)$

from (3.12) yields, with (3.16,17), the desired reorganization of logarithms due to (3.13).

For application to the setting of [2], we define

(4.12a) $\tilde{g} = -\left[\frac{\partial}{\partial p^2}\Gamma_B(p(-p))|_{p=0}\right]^{-2}\Gamma_B(000q),$

(4.12b) $\tilde{m}^2 = \left[\frac{\partial}{\partial p^2}\ln(-\Gamma_B(p(-p),))|_{p=0}\right]^{-1}.$

Here $\tilde{m}^2$ is the, for Euclidean computations convenient, second-moment definition of mass, and $\tilde{g}$ is a normalization-independent renormalized coupling constant. Inserting (4.10) into (4.12) and omitting at first the a^2 parts, indicating this by a zero subscript, we find from (3.8) that

(4.13a) $\tilde{g}_o = A(C_1, C_2),$

(4.13b) $\tilde{m}_o^2 = B(C_1, C_2)$

with

(4.14a) $C_1 = \bar{\rho}(g_B) + \ln a,$

(4.14b) $C_2 = \bar{\sigma}(g_B) - \ln\Delta m_B^2$

where

(4.15) $\bar{\sigma}(\bar{g}) = \int^{\bar{g}} d\bar{g}'\,\beta_o(\bar{g}')^{-1}\,\eta(\bar{g}';0).$

Solving (4.13b) for C_2 and inserting in (4.13a) gives, for dimensional reasons,

$$\tilde{g}_0 = \tilde{\rho}^{-1}(\bar{\rho}(g_B) + \ln(a\tilde{m}_0)) \tag{4.16}$$

It is not difficult to show that $\tilde{\rho}(\tilde{g}) = \int^{\tilde{g}} d\tilde{g}' \tilde{\beta}(\tilde{g}')^{-1}$, where $\tilde{\beta}$ is the β-function to "intermediate" renormalization specified by $\tilde{\Gamma}(00) = -\tilde{m}^2$, $[\partial/\partial p^2]\tilde{\Gamma}(p(-p))|_{p=0} = -1$, $\tilde{\Gamma}(0000) = -\tilde{g}$. Therefore, in (4.16) only the function $\bar{\rho}(g_B)$, or $g(g_B) = f_0^{-1}(g_B)$ of (4.5), depends on the regularization chosen.

Including also the a^2 corrections, one finds from (4.10) after an easy calculation that

$$\begin{aligned} \tilde{g}(g_B, a\tilde{m}) = {} & \tilde{\rho}^{-1}(\bar{\rho}(g_B) + \ln(a\tilde{m})) + \\ & + (a\tilde{m})^2 \sum_{i=4}^{10} \bar{c}_i(g_B) \sum_{j=4}^{10} W_{ij}(g(g_B)) \cdot \\ & \cdot F_j(\tilde{\rho}^{-1}(\bar{\rho}(g_B) + \ln(a\tilde{m}))) + \\ & + O((a\tilde{m})^4 (\ln a)^{\cdot}) \end{aligned} \tag{4.17}$$

where, besides $g(g_B)$ or $\rho(g_B)$, only the $\bar{c}_i(g_B)$ depend on the regularization chosen. Now observe that, with $\tilde{m}a = \xi^{-1}$,

$$\begin{aligned} \tilde{\rho}^{-1}(\bar{\rho}(g_B) - \ln\xi) \equiv \tilde{g}_{\log} = {} & b_0^{-1}\,\overline{\ln\xi}^{\,-1} + \\ & + b_0^{-1}\,\overline{\ln\xi}^{\,-2}\left[\bar{\rho}(g_B) - b_0^{-2}b_1 \ln(b_0\,\overline{\ln\xi}(\ln\xi)^{-1})\right] + \\ & + O(\overline{\ln\xi}^{\,-3}) \end{aligned} \tag{4.18a}$$

where

$$\overline{\ln\xi} = \ln\xi + b_0^{-2}b_1 \ln\ln\xi, \tag{4.18b}$$

such that it goes to zero as $a \searrow 0$. For the $\bar{c}_i$ and W^{-1}_{ij} in (4.17), only the power series in g_B are (in principle) available. From (4.9), (3.16), (3.17), and (4.12) we finally get

$$\tilde{g}(g_B, \xi) = \tilde{g}_{log} + \tag{4.19}$$

$$+ \xi^{-2}\left[\frac{1}{6}\tilde{g}_{log} - \frac{5}{84} g_B^{-3}\tilde{g}_{log}^{4} - \frac{3}{28} g_B^{-\frac{2}{3}} \tilde{g}_{log}^{5/3} + \dots\right] +$$

$$+ O(\xi^{-4}(\ln \xi)^{\cdot})$$

where in the square bracket only terms with at least one more factor g_B or $\tilde{g}_{log}$ relative to a kept one are omitted. Since the estimate (4.18) is only meaningful for $\xi \gg 1$ such that $\tilde{g}_{log} < g_B$, (4.19) is well behaved.

The interest of these corrections formulae is: Assume that $\tilde{g}$ is inside the region next to the origin where $\tilde{\beta}(\tilde{g})$ is positive ($\tilde{g}$ is always positive due to the Lebowitz inequality). Then $\tilde{\beta}^{-1}$ is a monotonic function of its argument. Therefore, if in (4.18) $\tilde{g}_{log}$ has a maximum at g_B' for some fixed ξ, it has a maximum at g_B' for any fixed ξ provided $\tilde{g}$ stays in the mentioned region. If the observed $\tilde{g}(g_B, \xi)$[2] does not behave so, this can only happen due to the correction terms in (4.17). (4.19) shows, however, that these corrections can be estimated quite well and are $O(\xi^{-2}(\ln \xi)^{-1})$ for $\xi \to \infty$.

If, on the other hand, $\tilde{g}$ is not monotonic in g_B and if this cannot be ascribed to corrections to $\tilde{g}_{log}$, then $\bar{\rho}(g_B) = \rho(g(g_B))$ is not monotonic in g_B. Due to

$$\bar{\beta}(g_B) = \left\{[\partial/\partial g_B]\,\bar{\rho}(g_B)\right\}^{-1} =$$

$$= \beta(g(g_B))\left[\frac{\partial}{\partial g_B} g(g_B)\right]^{-1}$$

this would mean, if the β-function is positive, that $\bar{\beta}(g_B)$ has at that g_B a pole of first (or higher odd) order if $\bar{\rho}(g_B)$ is differentiable there. Such behaviour was indeed already found by Wilson [14] on the basis of nine-terms high-temperature series, and this feature was confirmed by Baker and Kincaid [2] with ten-terms series. Neither authors found a Gell-Mann-Low eigenvalue g_B'' defined by $\lim_{\xi \to \infty} \tilde{g}(g_B'', \xi) = \tilde{g}_\infty > 0$, or $\bar{\rho}(g_B'') = +\infty$, and such g_B'' is needed for a nontrivial continuum ϕ_4^4 theory (or nontrivial continuum Ising$_4$ model, which would require $g_B'' = +\infty$ to be such fixed point) to exist, at least as a limit of the lattice-regularized theory (1.1). - However, Wilson ([14], and discussion remark at the lecture) has remarked that (1.1) may not be a suitable starting point for finding a fixed point $\tilde{g}(\xi=\infty)$ different from the vanishing Gaussian one.

5. DISCUSSION

Anyone familiar with the idea and practice of "asymptotic freedom" (AF) [15] will recognize that the assumptions that led to (4.17-19) are formally identical with the assumptions that underlie AF. AF of nonabelian gauge theory is not understood physically, however, but (so far) merely an outgrowth of the perturbation theoretical formalism. Thus, failure of (4.17-19) to account for non-perturbation theoretical results where these formulae should do so would indicate a serious flaw in the AF reasoning. Note, again, that (4.17-19) do not presuppose a nontrivial continuum ϕ_4^4 theory to exist.

The method of effective Lagrangeans applies directly to all renormalizable theories regularized on the lattice such as to have vertices and propagators. For Abelian gauge theory such form has been discussed by Sharatchandra [17]. An equivalent discussion for nonabelian gauge theory has not been given yet; however, the questions one is interested in there are not in the realm of perturbation theory, and the cutoff problem is supposed to be not acute due to AF.

ACKNOWLEDGMENT

I am much indebted to T.T. Wu for piercing questions, crucial suggestions, and collaboration in part of this work.

REFERENCES

1. K. Symanzik, in: "New Developments in Quantum Field Theory and Statistical Mechanics", Eds. M. Lévy, P. Mitter; New York: Plenum Press 1977, p. 265.
2. G.A. Baker, J.M. Kincaid, Phys. Rev. Letts. 42, 1431 (1979), and to be published
3. G. 't Hooft, M. Veltman, Nucl. Phys. B44, 189 (1972).
4. J.M. Drouffe, G. Parisi, N. Sourlas, Nucl. Phys. B161, 397 (1979)
5. G. 't Hooft, Nucl. Phys. B61, 455 (1973).
6. E. Brézin, J.C. Le Gouillou, J. Zinn-Justin, in: "Phase Transitions and Critical Phenomena", eds. C. Domb, M.S. Green, Vol. VI. London: Academic Press 1976
7. K. Symanzik, Comm. Math. Phys. 45, 79 (1975).
8. J.C. Collins, Nucl. Phys. B80, 341 (1974).
9. K. Hepp, "Théorie de la rénormalisation", Berlin: Springer 1969
W. Zimmermann, Comm. Math. Phys. 15, 208 (1969)

10. W. Zimmermann, Ann. Phys. (N.Y.) 77, 536, 570 (1973).
11. C.K. Lee, Nucl. Phys. B161, 171 (1979)
12. Y. Kazama, Y.-P. Yao, UM HE 79-16
13. Y. Taguchi, A. Tanaka, K. Yamamoto, Nucl. Phys. B132, 333 (1978)
14. K.G. Wilson, J. Kogut, Phys. Reps. 12C, 75 (1974) Sect. 13
15. D. Politzer, Phys. Reps. 14C, 129 (1974).
16. H.S. Sharatchandra, Phys. Rev. D18, 2042 (1978).

Postscript: Meanwhile, more detailed values of $\tilde{g}(g_B, \xi)$ have become available[2]. The values of ξ are 1, 2, 4, 8, 16, 32, 64, 128, and 1000. The g_B values range from 0.2 to ∞ (the Ising model). The function

$$\tilde{\rho}(\tilde{g}) + \ln\xi \equiv \hat{\rho}(\tilde{g}, \xi) = \bar{\rho}(g_B) + O(\xi^{-2})$$

has been computed using the first two terms of $\tilde{\rho}(\tilde{g})$ since the higher ones are not known for intermediate renormalization. $\hat{\rho}(\tilde{g}, \xi)$ is found always to decrease with increasing ξ at fixed g_B. For ξ not too small and $g_B \lesssim 1$, this effect seems to be compatible with RG prediction. For $g_B > 1$, $\hat{\rho}(\tilde{g}, \xi)$ is for $\xi = 1000$ far off on the negative (small $\tilde{g}$) side relative to smaller ξ, even allowing for the uncertainties given [2], and also the remaining decrease with increasing ξ seems to be difficult to be reconciled with RG prediction, although the g_B values are presumably too large for (4.19) to be applicable such that (4.17) ought to be used. While the uncertainties [2] are too large to allow to determine the coefficient functions $\sum_i \tau_i(g_B) W_{ij}(g_B) \equiv C_j(g_B)$ from the data, large values for these are implausible. - The conclusion is that the applicability of "AF" concepts is not yet disproven by the results of [2] so far but is certainly not verified either, and one may wonder why. Most desirable would be to employ also alternative ways of computation (e.g. Monte Carlo) to exclude the possibility of unexpectedly large systematic errors in the HTE analysis. -

The author thanks G.A. Baker Jr. for private communication of his and J.M. Kincaid's results prior to publication.

GAUGE CONCEPTS IN CONDENSED MATTER PHYSICS

Gérard Toulouse

Laboratoire de Physique de l'Ecole Normale Supérieure
24 rue Lhomond, 75231 Paris 05

Foreword

When I accepted, a year ago, to lecture at this School on "Gauge Concepts in Condensed Matter Physics", I thought that this was a valuable and original topic for an assessment. Only later, did I fully realise the difficulty of finding a proper level. I feel therefore that it is necessary to issue here a warning to the reader and an apology. No problem is solved here, no statement receives a proof : c'est une battue à travers la campagne.

References are casual, aiming to help the reader in finding his own way through the literature. For various mathematical aspects, we refer to the companion lecture of V. Poénaru : "Superalgebras and confinement in condensed matter physics".

Numerous discussions on various topics with my colleagues B. Derrida, B. Julia, V. Poénaru, J. Vannimenus, and stimulating conversations with Sir S.F. Edwards and F. Caron are very gratefully acknowledged. I wish to thank M. Brereton for explaining to me his work on "A Gauge Description of Topological Entanglements" and allowing me to mention it prior to publication.

I. INTRODUCTION

The description and classification of phase transitions is an important task in condensed matter physics. This has led to varying definitions in the past. Two trends are clearly visible : there is a proliferating trend, with an increasing number of types of phase transitions being brought to the scene, and there is a unifying trend, establishing correspondences and equivalences between apparently remote systems. It provides an amazing show, this domain of physics constantly proliferating and constantly bending back on itself.

Introduced long ago, the Ehrenfest classification of phase transitions considers which derivative of the free energy exhibits a discontinuity at the transition : first order transition if a first derivative is discontinuous, second order if first derivatives are continuous but a second derivative is discontinuous, etc... Let us just make the remark that, within this scheme, the ideal Bose condensation at constant volume appears a bit odd, as a third order transition.

When the Landau theory of phase transitions was introduced, the emphasis was shifted to considerations of symmetry : symmetry breaking in the ordered phase, order parameter related to the symmetry breaking. This was a much more powerful theory, since it was not a mere classification, but a theory allowing one to make predictions on the nature of phase transitions from symmetry-only arguments. As a consequence of the symmetry breaking ideas, the (low temperature) ordered phase appears as profoundly different from the (high temperature) disordered phase. Also the order of the transition is now related to the behaviour of the order parameter : first order if the order parameter is discontinuous at the transition, second order if it is continuous. This classification differs from Ehrenfest's in a number of cases : for instance, the ideal Bose condensation at constant volume appears as a standard second order phase transition, within the Landau classification. On the other hand, the liquid-gas transition (and also several other transitions, such as various metal-insulator transitions, or the α-γ transition in Cerium, etc...) now appears a bit strange, because it does not display any clear symmetry breaking. Reconciliation with the general pattern results from a mapping into the Ising model (lattice-gas model), which is a first example (many more to come) of an oddity reconciled by a formal equivalence.

At this stage, it is possible to give a picture of "standard", normal, phase transitions, which will serve as a convenient background for later eventual modifications.

A "true" phase transition occurs only in the thermodynamic limit of infinite volume. A magnetic model, with discrete (Ising) or continuous (planar, Heisenberg,...) spins illustrates most clearly what is involved in a phase transition. The phase transition point separates a disordered (high temperature) phase from an ordered (low temperature) phase. There is a degeneracy of ordered phases : a symmetry breaking at low temperature, with infinite barriers separating states of equal energy, leads to ergodicity breaking. The order is characterized by an order parameter : the average of the order parameter (which is defined as the local magnetization in a ferromagnet) is non zero in an ordered phase and vanishes in the disordered phase. The order parameter two-point correlation function tends toward a constant for large separation in an ordered phase, whereas it decreases exponentially in the disordered phase. At low temperatures, the free energy of an ordered phase is lower than the free energy of the disordered phase.

In this same background picture, examples of "false" phase transitions can be given : transitions in small systems, transitions in heterogeneous systems, glassy transitions, where effects of finite size, of dispersion of local T_c values, or of non equilibrium lead to a rounded, smooth crossover region instead of a sharp transition point.

II. PROLIFERATION OF TRANSITIONS

In their quest for novelty, physicists have invented or stumbled upon transitions which appeared not to fit into the previous classification. Quite frequently, it has occurred that subsequent effort was able to establish mappings with previously known transitions. Often the quest for novelty is thus apparently defeated ; but it can also be said that this process leaves no concept unturned and that "hidden faces" of phase transitions are thereby revealed.

1. Percolation

The percolation problem, initiated by applied mathematicians (S.R. Broadbent, J.M. Hammersley, 1957), has spread into many fields and countless review papers are now being devoted to it. For short, a single recent reference is given, the Proceedings of the 1978 Les Houches Summer School[1], which will be useful also for other topics.

The percolation problem is most easily defined on a lattice. Sites can be black with probability p or white with probability (1-p). When p increases, the black sites form clusters of increasing sizes, until at p_c an infinite cluster appears. In another analogous formulation (random resistor network), bonds (links) are randomly conducting or non conducting, and the network is non conducting until an infinite cluster of conducting links appears. A priori, the percolation transition appears to differ radically from standard phase transitions, for it is a purely geometrical problem (no temperature, no partition function are involved). Besides, no symmetry breaking is apparent : actually, it belongs to the category of random or heterogeneous problems, of which we shall see several other examples (it is the simplest of them all). Note also, anecdotically, that for a finite resistor network, a sudden insulator-conductor transition occurs, when the number of conducting links is increased.

There was however an obliged point of contact with thermal phase transitions. Suppose that the black sites carry a magnetic moment while the white sites are non-magnetic, and that magnetic interactions are nearest-neighbour. A magnetic phase transition can occur only if an infinite cluster of magnetic sites exist, so that the percolation transition on the p-axis ($T=0$) appears as a limit point of the magnetic transition line in a (temperature T, concentration p) phase diagram.

The decisive "bending back" step was provided by C.M. Fortuin and P.W. Kasteleyn (1972), who established a formal correspondence between the percolation problem and the $N \to 1$ limit of the Potts model. The Potts model is a generalization of the Ising model, where each site has N internal states and neighbouring sites interact when they are in the same internal state. Thus by virtue of an analytic continuation procedure ($N \to 1$), the percolation problem was brought back to standard procedures of phase transition physics (partition function, etc...). Moreover, the even more formal limit $N \to 0$ was shown to describe the statistics of trees on a lattice.

Since then, the set of correspondences found between thermal and geometrical problems, via mappings of generating functionals and analytic continuations, has much expanded (Lubensky[1]), other instances being mentioned below.

2. Localization

The localization problem[1] was initiated by P.W. Anderson (1958) and it may be described as a quantum percolation problem. Consider the Schroedinger equation with a random potential. The eigenstates may be localized or extended (eigenfunctions square integrable or not). As a function of eigenenergy E, a transition may occur between localized behaviour ($E<E_c$) and extended behaviour ($E>E_c$), where E_c is a mobility edge.

Again, this problem appears quite different from thermal phase transition problems.

Many attempts and conflicting guesses have been made to establish correspondences with other problems (including the self avoiding walk and spin glass problems, to be introduced later), and it is not yet clear whether the localization problem will maintain an independent status. A presently debated question is whether, in two dimensions, all eigenstates are localized or whether a mobility edge may exist.

In the shade behind the localization problem, stands the Fermi glass problem (Anderson[1]), where interacting electrons move in a random potential.

3. Polymer Statistics

This is a field which has considerably developed[1], since the early works of P.J. Flory and other chemical physicists. The simplest problem has to do with the statistics of one linear polymer, taking into account excluded volume effects, which can be reformulated as the statistics of self-avoiding random walks on a lattice. Concentrated solutions are also of interest. Linear polymers are made with bi-functional monomers : with poly-functional monomers, one can get branched polymers (with or without loops). Besides, the phenomena of gelation and vulcanization[1] appear .

The work of S.F. Edwards initiated the "bending back" process which has been flourishing. de Gennes (1972) established a correspondence between the self-avoiding walk and the $n \to 0$ limit of the phase transition problem for n-component spins. The correspondence with phase transitions was later deepened and extended to finite polymer concentrations by des Cloizeaux (1974). Correspondences between the problems of gelation and vulcanization (with or without loops) and the percolation problem (or the general Potts model) have also been established (Lubensky[1]).

The problem of topological entanglements (links or knots) of ideal polymer rings is presently receiving increased attention. We shall consider it more closely below, because of its connections with spin glasses and gauge fields.

4. The Two-Dimensional Planar Spin Model

The systems considered in this paragraph have created puzzles of a different nature than the previous ones (there, the systems had a feature of geometrical randomness, with no obvious similarity with standard thermal phase transitions and the progress involved the establishment of formal equivalences). Here, the problems (of which the best representative is the two-dimensional planar spin model, named hereafter for short the 2D planar model) are formulated in a standard way but the physical properties are found to be unusual.

The spontaneous magnetization vanishes at any finite temperature in the 2D planar model (two-dimensional spins on the sites of a two-dimensional lattice, with usual exchange interactions) because of spin wave excitations : therefore there is no long-range order, no symmetry breaking of the usual type in ferromagnetic systems. However, the magnetic susceptibility diverges at a temperature T^*, and remains infinite below it (H.E. Stanley, T.A. Kaplan, 1966). Is there a phase transition or not at T^*? This question may appear semantical, but it is not. It is easy to construct a model (such as the Gaussian planar model, or the Harmonic lattice model) where the two previous features (no long-range order at any finite temperature, divergence of a response function below some temperature) are found, but which everyone agrees to consider "trivial", because the free energy (exactly calculable in these cases) is perfectly analytic as a function of temperature.

The question whether the 2D planar model exhibits trivial behaviour (as in the Gaussian planar model) or not has prompted much study and there is still no general consensus. However, an interesting picture has emerged (particularly developed by J.M.Kosterlitz and D.J. Thouless [2]), where the usual role of symmetry breaking (to bring about ergodicity breaking in the low temperature phase) is somehow taken up by topological excitations. More precisely, the topologically stable defects for two-dimensional spins in two-dimensional space are point vortices which are classified by an integer. The free energy for creation of a vortex diverges, at low temperatures, as $\lambda(T)$. Log R, where R is the size of the sample, but $\lambda(T)$ vanishes at T^*. Therefore the transition at T^* appears as an insulator-metal or bound-free transition, with bound vortex-antivortex pairs at low temperatures and free vortices at high temperatures.

Other problems of a similar nature, where topological defects are expected to play an essential role, include the problems of two-dimensional melting, transitions in adsorbed layers and superconducting films. Let us remark, en passant, that in superconducting films (or in the Smectic A-Nematic transition in three dimensions) the gauge invariance (or approximate gauge invariance) introduces special features.

As a final comment, the surface roughening model, which had been introduced to describe crystal growth, appears to be related to a variant (the periodic Gaussian model of Villain) of the 2D planar model, via duality relations.

5. Gauge Field Transitions

Retrospectively, it appears that several paths lead naturally from standard phase transitions to gauge field transitions. Let us just list three such paths :

i) duality correspondences,
ii) random magnetic systems (spin glass problems),
iii) topological defects and the description of ordered media with a finite concentration of them.

Let us first introduce briefly the gauge field problem, in its lattice formulation[3]. A standard magnetic problem is obtained by defining variables (the spins $\vec{S}_i$) on the sites of a lattice, interactions J_{ij} on the bonds, and a Hamiltonian :

$$H = - \sum_{<ij>} J_{ij} \vec{S}_i . \vec{S}_j \quad .$$

In the lattice formulation of a gauge field problem, the variables J_{ij} are defined on the bonds of a lattice, the interactions I_p on the elementary loops or plaquettes of the lattice and the Hamiltonian (for simplicity, in the Ising case) has the form

$$H = - \sum_p I_p (JJ...J)_p \quad ,$$

where the product of J, indexed by p, is taken along the perimeter of plaquette p (this product of J along a closed loop is the Wilson loop function or frustration function for the loop; a plaquette is said frustrated if its product is positive and unfrustrated if it is negative). The Hamiltonian has a local (gauge) invariance property (H is invariant if the J values on all bonds adjacent to an arbitrary site are simultaneously flipped).

i) Duality. The duality path to gauge fields was first explored by L.P. Kadanoff and H. Ceva[4] and by F.J. Wegner[5], although curiously the term gauge does not appear in their papers and they seem not to have been bothered with that connection. Duality relations in statistical physics establish a correspondence between the high-temperature phase of some model and the low-temperature phase of some other model. Sometimes, the other model is the same model : this is a case of self-duality, and it occurs for the square two-dimensional Ising model (H.A. Kramers, G.H. Wannier, 1941), allowing in that case the exact determination of T_c (because there is only one phase transition). Such a self-duality relation is somewhat puzzling because it establishes a correspondence between a high-temperature phase and a low-temperature phase, which the "background" picture of standard phase transitions presents as qualitatively different. More precisely, the order parameter, for the square 2D Ising model, has a dual : the "disorder" parameter, which is obviously worth examining (and it is strange that thirty years elapsed before it was examined). The definition of the disorder parameter in the 2D Ising model (we assume that all interactions are nearest-neighbour, ferromagnetic and of uniform value) involves the consideration of a modified system which differs from the original one by having one frustrated plaquette (all plaquettes are unfrustrated in the original system). The average value of the disordered parameter is the ratio of partition functions for the modified system and the original system. The two-point correlation function of the disorder parameter is the ratio of partition functions for a modified system, with two frustrated plaquettes, and the original system. It is easily checked that the disorder parameter behaves as it should and that it characterizes the phase transition just as well as the order parameter. Moreover, it appears clearly that the disordered parameter is directly coupled to the topological defects (which are lines in the 2D Ising model), a frustrated plaquette acting as source or sink of a line defect. Thus, for this model, the concept of disorder parameter is both a neat way of describing the physics of defects and a concept as rich as its dual, the order parameter.

This raises the question whether it is possible to define a disorder parameter, for any phase transition, even when there is no self-duality guide (or self-duality, but no unique transition). A question to be re-examined later

In three dimensions, duality leads from the Ising problem on a cubic lattice to the gauge field problem on the same lattice. A characterization of the gauge phase transition is obtained from the asymptotic behaviour of the Wilson loop function (e^{-L}, where L is the loop perimeter, in the low temperature phase and e^{-S}, where S is the area enclosed by the loop, in the high temperature phase). Indeed, duality relates the Wilson loop function to what appears to be a natural generalization of the Ising disorder parameter correlation function in going from 2 to 3 dimensions. Remember that in $d = 2$, this function was obtained by considering a modified Ising system with two frustrated plaquettes. In $d = 3$, an isolated frustrated plaquette cannot exist because of a geometrical conservation law for the frustration, and one is led to consider a ring of frustrated plaquettes. The disorder parameter correlation function for the 3-dimensional Ising model is therefore defined as a loop function, the ratio of partition functions for the modified system, with one loop of frustrated plaquettes, and the original system.

The connection between usual phase transitions and gauge field transitions, because it is iterable, reveals the existence of a whole hierarchy of problems (Table I). In order to make this clearer, let us introduce some vocabulary : a usual phase transition problem will be called a matter problem, a gauge field transition problem will be called a gauge problem. Now, in the Ising case (on cubic lattices), we see that there exists a hierarchy of problems, that we shall call the p-hierarchy, the first of which are the matter problem ($p = 1$) and the gauge problem ($p = 2$). The disorder parameter of the p-problem is defined by introducing into the system the defects of the (p+1)-problem, which act as sources or sinks (boundaries) of the defects of the p-problem. For cubic lattices, duality brings relations between a (d,p) problem and a (d,d-p) problem (for lattices which are not self-dual, duality relates a (d,p) problem on one lattice and a (d,d-p) problem on the dual lattice).

TABLE I - The p-hierarchy of problems in the Ising case

Type p of problem	Variables defined on	Interactions defined on	Hamiltonian	Exact symmetry	Mixed symmetry	Order parameter correlation function defined on	Asymptotic behaviours low temp.	Asymptotic behaviours high temp.	Dimension of defects	**Disorder parameter** correlation function defined on	**Asymptotic behaviours low temp.**	**Asymptotic behaviours high temp.**
p = 1 matter problem (usual phase transitions)	sites	bonds	$H = \sum_{\text{bonds}} J_{ij}S_iS_j$	global $S_i \to - S_i$ on all sites	gauge $S_i \to - S_i$ $J_{ij} \to - J_{ij}$ i arbitrary j neighbour	2 points sphere Σ_o	$\to C$ when $R = \lvert \vec{r}_1 - \vec{r}_2 \rvert \to \infty$	e^{-R}	d - 1	minimal ensemble of frustrated plaquettes sphere Σ_{d-2}	$e^{-V_{d-1}}$	$e^{-V_{d-2}}$
p = 2 gauge problem	bonds	plaquettes	$H = \Sigma I_p (JJ..J)_p$ plaquettes	gauge $J_{ij} \to -J_{ij}$ i arbitrary j neighbour	hyper gauge $J_{ij} \to -J_{ij}$ $I_p \to -I_p$ <ij>arbitrary p adjacent	loop sphere Σ_1	e^{-L}	e^{-S}	d - 2	minimal ensemble of frustrated plaquettes sphere Σ_{d-3}	$e^{-V_{d-2}}$	$e^{-V_{d-3}}$
p (p<d)	$S_{(p-1)}$	$I_{(p)}$	$H = \Sigma I(SS..S)$	$S_{(p-1)} \to -S_{(p-1)}$ adjacent to a (p-2) object	$S_{(p-1)} \to -S_{(p-1)}$ $I_{(p)} \to -I_{(p)}$ all(p) adjacent to the (p-1)	Σ_{p-1}	$e^{-V_{p-1}}$	e^{-V_p}	d - p	Σ_{d-p-1}	$e^{-V_{d-p}}$	$e^{-V_{d-p-1}}$

velocity was not anymore a gradient, its three components should be taken as independent hydrodynamic variables. I.E.Dzyaloshinskii and G.E. Volovik noticed recently[9] that if the superfluid velocity is considered as a gauge field (much like the electromagnetic vector potential), then topological conservation laws for the vortices are automatically satisfied, which shows that the gauge language is indeed a natural language to describe these materials[10].

The case of superfluid Helium 4 (as well as the case of Ising models discussed above) has a simple feature : the symmetry group involved is an Abelian group, U(1) (in the Ising case, remember, it is the two-element group Z_2). This property allows simple derivation of duality relations[11], and thereby some guidance for the definition of disorder parameters. Another simple feature is the fact that the vortices have a topologically additive index (in other terms, $\pi_1[U(1)] = Z$). This property (not extant in the Ising case, since $\pi_0[Z_2] = Z_2$) allows simple transfer from homotopy to homology and the use of differential forms[10]. Within this scheme, the p-hierarchy for U(1) problems appears quite naturally. Conventional electrodynamics appears as a p = 2 system without defects much like irrotational superfluid Helium 4 is a p = 1 system without defects (the natural defects of electrodynamics being points in space : magnetic monopoles).

Some confusion has arisen in the past (in the context of spin glasses) between the introduction of gauge fields as static variables to describe quenched disorder (item ii) and the introduction of gauge fields as hydrodynamic variables to describe the topological excitations of a medium which, besides, may be homogeneous or random (item iii).

6. Spin Glasses

The term spin glass (B.R. Coles, 1965) has been given to a class of random magnets which exhibit several characteristic experimental properties (behaviour of the magnetic susceptibility $\chi(T)$ and of the specific heat, remanence effects,...)[1]. The name itself (glass) reveals that these systems were considered to undergo a glassy transition, that is a "false" (rounded, non equilibrium) transition. Though it is true that these systems share common properties with glasses (remanence effects), it was found later on (V. Cannella, J.A. Mydosh, 1972) in their low-field magnetic susceptibility a sharp peak at a definite temperature T_{SG}.

This raised considerable interest because it appeared that spin glass phase transitions were standing in an intermediate position between usual phase transitions and glassy transitions, which opened the possibility of a new type of phase transition and of an eventual new approach to glassy transitions (could the glassy

Incidentally, this parameter p appears as a nice candidate for the role of "hidden parameter" (a role played in other problems by the space dimension d, or n or N in systems with O(n) or SU(N) symmetry,...).

ii) Random magnets. The problems in random magnets can be divided into two classes[1]. One is concerned with the modification of critical properties of homogeneous phases when some heterogeneity is introduced (weak randomness). The other is concerned with the appearance of new phases (so-called spin glass phases) and their characterization (strong randomness).

Although the physical systems of interest may present a variety of characteristics (impurities on random sites, long range forces, etc...), the attention has been largely focused on the following random-bond Hamiltonian :

$$H = - \sum_{<ij>} J_{ij} \vec{S_i} . \vec{S_j} \quad ,$$

where the J_{ij} interactions can take positive, zero or negative values, according to some random probability law. If the J_{ij} take positive or zero values, this is a model of dilution randomness[1], which is related to the percolation problem, and does not lead to the appearance of new phases. However, if the J_{ij} can take positive or negative values, opposition effects between the bonds result, for which it is natural to introduce the gauge concepts of plaquettes, frustration functions, etc...[6]. An important fact to keep in mind is that, in the random magnet problem, the J_{ij} are quenched random variables, whereas in a gauge field problem the J_{ij} are thermodynamic (annealed) variables. However, it is generally quite instructive to compare the phase diagrams of matter and gauge field problems and of random magnets problems.

An important question is whether at the pure gauge field transition the energy exhibits a discontinuity (first order transition) or not (second order)[3,7]. A similar question is asked in the random magnet case and for fully frustrated lattices (see below).

iii) Ordered media containing a density of topological defects. For various materials, it is possible to obtain ordered phases with a density of topological defects. The best examples are crystals for which metallurgical techniques allow the control of the density of dislocation lines (a topic of much technical importance) and superfluid Helium 4 for which a varying density of vortices can be obtained either by putting the liquid in a rotating vessel or by reaching a turbulent stage in a non-equilibrium flow.

In this last case of superfluid Helium 4 with vortices, it was recognised empirically long ago[8] that, since the superfluid transition be close to some "true" transition ?).

Before reviewing some of the work motivated by the spin glass problem, it seems fair to say that it has opened a Pandora box of surprises, paradoxes and mysteries, which make the topic a presently very fascinating one, even if the final physical outcome turns out to be dull, as predicted by some reluctant experimentalists.

One direction of study has been the search for a mean-field theory of the spin-glass transition. The use of the replica trick ($n\to0$), appropriate for quenched randomness (S.F. Edwards, P.W. Anderson, 1975), allows in principle to solve simply the limit case where each spin interacts with every other spin (this is the Sherrington-Kirkpatrick model, 1975, where moreover the bond distribution probability is taken Gaussian). This is in effect a limit of infinite dimensionality for which mean-field theory should apply. But, much to the surprise of everyone, an unacceptable negative zero-point entropy has been found, unless the replica symmetry is broken (which is mysterious enough) and moreover in a very strange way (A.J. Bray and M.A. Moore ; A. Blandin ; G. Parisi). A breakdown of the classical fluctuation-dissipation theorem (a bit like in quantum systems ?) has also been evoked (Bray and Moore). In brief, the present situation is intolerable and therefore promising.

Outside of the replica scheme, there are also doubts on the general validity of the mean-field equations (defined again as equations correct in the limit of infinite dimensionality) which have been proposed (Thouless, Anderson, Palmer), even within the simple framework of spin exchange Hamiltonians (B. Derrida, 1979). Besides, the free energy of the spin glass phase is found to be higher than the continuation of the free energy of the paramagnetic phase (an unusual feature, which however is not totally unprecedented since it is found also in the ideal Bose condensation at constant volume, as noted by P.W. Anderson[1]).

Another line of work has tried to tackle spin glass models in more realistic dimensions. Numerical simulations[1] have brought a great deal of information, although simple questions like the determination of the range of the ferromagnetic phase in the phase diagram, or the value of the lower critical dimensionality of the spin glass phase (for Ising, planar or Heisenberg spins), and the existence of ergodicity breaking, have not yet been fully answered. Indeed it is even questioned whether there may be only one or several sharply different types of spin glass phases in the same phase diagram. In particular, the existence of a mixed ferromagnetic-spin glass phase, a sort of continuation of the spin glass phase within the ferromagnetic phase, has been suggested in various contexts.

Renormalization group expansions around the upper critical dimensionality (presumed to be d = 6) have also produced oddities, such as complex critical exponents[1].

The gauge concepts of frustration have been used, in order to describe the quenched disorder of the interactions, and in view of characterizing the ground states and the elementary excitations for Ising and continuous spins. J.A. Hertz[12] has studied a continuum Landau-Ginzburg model, for continuous spins, which incorporates the concept of frustration. His renormalization calculations (around d = 4) seem to indicate that the unfrustrated fixed point is unstable against the introduction of a small amount of frustration, and that, no new fixed point appearing, the system renormalizes towards maximum frustration.

7. Fully Frustrated Lattices

The spin glass models having raised so many riddles, it is natural to examine models somewhat intermediate between those and usual phase transition models. Such appear to be the fully frustrated models, where the mixture of frustration and randomness has been disentangled by deleting the latter[13]. Moreover, there exists a number of homogeneous (periodic) materials which have a high degree of competition between nearest-neighbour interactions of various range[14]. Incidentally, fully frustrated systems are also a natural limit of the dislocated antiferromagnet considered by I. Dzyaloshinskii (1977) where the dislocation lines of the lattice constitute a frustration network for the spins.

Strictly speaking, fully frustrated models are such that interactions are nearest-neighbour only, of fixed magnitude, and so distributed that every plaquette of the lattice is frustrated. This is always realizable with a regular, periodic, configuration of bonds and the fully frustrated models are therefore the simplest models for homogeneous systems with a high degree of bond competition. Their systematic study has begun only recently but indeed they do present, in a simpler framework, some of the oddities encountered in spin glass models. Besides they correspond to one limit boundary in the phase diagrams of lattice gauge theories.

III. UNIFYING TOOLS

1. Equivalences Between Generating Functions

Such equivalences have been mentioned about percolation and polymer statistics. A good example is the equivalence between the excluded volume problem in linear polymer statistics and the $n \to 0$ limit of the n-vector model for phase transitions (de Gennes, 1972). In the latter case, a diagrammatic expansion of the two-point order parameter correlation function exhibits diagrams which look precisely like polymers with monomer interactions, provided some diagrams (those with loops) are discarded, a goal which is achieved by taking the limit $n \to 0$.

2. Renormalization Group

The introduction of renormalization techniques in the physics of critical phenomena has given a basis to the notions of universality and irrelevant variables, which are powerful unifying concepts indeed.

A case which has been well underlined by T. Lubensky[1] concerns the phase transitions in heterogeneous materials (weak randomness). Prior to renormalization analyses, it was generally believed that phase transitions would always be rounded in heterogeneous materials on the basis that the material would behave as a collection of independent samples with a distribution of T_c values. Renormalization studies showed that the fixed point of the homogeneous system could indeed be stable with respect to weak randomness, in this case an irrelevant variable ; in other cases, weak randomness was found to be a relevant variable leading to another fixed point, with different exponents. So, in many cases, the phase transition would indeed remain sharp.

In a different vein, the Migdal (1975) approximate recursion relations for matter field (n-vector model, $n \geqslant 2$) transitions in $d \sim 2$ have revealed a suggestive analogy with the corresponding gauge field transitions in 2D.

3. Topological Excitations

The classification of topologically stable excitations (generally called defects and textures, or non singular configurations, in the matter problems ; monopoles, or strings, and instantons in the gauge problems;...) has been shown to be provided by the homotopy groups of some manifold (the manifold of internal states in matter problems, the gauge group in gauge problems, ...). Therefore their existence is related to very general considerations of symmetry, with a high degree of universality.

Moreover we have seen that, in many cases, a phase transition could be described via a disorder parameter, which reflects the physics of the topological excitations. As a consequence, two transitions which have topological defects of a similar nature (such as points, or lines,...) will share some analogies. Such an analogy has been repeatedly noted, for instance, between a matter problem in dimension d and the corresponding gauge problem in dimension (d+1), for Abelian symmetry groups.

Indeed, some of the difficulties with non-Abelian symmetry groups appear clearly within this topological analysis : the symmetry breaking for matter problems can occur in different ways (in the case of SO(3), one may have ferromagnets, or nematics, or non-colinear antiferromagnets), for one matter phase there may exist several types of topologically stable defects (in the case of nematics, there are lines and points). This introduces ambiguities in the choice of disorder parameters and breaks the simple p-hierarchy of Abelian problems.

Behind these difficulties, there is presumably a richer physics. Thus in the case of 2D crystals, the existence of two distinct kinds of topological defects : point dislocations and point disclinations, has led to predict the existence of two melting transitions (B.I.. Halperin, D.R. Nelson, 1978), with distinct order and disorder parameters.

4. Duality Relations

The duality relations described above in the Ising case (Wegner, 1971) have a direct generalization for any Abelian symmetry group (Drouffe, 1978). To see this, it is necessary to use the notion of duality for groups. The dual of a group G is the ensemble of its irreducible representations. If the group G is Abelian, its dual ensemble has a natural group structure, and the dual of its dual is the group G itself.

There are duality relations between a (d,p) problem with symmetry group G on a lattice L and a (d,d-p) problem with dual symmetry group G^* on the dual lattice L^*. These relations are direct generalizations of the Ising case, and the high temperature phase of one problem is mapped into the low temperature phase of the dual problem.

The discrete groups, Z_2 (Ising) and more generally Z_n, are self-dual. The dual of U(1) is the group of integers Z. This is the basis for the relations between U(1) problems and gases of Coulomb particles (if d = p+1), or Coulomb vortices (d = p+2), etc..

In practice, the game of duality has been brought, in particular in the case of U(1) problems, to a high degree of sophistication, establishing through various algebraic transformations equivalences between one given problem and a multiplicity of others[15].

A starting trick consists in using the Villain formulation (J. Villain, 1975) for any (d,p) lattice U(1) problems : in the case of matter problems (p=1), this is the Periodic Gaussian model which incorporates conveniently a manifest U(1) symmetry. A simplified model is the Gaussian model, where a low temperature approximation of the Periodic Gaussian model is assumed to hold at all temperatures : this, in effect, eliminates the topological defects associated with the U(1) symmetry (one is playing with the fact that U(1) and R have same Lie algebra). At the gauge level (p=2), one has a similar formulation for Periodic Electrodynamics, a mutilation of which corresponds to Conventional Electrodynamics. Both versions have been found subtly related via duality (M.Peskin, 1978).

Another feature of interest is the natural appearance of electric-magnetic duality relations, in d = 3+1, and thus for instance the justification by duality of the "magnetic superconductor" picture for confining phases in Abelian (M. Peskin) or non-Abelian (G. 't Hooft, 1978) gauge theories.

There is a scheme for establishing duality relations which starts from a Hamiltonian (or Transfer Matrix) formulation of the statistical physics problem. With the advantage of reducing the dimensionality of the problem by one unit and the difficulties of doubling the number of variables, it brings an interesting other viewpoint[16].

To my knowledge, no way has yet been found to extend the duality ideas to non-Abelian problems (note however the short paper of J.M. Drouffe, C. Itzykson, J.B. Zuber, Nucl. Phys., B 147, 132 (1979) on the special case of solvable groups).

IV. GAUGE CONCEPTS IN CONDENSED MATTER PHYSICS

1. Matter Field Coupled to a Gauge Field (Higgs-like phenomena).

By Higgs-like phenomenon is meant a mechanism by which a gauge field acquires a mass, through coupling to a matter field. In the case of electrodynamics in condensed matter, this can occur in two characteristic ways symbolized in the following lines :

$$\vec{j}.\vec{A} \rightarrow m^2\vec{A}^2 \quad .$$

or $$\rho V \rightarrow \mu^2 V^2 \quad .$$

Before discussing the archfamous case of the superconductor, it is worth mentioning briefly the metal-insulator transition. In a metal, there are three branches of electromagnetic modes (one longitudinal, two transverse) which go to finite frequency, in the limit of small wavevectors q→0. This is the famous plasma frequency ω_p, given by

$$\omega_p^2 = \frac{4\pi ne^2}{m} \quad ,$$

where e,m,n are the electron charge, mass and density. This finite plasma frequency can be interpreted (P.W. Anderson, 1963) as a mass acquired by the photon in a metal. It is related to the phenomenon of charge screening : an extra charge inserted in a metal is screened by a modification of the electron density, so that the charge appears to be effectively expelled toward the surface. Actually, the plasma frequency ω_p can be written as :

$$\omega_p = V_F . \lambda_{T.F.} \quad ,$$

where V_F is the Fermi velocity and $\lambda_{T.F.}$ is the Thomas-Fermi screening length. In a metal, the Poisson equation :

$$\Delta V = \rho \quad ,$$

because $\rho \sim \frac{V}{\lambda_{TF}^2}$, takes the following form :

$$\Delta V \sim \frac{V}{\lambda_{TF}^2} \quad ,$$

implying that an external charge is screened over distances of the order λ_{TF} . Note that a metal-insulator transition is generally similar to a liquid-gas transition for a fluid, in the sense that it is possible to go in a completely smooth way from one phase to the other.

In a superconductor, the (London) equations are analogous, up to an electric-magnetic interchange. Ampère's equation :

$$\overrightarrow{\mathrm{rot}}\, H = \vec{j} \quad ,$$

because $\vec{j} \sim \frac{\vec{A}}{\lambda_L^2}$, takes the form :

$$\mathrm{rot}\, \overrightarrow{\mathrm{rot}}\, A \sim \frac{\vec{A}}{\lambda_L^2} \quad ,$$

implying that an external magnetic field is confined over distances of the order of the London penetration length λ_L (Meissner effect). Note that,in contradistinction to the previous case, a superconducting-normal transition is rather similar to a solid-liquid transition, in the sense that is not possible to transform smoothly one phase into the other.

If one compares superfluid Helium 4 (a neutral superfluid) with superconductors (charged superfluids coupled to the electromagnetic field), one observes some qualitative differences :

i) there are Type I and Type II superconductors (with different phase diagrams in the presence of a magnetic field),
ii) the superconductor-normal transition is predicted[17] to be

first order, in dimensions $2<d\leqslant 4$, because the fluctuations of the electromagnetic field, when summed over, add to the Ginzburg-Landau free energy a term proportional to

$$|\psi|^d \quad ,$$

where ψ is the superfluid order parameter (a complex number). Note that this problem (sometimes called scalar electrodynamics) has conventional electrodynamics in it and therefore differs from the matter gauge problem in lattice U(1) theories where it is periodic electrodynamics which enters.

iii) the energy of the topological defects (vortices) is reduced, as a direct consequence of the gauge invariance. Thus the energy of a vortex ring of length L (in a 3 dimensional sample), which is of order $\sim$ L Log L for a superfluid, is of order $\sim$ L in a superconductor, and (in a 2 dimensional sample) the energy of a vortex point which diverges like Log R (where R is the size of the sample) in a superfluid becomes finite in a superconductor. As a consequence of what has been said earlier, this seems to imply that whereas a two-dimensional superfluid sample may have a phase transition (of the type discussed for the 2D planar model) a strictly two-dimensional superconductor sample should not have any. Actually, it has been recently predicted that for actual superconductor films (of finite thickness), a quite rich phase diagram could be observed, with fascinating phenomena due to the properties of vortices (S. Doniach, B. Huberman, 1979).

There exists, in the realm of liquid crystals, a transition, the smectic A-nematic transition, which has a partial analogy with the superconductor-normal transition, as first noted by de Gennes (1972). The smectic A order parameter describes the lamellar order (a matter density wave) and it is a complex number with an amplitude and a phase ϕ. In the nematic phase, the elongated molecules are aligned and the order parameter is a director n. Now, the Ginzburg-Landau free energy for the smectic A contains the following elasticity terms :

$$F \sim (\nabla_z \phi)^2 + (\vec{\nabla}_\perp \phi - \vec{\delta n})^2 + O\left[(\nabla n)^2\right] \quad ,$$

where z is the direction normal to the smectic layers. The middle term has a typical gauge invariant form and expresses the physical fact that a transverse fluctuation of the phase ϕ can be compensated by a rotation of the alignment axis of the molecules. As a consequence, the thermal fluctuations of the phase ϕ are such that

$$< |\phi_k|^2 > \sim \frac{T}{k_z^2 + k_\perp^4} \quad ,$$

and, when integrated over k-space, they are seen to give an infrared divergent contribution below d=3 (whereas in usual systems, with continuous order parameters, this occurs below d = 2). Therefore, this quasi-gauge invariance has the effect of pushing up the lower critical dimensionality from d = 2 to d = 3 (a situation intermediate between pure matter and pure gauge problems). It leads also to an anisotropic screening of the topological defects (dislocation lines) in the smectic A phase. Recently, it has been found possible (B. Malraison, Y. Poggi, E. Guyon, 1979) to measure directly the transverse fluctuations of the nematic order parameter (our "photon" field) in the nematic phase and to observe how they become quenched (a precursor of the Higgs effect) when one approaches the Sm A-Nem transition from above. Note that the nematic "photon" field admits defects, a feature with some analogy with periodic electrodynamics.

2. Spin Glass Models

We consider magnetic systems described, typically, by the following exchange Hamiltonian :

$$H = - \sum_{<ij>} J_{ij} \vec{S_i} . \vec{S_j} \quad .$$

The spin glass problem is then most simply summarized as the elucidation of a phase diagram (T,c), where T is the temperature and c is a measure of randomness or frustration in the interactions of the above Hamiltonian. For instance, c might be the concentration of antiferromagnetic bonds.

A model was proposed by D.C. Mattis (1976) where

$$J_{ij} = J \varepsilon_i \varepsilon_j \quad ,$$

and $\varepsilon_i = \pm 1$ is a site random variable. Such a model is trivially soluble in terms of properties of the pure homogeneous system and this made it clear that the serious properties due to randomness were related to frustration, that is deviations from pure gauge transformations.

Assuming thus that the bond interactions are random variables (taking values ± 1), we see that the phase diagram (T,c) will be qualitatively different on lattices with even or odd number of bonds per plaquette. Indeed, in the antiferromagnetic limit $c \to 1$, the square lattice is not frustrated whereas the triangular lattice is fully frustrated.

Consider now the phase diagram (β_m, β_g) for a typical (say Ising) matter gauge lattice problem. The action is taken as

$$S = \beta_m \sum_{<ij>} J_{ij} S_i S_j + \beta_g \sum_p (JJ...J)_p \quad .$$

In the limit $\beta_g \to \infty$, the frustration is frozen and one recovers the pure matter problem (that is the limit $c \to 0$ of the previous problem). When β_g decreases (or when c increases) the frustration increases gradually but there is a difference between the two problems, because in one problem the J_{ij} are quenched variables, whereas in the other the J_{ij} are thermodynamic (annealed) variables. Indeed in the limit $\beta_g \to 0$, of the matter + gauge problem, which corresponds to a density $\frac{1}{2}$ of frustrated plaquettes, there is a trivial solution and no transition as a function of β_m. Whereas, in the disordered magnet, this density of frustrated plaquettes is achieved for $c = \frac{1}{2}$ (on lattices with even number of bonds per plaquette) and it is expected that a spin glass-paramagnet transition occurs as a function of T.

There is however another limit which is of interest in both problems and this is the limit of full frustration. In the matter + gauge problem, this is obtained in the limit $\beta_g \to -\infty$ for lattices with even number of bonds per plaquette (there is a symmetry $\beta_m \to -\beta_m$ in the phase diagram) or in the limit ($\beta_g \to +\infty$, $\beta_m < 0$) and ($\beta_g \to -\infty$, $\beta_m > 0$) for lattices with odd number of bonds per plaquette (there is a symmetry $\beta_m, \beta_g \to -\beta_m, -\beta_g$ in the phase diagram).

The closest connection between the Z_2 matter + gauge field diagram and the Ising spin glass problem appears now to be the following. The line ($\beta_g = 0$) corresponds to the annealed version of the spin-glass problem ; in this case, the interactions J_{ij} are thermodynamic variables and develop correlations which are best measured by the frustration function W taken around one plaquette. Now, in the diagram (β_m,β_g) one can draw the lines of constant W. In particular, there is an iso-W line for which W = 0, passing by the points ($\beta_m = \beta_g = 0$) and ($\beta_m = +\infty$, $\beta_g = -\infty$) : this last information comes from the knowledge gained in the study of fully frustrated systems. On this line, the matter + gauge system mimics best a quenched spin glass, because the plaquette frustration function is temperature independent, as it is, obviously, when the bonds are frozen. Up to now, there has been no numerical study of the relevant quadrant of the diagram (β_m,β_g). However, if it bears any similarity with the ($\beta_m,\beta_g > 0$) quadrant, there may exist in it a transition line terminating at a critical point. This would offer the possibility for the W = 0 line either to cross this transition line (with a true transition as a function of temperature) or to pass in the vicinity of the critical point (with a rounded transition). This promising approach is presently under investigation.

The frozen frustration model gives a particularly simple picture of Ising disordered magnets, with the frustration network acting as sources or sinks of defects, and giving rise to a large ground state degeneracy. In two dimensions, on a square lattice, the partition function of a disordered magnet with N frustrated plaquettes is related by duality to the N-point correlation function of the pure magnet[18]. This is not a priori very helpful because N has to be infinite for finite concentration in the thermodynamic limit. However, H.G. Schuster[19] has predicted on such grounds a transition between two spin glass phases which, at zero temperature, would occur for a concentration $c \sim 30\ \%$ of negative bonds. Note that there is still uncertainty in the determination of the critical concentration for disappearance of ferromagnetism, values ranging from 9 % to 16 % having been obtained by various criteria[20].

In the case of continuous spins, the difference between the random magnet problem and the corresponding matter + gauge problem increases. The reason is that in the random magnet, the bonds keep taking real values, whereas in the gauge problem, for continuous symmetries, the bonds take values within the symmetry group. Thus, in three dimensions, for planar spins (symmetry U(1)), a ring of frustrated plaquettes in the random magnet problem gives rise to a half-vortex ring for the spins (J. Villain, 1977), whereas a topological defect of the gauge field is a monopole which acts as source of a vortex of integer strength.

This somewhat mixed situation where the frustration network produces canting of the spins, without however providing natural sources and sinks of topological defects, justifies an analogy between the frustration network for planar spins and a conventional (non periodic) electrodynamic gauge field. This is actually the content of the Hertz model (1978), where a Ginzburg-Landau free energy is written as :

$$F = \frac{1}{2} \int d^dx \left[r_o |\phi(x)|^2 + \frac{u}{4} |\phi(x)|^4 + |(\vec{\nabla}-i\vec{Q})\phi(x)|^2\right] ,$$

where $\phi(x)$ is a complex number representing the magnetization at point x and $\vec{Q}(x)$ is a quenched random variable with probability distribution :

$$P|\vec{Q}| \sim \exp(-\frac{1}{2f} \int d^dx \sum_{\mu\nu} F^2_{\mu\nu}(x)) ,$$

where

$$F_{\mu\nu}(x) = \partial_\mu Q_\nu(x) - \partial_\nu Q_\mu(x)$$

This continuum formulation of the problem is obviously very close to the problem of scalar electrodynamics, discussed earlier in another context, with still the difference that the "gauge" field is here quenched instead of being a thermodynamic variable. In his renormalization calculations around dimension 4, Hertz found that, for $d > 4$, f renormalizes to zero, i.e. the frustration is an irrelevant variable. However, for $d > 4$, f grows without limit, a runaway interpretable either as a first-order transition or as no transition at all. Further study of this model, maybe with other techniques, appears desirable.

There is an obvious formal generalization of this model for the case of Heisenberg (three-component) spins, which involves the introduction of a (quenched) Yang-Mills "gauge" field of the group SO(3). (Note however that there is no equivalent of conventional electrodynamics for a non Abelian group like SO(3) because its Lie algebra is not compatible with a non compact group).

Such an introduction had been earlier suggested by Dzyaloshinskii and Volovik (1978), but in a different formulation where the gauge field appeared more like a thermodynamic variable. It seems that the justification for the introduction of a gauge field, in one formulation or another, must come from a deeper study of how the frustration network distorts the spins and how it effects their topological excitations.

One obvious effect[21] is to create packets of hirsute spins so that the effective manifold of internal states is (at least) enlarged from S_2 to O(3) (in the case of planar spins, the enlargement was less spectacular, from SO(2) to O(2)). This suggests that the number of local degrees of freedom has increased and is a justification for the introduction of new thermodynamic variables. Actually, it is not a priori decidable whether these packets of spins have themselves internal degrees of freedom, the total number of which would add up to the number of components of an SO(3) gauge field. This has probably to be decided by "experiments", that is by careful numerical simulation studies of the topological excitations in various parts of the phase diagram of the Heisenberg random magnet.

Indeed, this problem of formation of hirsute packets and of their internal degrees of freedom exists also on fully frustrated periodic lattices. These systems, where randomness has been eliminated, deserve more study. As a very simple example, it is not yet known whether the triangular antiferromagnetic lattice, with planar spins, exhibits zero, one or two phase transitions (with Heisenberg spins, there is probably none). In three dimensions, it is not yet known whether the fully frustrated simple cubic lattice with Ising spins has a phase transition (on the fcc lattice, there is a first order transition[22]), and the case of continuous spins has not yet been examined.

Note finally some first studies of percolation on fully frustrated lattices, which is a simple way of introducing randomness[23].

3. Entanglements of Polymer Rings

By itself, the problem of entanglements of polymer rings incorporates enough physics to deserve interest. The proof[24] that linear topological defects in some three-dimensional materials may encounter topological obstructions for crossing (and generalizations thereof in higher dimensions) adds another motivation. Moreover, a formal connection with a spin glass model has recently been discovered[25]. And finally, it carries the charm of topology.

Entanglements may be links (two rings, or more) or knots (one ring alone). Knots raise more mathematical difficulties than links, and we shall begin with the latter.

The topological classification of links of two rings leads to the definition of an infinite list of integer invariants, the first of which is the so-called (Gauss) linking number. This linking number can be expressed as a double loop integral :

$$\ell k(C_1,C_2) = \frac{1}{4\pi} \int_{C_1} \int_{C_2} (\overrightarrow{dr_1} \times \overrightarrow{dr_2}) . \vec{\nabla} \frac{1}{r_{12}} ,$$

and actually mesures the flux through one ring of a unit electric current flowing along the other (or equivalently the magnetic flux of one vortex through another in a superconductor). This language immediately suggests a connection with electromagnetism, that is with U(1) gauge theory. The introduction of gauge field variables in the generating functional for polymer statistics was first done by S.F. Edwards[26], in the context of two-dimensional polymers entangled with a fixed point obstacle.

This problem has been taken up recently, in a three-dimensional context, by M. Brereton[25] who studied the statistics of polymer rings, in a box, entangled with a fixed polymer ring, an average over the position of the latter in the box being taken subsequently. The entanglement constraint is defined by a given linking number, which is a convenient starting hypothesis. Algebraic manipulation of the generating functional leads then to the appearance of an effective free energy which is exactly similar to the Hertz free energy for the planar spin glass, presented above. Quite interestingly, in this approach, the frustration parameter $f^{1/2}$, or "charge" coupling constant, appears as the variable conjugate to the linking number so that, in the limit where the latter is indeterminate, one recovers neatly the free (unconstrained) polymer problem. Physically, a topological constraint of zero linking number leads to a swelling of the polymer ring, only partially analogous to an excluded volume effect (more like the shape of a confined loop).

In the case of knots of one single ring (for which a no-knot constraint leads also to a swelling), there is a difficulty for finding a proper generalization of the linking number, because for two coinciding loops the double integral above is not well defined. This has led to some controversies in the past[27] (leading some authors to devise alternate methods of knot analysis[28]) but I have been told recently (S.F. Edwards, private communication) that a well-defined limiting procedure for the double integral still affords a distinction between classes of configurations of one ring (thesis of Robin Ball, Cambridge, U.K.).

Actually, as is often the case, this difficulty is the sign of a rich underlying structure, with application to the topology of ribbons, which has been put to use in the description of the spatial conformations of DNA (superhelix structure, the role of nicking enzymes,...)[29].

REFERENCES

1. Proceedings of the 1978 Les Houches Summer School on "Ill-Condensed Matter", Eds. R.Balian, R. Maynard, G. Toulouse (North-Holland, 1979).
2. J.M. Kosterlitz, D.J. Thouless, Progr. Low. Temp. Phys., 7 b (1978).
3. K.G. Wilson, Phys. Rev. D 14, 2455 (1974) ;
R. Balian, J.M. Drouffe, C. Itzykson, Phys. Rev. D 11,2098 (1975).
4. L.P. Kadanoff, H. Ceva, Phys. Rev. B 3, 3918 (1971).
5. F.J. Wegner, J. Math. Phys., 12, 2259 (1971).
6. G. Toulouse, Comm. on Phys., 2, 115 (1977) ; Proc. of the XVI Karpacz Winter School 1979 (Springer, to be published).
7. M. Creutz, L. Jacobs, C. Rebbi, Phys. Rev. Letters, 42, 1390 (1979) ;
D. Horn, S. Yankielowicz, preprint.
8. I.M. Khalatnikov, Introduction to the Theory of Superfluidity (Benjamin, 1965).
9. I. Dzyaloshinskii, G. Volovik, J. Physique, 39, 693 (1978); Zh. ETF, 1102 (1978) ; Phys. Rev., to be published.
10. B. Julia, G. Toulouse, J. Physique Lettres, 16, L395 (1979).
11. J.M. Drouffe, Phys. Rev. D 18, 1174 (1978).
12. J.A. Hertz, Phys. Rev. B 18, 4875 (1978).
13. S. Alexander, P. Pincus (1978), to appear in J. Phys. A ;
B. Derrida, Y. Pomeau, G. Toulouse, J. Vannimenus, J. Physique, 40, 617 (1979) ;
B. Derrida, G. Toulouse, Proc. 8th Int. Coll. on Group Theor. Methods in Phys., 1979 (Springer, to be published).
14. J. Villain, Z. f. Physik, B 33, 31 (1979).
15. J.V. José, L.P. Kadanoff, S. Kirkpatrick, D.R. Nelson, Phys. Rev. B 16, 1217 (1977) ;

L.P. Kadanoff J. Phys. A, 11, 1399 (1978) ;
A.M. Polyakov, Phys. Letters, B 59, 79 (1975) ;
T. Banks, R. Myerson, J. Kogut, Nucl. Phys. B 129, 493 (1977);
R. Savit, Phys. Rev. Letters, 39, 55 (1977) ;
M.E. Peskin, Ann. Phys., 113, 122 (1978).

16. E. Fradkin, L. Susskind, Phys. Rev. D 17, 2637 (1978) ;
J. Kogut, Introduction to Lattice Gauge Theories, Urbana Lecture notes, 1979.
D. Horn, M. Weinstein, S. Yankielowicz, preprints.
17. B.I. Halperin, T.C. Lubensky, S.K. Ma, Phys. Rev. Letters, 32, 292 (1974) ;
J.H. Chen, T.C. Lubensky, D.R. Nelson, Phys. Rev. B 17, 4274 (1978).
18. E. Fradkin, B.A. Huberman, S.H. Shenker, Phys. Rev. B 18, 4789 (1978).
19. H.G. Schuster, preprint.
20. J. Vannimenus, G. Toulouse, J. Phys. C 10, 537 (1977) ;
J. Vannimenus, J.M. Maillard, L. de Sèze, to be published ;
S. Kirkpatrick, Phys. Rev. B 16, 4630 (1977) ;
G. Grinstein, C. Jayaprakash, M. Wortis, Phys. Rev. B 19, 260 (1979) ;
R. Rammal, R. Maynard, private communication.
21. G. Toulouse, Phys. Reports, 49, 267 (1979).
22. M.K. Phani, J.L. Lebowitz, M.H. Kalos, C.C. Tsai, Phys. Rev. Lett., 42, 577 (1979).
23. L. de Sèze, J. Phys. C 10, L 353 (1977).
G.S. Grest, preprint.
24. V. Poénaru, G. Toulouse, J. Physique 38, 887 (1977) ;
J. Math. Phys. 20, 13 (1979).
25. M. Brereton, to be published.
26. S.F. Edwards, Proc. Phys. Soc. 91, 513 (1967).
27. S.F. Edwards, J. Phys. A 1, 15 (1968) ;
A.V. Vologodski, A.V. Lukashin, M.D. Frank-Kamenetskii, V.V. Anshelevitch, Sov. Phys. J.E.T.P. 39, 1059 (1974).
28. M.D. Frank-Kamenetskii, A.V. Lukashin, A.V. Vologodskii, Nature, 258, 398 (1975).
J. des Cloizeaux, M.L. Mehta, J. Physique, 40, 665 (1979).
29. F. Brock Fuller, Proc. Nat. Acad. Sci. USA, 68, 815 (1971).
F.H.C. Crick, Proc. Nat. Acad. Sci. USA, 73, 2639 (1976).

APPENDIX I - REMARKS ON ORDER AND DISORDER PARAMETERS

a) Order Parameters

The concept of order parameter was introduced by Landau as a measure of symmetry breaking. Since we have encountered phases where there exists an ergodicity breaking, which does not appear, in any simple way, to be due to symmetry breaking, it is natural to look for an extension of the notion of order parameter, suitable for describing such situations.

For random magnets, S.F. Edwards and P.W. Anderson (1975) have defined an order parameter q as

$$q = \overline{< \vec{S}_i >^2} \quad ,$$

where < > means a thermodynamic average for one configuration of interactions, and the upper bar means an average over the probability law for these configurations. Presumably (always ?), it can also be written as

$$q = \frac{1}{N} \sum_{i=1}^{N} < \vec{S}_i >^2 , \; N \to \infty \quad .$$

Since we wish to consider here the Edwards-Anderson order parameter in the simple context of homogeneous systems, we shall prefer the second expression. The quantity q is non-zero for any phase with broken symmetry, and this makes it useful, for instance in numerical solutions, when one wants to know whether there is some phase transition without any preconception about the symmetry breaking. But it is apparent that this quantity q is of little help in more general situations (ergodicity breaking without symmetry breaking).

Consider the case of the 2D planar model. At low temperatures, the parameter q still vanishes, despite the existence of a phase transition, and despite the ergodicity breaking in phase space between valleys with different global topological indices. (Note that in the context of gauge theories, Elitzur's theorem (S. Elitzur, 1975) says that the corresponding quantity $q = \sum_{<ij>} < J_{ij} >^2$ is always vanishing).

Therefore, the Edwards-Anderson order parameter, although a valuable addition to the panoply of the theoretician, is no universal panacea, even within the context of homogeneous matter problems.

For the 2D planar model, an order parameter can be defined as a loop function

$$< e^{i \int_c \vec{\nabla}\phi . d\vec{\ell}} > \quad ,$$

where ϕ is the phase of the two-component spins[2]. This loop function is supposed to have an asymptotic behaviour

$\sim e^{-L}$, in the low temperature phase,

$\sim e^{-S}$, in the high temperature phase.

This shows that the deficiency of the quantity q, here, is its strictly local character. (Notice that the loop function above is a sort of Wilson loop integral defined for a matter system : more generally, a "topological order parameter" is defined on the surrounding subspaces of the topological defects).

Another sort of difficulty is met with the order parameter in systems where the translation symmetry is broken. In these cases, it does not appear to be possible to define a strictly local order parameter. This difficulty shows up in the classification of topological defects, where metric considerations spoil the purely homotopic classification (Mineyev, Mermin, Poénaru), a complication which does not exist in systems where the rotation symmetry only is broken.

b) Disorder Parameters

It has been seen, in the case of Ising models (matter, gauge, or more generally p-problems), that the disorder parameter correlation function is obtained from a modification of the system which involves the creation of sources and sinks which act as boundaries of the natural topological defects. Thus, for an Ising matter problem, the disorder parameter correlation function is a two-point function in dimension 2, a loop function in dimension 3, etc... whereas for an Ising gauge problem, it is a two-point function in dimension 3, a loop function in dimension 4, etc...

This is generalizable for other Abelian problems. Thus, for a U(1) matter problem, the disorder parameter correlation function is a two-point function in dimension 3, a loop function in dimension 4, etc... whereas for a U(1) gauge problem, it is a two-point function in dimension 4, a loop function in dimension 5, etc...

This is generally satisfactory, but does not tell what to do in some special lower dimensions such as d = 1 (for Ising matter problem), d = 2 (for Ising gauge problem, or U(1) matter problem),.. In many cases, there is no phase transition in such low dimensions, but there are exceptions, such as the U(1) matter problem (or 2D planar model), for which the general definition above is not sufficient.

Another question concerns the way in which sources and sinks are going to be introduced, that is how practically the system is to be modified. In the case of Ising systems, it has been seen that this involved the introduction of frustrated plaquettes, or hyper-frustrated blocks, etc... a result which is achieved by a simple change of sign of some interactions. However, this straightforward procedure, for continuous spins, does not necessarily create sources or sinks of the appropriate dimensionality and, in the case of planar spins for instance, it tends to create topological defects of half-integer strength. Therefore, one is left with the alternative, either of creating a rather complex modification of the interactions, or of imposing suitably chosen boundary constraints. Besides, the reason why definition of the disorder parameter in Ising systems appears simple, is because one has in mind for instance a ferromagnetic matter model (more generally, an unfrustrated model). But if one considers a random Ising system, or even, more simply, a fully frustrated matter system, it is not clear whether the recipe above still works. Thus, in the triangular antiferromagnetic lattice dis-frustrating two plaquettes gives correctly a constant asymptotic behaviour at all temperatures, but the energy of the system is being lowered instead of increased (B. Derrida).

From another standpoint, one may notice that such a modification of the system, introducing sources and sinks, is a generalization of the notion of susceptibility or response function. And the Wilson loop function itself, the order parameter function for gauge problems, can be viewed as the response of the system to the introduction of a closed electric current loop.

In Hamiltonian formulations of matter and gauge problems, the disorder parameter becomes an operator (which is a point-function, or loop-function, etc...) and can be defined by its commutation relations with the order parameter operator[16] ('t Hooft, 1978).

APPENDIX II - TOPOLOGICAL EXCITATIONS

This appendix will be short because its content has been presented at the Lausanne International Conference on Mathematical Physics (August 1979) and will appear in the Proceedings (Springer Lecture Notes in Physics).

The reader is referred to the companion lecture of V. Poénaru. Here is also a list of review articles on this topic :

L. Michel, Proc. of the Tübingen Conf. on Group Theoretical Methods in Physics, 79, (Springer, 1978); Rev. Mod. Phys.(1979).
N.D. Mermin, Rev. Mod. Phys., 51, 591 (1979).
V.P. Mineyev, Sov. Phys. Uspekhi (1979).

APPENDIX III - FULLY FRUSTRATED LATTICES

In a fully frustrated lattice, each plaquette has at least one frustrated bond in any spin configuration (a frustrated bond is a bond for which the product $J_{ij}S_iS_j$ is negative ; we consider for the moment Ising spins). In a ground state configuration, can each plaquette be in its minimal state ? It has been found, for face centered cubic and simple cubic lattices, that the answer is positive for $d \leq 4$, and negative for $d > 4$[13]. Thus, for both families of lattices, the ground state energy per spin is proportional to the coordination number z, for $d < 4$, and becomes asymptotically proportional to the square root of z, for $d > 4$:

$$E_o \sim \sqrt{-\frac{z}{2}} \;.$$

This is a behaviour reminiscent of spin glasses.

The determination of the critical temperature T_c is more interesting but difficult. However, if one considers the spherical model on a simple cubic lattice, it is possible to obtain a relation between the fully frustrated T_c and the fully unfrustrated T_c' . This relation

$$T_c = \frac{T_c'}{2\sqrt{d}}$$

is valid in any dimension. The limit of high dimension $d \to \infty$ is interesting to consider because some sort of mean field theory should be correct in that limit, and comparison with an exact value provides a stringent test. Standard mean field theory is found to be wrong by a factor 2 (over-estimation of T_c). It was also wrong by this same factor 2 in spin glass models, where an improved mean field theory (the Bethe approximation which considers not only one spin but also its first shell of neighbours) was found to give the correct answer. However, this improved mean field theory predicts no transition in the spherical model limit for the fully frustrated sc lattice.

For ordinary (say, ferromagnetic) phase transitions, standard mean field theory becomes correct in the limit of high dimensions because local field corrections are small with respect to the mean field (which is of order z). However, for the class of systems where the mean field (or ground state energy, or transition temperature) is much smaller and of order $\sqrt{z}$, the local field corrections become quite important. Apparently, we do not yet have a mean field theory which would be generally valid for this class of systems.

APPENDIX - ENTANGLEMENTS

The linking number of two closed loops can be seen as the first invariant number in a series, which characterises in more and more detail the possible states of linkage of two loops. Deeper mathematical concepts are introduced in the companion lecture of V. Poénaru.

The linking number can be computed in a variety of ways, of which we list here three.

1) The linking number from a double integral.

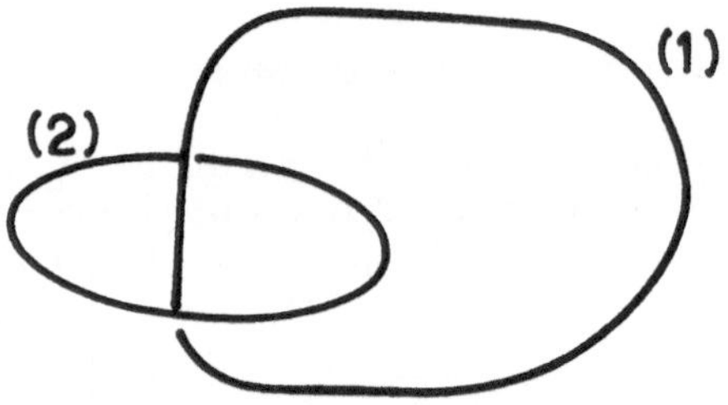

One wants to measure the flux through loop (2) of an electric current carried by loop (1).

$$\ell k(C_1,C_2) \sim \iint_{(2)} \vec{J_1}.\vec{dS_2}$$

from Ampère's law

$$\overrightarrow{\text{rot } H_1} \sim \vec{J_1}$$

$$\iint_{(2)} \vec{J_1}.\vec{dS_2} \sim \int_{C_2} \vec{H_1}.\vec{dr_2}$$

since

$$\vec{H_1} = \overrightarrow{\text{rot } A_1} \qquad \text{with} \qquad \vec{A_1}(\vec{r}) \sim I_1 \int \frac{\vec{dr_1}}{|\vec{r}-\vec{r_1}|}$$

$$\vec{H_1}(\vec{r}) \sim I \int_{C_1} \frac{\vec{dr_1} \times \vec{r}}{|\vec{r}-\vec{r_1}|^3}$$

So, re-establishing numerical coefficients,

$$\ell k(C_1,C_2) = \frac{1}{4\pi} \int_{C_1} \int_{C_2} (\vec{dr_1} \times \vec{dr_2}).\frac{\vec{\nabla 1}}{r_{12}} \quad .$$

To see that this double integral gives an integer number, just recall that

$$\vec{H}_1(\vec{r}) = \vec{\nabla}_{\vec{r}} \, \Omega_1 \quad ,$$

where Ω_1 is the solid angle through which loop (1) is seen from point $\vec{r}$.

2) The linking number from Seifert surfaces

A Seifert surface of a closed loop is an orientable surface whose edge is the loop. A Seifert surface for one of the two rings is punctured by the other ring. The algebraic sum of these crossings is the linking number.

3) The linking number from plane projection

Consider a non singular projection of the two rings on a plane. The rings are oriented and, on the projection, a double point is counted as + 1 if the priority of overpass from the right is observed and (-1) if it is not. The linking number is half of this algebraic sum of intersections[28].

Moving from links of two rings to knots of one single ring, it is tempting to look for a "knot number" which would be somehow the equivalent of the linking number. There have been attempts to make such a derivation, starting from the double integral formulation[27]. The nature of the difficulties which have been met is best understood from the third formulation. The algebraic sum on a plane projection for one single ring depends on the projection axis $\vec{\theta}$; actually it is the linking number of the ring with another ring, shifted infinitesimally in the uniform direction $\vec{\theta}$. Let us call this number :

$$\ell k(C,\vec{\theta}) = Wr\,(C,\vec{\theta}) \quad ,$$

where Wr comes from the word writhing. This number is well defined if $\vec{\theta}$ is nowhere tangent to the ring C. It is natural to consider the average of $Wr(C,\vec{\theta})$ over the sphere of all directions $\vec{\theta}$; this yields a number, called the writhing number,

$$Wr(C) \quad ,$$

which is not in general an integer, and which is not invariant under arbitrary deformations of C. The writhing number of a closed loop has interesting properties[29].

The consideration of a shifted loop leads naturally to a consideration of ribbons (of interest, for instance, for DNA molecules). Consider a closed loop C ; in any of its points, is defined a transversal vector $\vec{U}$, in a continuous fashion. The loop, shifted from C along $\vec{U}$, has with C a linking number denoted as

$$\ell k(C,\vec{U}) \quad .$$

Now there is a theorem which says that :

$$Wr(C) = \ell k(C,\vec{U}) - Tw(C,\vec{U}) \quad .$$

We have to define the last term, a twisting number. Consider locally, on one point of C, the frame $(\vec{T},\vec{U},\vec{T}\times\vec{U})$ where $\vec{T}$ is the tangent to C. Consider the rotation vector $\vec{\Omega}$ (which defines the rate of rotation of this frame with respect to arc length s along C)

$$\frac{d\vec{T}}{ds} = \vec{\Omega}\times\vec{T} \qquad , \text{etc...}$$

The rotation vector $\vec{\Omega}$ can be expressed as

$$\vec{\Omega} = \omega_1\,\vec{T} + \omega_2\,\vec{U} + \omega_3\,\vec{T}\times\vec{U} \quad ,$$

where ω_1, by definition, is the twist. Now the twisting number is the integral of the twist

$$Tw(C,\vec{U}) = \frac{1}{2\pi}\int_C \omega_1 ds \quad ,$$

not an integer in general.

The twisting number is related to the elastic energy. For a given $\ell k(C,\vec{U})$, there is an optimal Wr(C), which defines a class of optimal configurations of the ring.

In biological cells, there is a nicking enzyme which is able to cut one strand of DNA and thus is able to change the linking number of a double-stranded DNA ring by units. Such unit change entails changes of the average size of the ring and this is observed by chromatography because the migration rates of the molecules depend sensitively on their sizes. One has observed succession of dark lines, each line corresponding to rings with a given linking number, and two successive lines corresponding to rings differing by one unit of linking number.

MONTE-CARLO CALCULATIONS FOR THE LATTICE GAUGE THEORY*

Kenneth G. Wilson

Laboratory for Nuclear Studies
Cornell University
Ithaca, New York 14850, USA

ABSTRACT

Monte-Carlo renormalization-group methods are formulated for the lattice gauge theory following earlier ideas of Ma and Swendson in statistical mechanics. Expectation values of gauge loops and "block gauge loops" have been computed numerically by Monte-Carlo methods on lattices of maximum size $8 \times 8 \times 8 \times 8$. The expectation values of block gauge loops are used to determine the dependence of the bare coupling of the lattice theory on the lattice spacing. The lattice Lagrangian used in the renormalization calculations contained three different loops with coefficients in a fixed ratio. The calculations reported here are preliminary and the renormalization calculation is complete only for one value of the coupling constant. Some simple loop expectation values for the simple plaquette Lagrangian are also reported; similar results have previously been reported by Creutz.

The purpose of this paper is to report the present state of calculations on the renormalization of the SU(2) gauge theory (without quarks) in the intermediate coupling region. The calculations are Monte-Carlo integrations of a kind used in statistical mechanics. The Monte-Carlo methods used here are very similar to those used by Creutz *et al.*[1,2] in gauge theory calculations (which will be reviewed later); the numerical method of studying renormalization described here is similar to the method proposed by Swendson[3] for statistical mechanical problems.

*Supported in part by the National Science Foundation

Renormalization theory in its original form was a method for removing the infinities in the perturbation expansion of Quantum Electrodynamics. The infinities arose due to the singular small distance behavior of the continuum quantum field theory. The problem of renormalization seems at first sight to be an unlikely candidate for study by purely numerical methods, especially the very mundane procedures of Monte-Carlo calculations. For example, to see the problem of renormalization one must be close to the continuum limit. However, for numerical work (as will be seen later) a very strong ultra-violet cutoff is required.

Nevertheless, a numerical approach to renormalization is possible; actual calculations have been performed and the preliminary results are reported here. In order to proceed from the formal ideas of removing infinities to a serious calculation, a clear understanding of the physics of renormalization is necessary, i.e., to appreciate that one can study the problem of renormalization with considerable precision without ever seeing a large number. The necessary understanding has been provided largely as a consequence of modern renormalization-group theory. In the first part of this report I give an elementary discussion of lowest-order perturbative renormalization theory slanted to bring out the formal features needed as background for the numerical methods. The second stage is to ruthlessly truncate the formal continuum theory to a lattice theory of manageable size (an $8 \times 8 \times 8 \times 8$ lattice). The truncation procedure involves both infrared and ultraviolet cutoffs and associated errors; these errors must be estimated and various tricks are introduced to minimize their effects. At the end of this stage it will be seen that the renormalization problem is a uniquely *favorable* one for numerical treatment! The third part of this paper involves the practical details of actual calculations - how the Monte-Carlo method works and what is calculated using it, and how the results relate to renormalization.

To lead into the numerical calculations some general features of renormalization of field theory will be reviewed. The ϕ^4 theory (interaction $g_0\phi^4$, before renormalization) will be used as an example for the general discussion, with the complications of non-Abelian gauge theories referred to when necessary.

Look at some of the simplest diagrams for the four-point function of the ϕ^4 theory (shown in Fig. 1). The first diagram contributes the constant g_0 to the four-point function. The second diagram contains the integral

$$g_0^2 \int d^4k \frac{1}{(k^2+m^2)\,[(k+q)^2+m^2]} \tag{1}$$

(an Euclidean metric will be used throughout this report), where $q = q_1 + q_3$ and m is the mass of the field. This integral

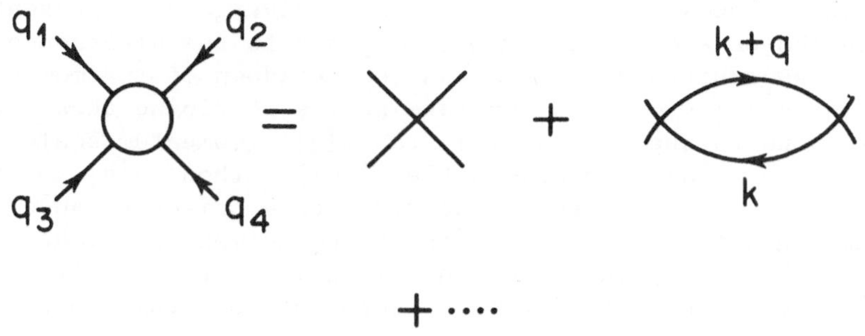

Fig. 1. Diagrams in the expansion of the four-point function of ϕ^4 theory.

contains a logarithmic divergence, coming from the region of k where $|k|$ is much larger than either m or $|q|$. For this range of k, the integral reduces to

$$\int^{\infty} \frac{d^4k}{k^4}$$

which is logarithmically divergent.

What is the origin of the divergence? Write the divergent integral as a sum of finite pieces. Assume for convenience, that $|q| >> m$. Then, roughly, the divergent integral starts from $|k| = |q|$ and is a sum of pieces

$$\int_{|q|}^{\infty} \frac{d^4k}{k^4} \sim \int_{|q|}^{\infty} \frac{dk}{k} = \int_{|q|}^{2|q|} \frac{dk}{k} + \int_{2|q|}^{4|q|} \frac{dk}{k} + \int_{4|q|}^{8|q|} \frac{dk}{k} + \ldots \quad (2)$$

Each subintegral is finite, in fact each subintegral integrates to $\ell n2$. The divergence is due to the existence of an infinite number of subintervals between $|q|$ and ∞.

Each subintegral in the logarithmic divergence represents the contribution of a different *momentum scale* to the divergence. We have made an arbitrary but reasonable breakdown of the momentum range $|q| < |k| < \infty$ into separate scales, defining these scales to correspond to the ranges $|q|$ to $2|q|$ (momentum scale $|q|$) then $2|q|$ to $4|q|$ (momentum scale $2|q|$), then $4|q|$ to $8|q|$ (momentum scale $4|q|$), etc. The divergence arises because *every* momentum scale from $|q|$ to ∞ is trying to make an *equal* contribution to the four-point function. This leads to an infinite result since there is an infinite number of momentum scales above $|q|$.

Rather than cover the usual procedures of renormalization with all the technical complications involved (plus making the necessary apologies for manipulating infinite quantities), I will describe a simple analogy which not only shows that this divergence is not suprising but also how it should be understood and removed. The analogy is the hydrodynamics of water. The motions of water on a macroscopic scale, for example water waves with wavelengths of order meters, are described by hydrodynamics. On a much smaller length scale, namely atomic scales (of order 10^{-8} cm) one sees the atomic structure of water, and water must be described by the Schrödinger equation for the electrons making up the atoms of water. In hydrodynamics, there is no reference to atomic scales except through some of the parameters in the hydrodynamic equations, such as the density and viscosity of water. The density and viscosity can only be determined by solving the atomic problem. In other words, the description of water on a scale of meters is independent of the description of water on much smaller scales, except for a few constants. These constants are dependent on smaller scales but can be determined by solving the theory at the next smaller scale of importance, namely atomic scales.

In the same fashion, there are parameters in the description of the atomic scales which depend on the details of even smaller scales. For example, the masses of the nucleii of the atoms of water depend on nuclear or subnuclear physics at scales of 10^{-13}cm.

Generalizing from this analogy, a physical system generally has a number of length scales, each of which is described by a different set of laws (hydrodynamics or quantum mechanics or whatever). For each separate length scale there is a separate set of parameters needed to describe the physics - the density and viscosity for hydrodynamics, electric charge and various masses for atomic scales, etc. The parameters for a given length scale are determined from the parameters for the *next smaller* length scale. Note that it is not necessary to know all the parameters governing yet smaller length scales. For example, one does not have to solve the nuclear-physics problem to determine the hydrodynamic parameters, as long as the parameters of the atomic physics are known. In other words, the effects of the nuclear physics on all larger scales are already contained in the parameters of the atomic physics.

Revert now to the ϕ^4 field theory. What is seen from the divergent integral discussed above, is that *every* momentum scale from the mass m up to ∞ must be considered important. Translated into position space, these momentum scales correspond to every length scale from $1/m$ down to 0. These scales are no longer represented by length parameters. In the hydrodynamical analogy each important length scale was characterized by a length parameter: the Bohr radius characterizing atomic physics and the nuclear radius characterizing nuclear physics. But in the ϕ^4 theory there are no obvious length parameters corresponding to the length scales $1/2m$, $1/4m$, $1/8m$, etc. Nevertheless, it is clear from the perturbation theoretic example that all these length scales must be taken into account. The other difference between the hydrodynamic example and the ϕ^4 theory is that the physical laws describing the different length scales do not change from scale to scale: the same ϕ^4 theory applies to all of them. However, the parameters appearing in the ϕ^4 theory *do* change from scale to scale: this is the basic result of the theory of renormalization, as will be discussed below.

Thus we cannot expect to discuss the ϕ^4 theory in terms of a single coupling constant g_0; instead there will be a momentum-dependent effective coupling $g_{eff}(q)$ which is used to compute vacuum expectation values with external momenta of order q.

(The case where the external momenta cover a range of momentum scales instead of a single scale has to be treated separately.) The parameter $g_{eff}(q)$ has no unique precise definition, it is usually arbitrarily defined to be the four-point function for some fixed choice of the ratios of external momenta with the magnitude of the external momenta proportional to q.

The example discussed at the beginning showed that the four-point function for momentum q gets contributions from all larger momentum scales. Now some examples will be given to show that the contributions from larger scales are actually contributions to the effective coupling $g_{eff}(q)$. Apart from contributions to $g_{eff}(q)$ (and to the renormalized mass, which will not be discussed here) the higher momentum scales do not affect vacuum expectation values with momentum q. In the case of the four-point function, the divergent part of the second-order diagram discussed earlier is a constant independent of the external momenta, like the lowest-order diagram. Thus the divergence can be combined with g_0 to form an effective coupling parameter. Some typical diagrams for the six-point function are shown in Fig. 2. The first two diagrams are convergent, which means the contributions of internal momenta much larger than q are negligible. The last diagram shown has a divergent loop, but it is easily seen that the divergent part of the last diagram combines with the first diagram to replace g_0 at one of the vertices of the first diagram by $g_{eff}(q)$ (assuming all the external momenta and some sums of external momenta are of

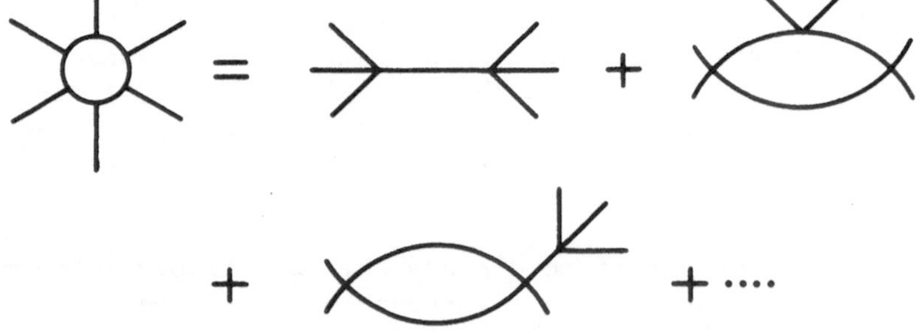

Fig. 2. Diagrams in the expansion of the six-point function of ϕ^4 theory.

order q). In general, the divergences in higher-order diagrams combine with lower orders to replace every coupling constant g_0 by the effective coupling $g_{eff}(q)$, so that g_0 and the divergences both disappear from the final result.

A procedure is required to compute $g_{eff}(q)$. According to the analogy with hydrodynamics, one would expect to determine $g_{eff}(q)$ knowing the coupling parameter for the next larger scale, say $g_{eff}(2q)$. However, the diagrammatic expansion gives $g_{eff}(q)$ in terms of g_0, and we must convert the result by hand to an equation expressing $g_{eff}(q)$ in terms of $g_{eff}(2q)$. For example, if only the diagrams of Fig. 1 are used, the equation resulting from the diagrams is, roughly,

$$g_{eff}(q) = g_0 + c\, g_0^2 \int_q^\infty \frac{dk}{k} \tag{3}$$

(where c is a negative constant) in terms of g_0. Most of the divergence is used to define $g_{eff}(2q)$, leaving

$$g_{eff}(q) = g_{eff}(2q) + c g_0^2 \int_q^{2q} \frac{dk}{k} . \tag{4}$$

This equation still involves g_0 but no explicit divergences. However, this equation neglects terms of order g_0^3 and higher; neglecting order g_0^3, we can replace g_0^2 by $g_{eff}^2(2q)$, giving

$$g_{eff}(q) = g_{eff}(2q) + c\ g_{eff}^2(2q)\ \ell n2. \tag{5}$$

This equation for $g_{eff}(q)$ expresses $g_{eff}(q)$ in terms of $g_{eff}(2q)$; all divergences have disappeared, meaning momentum scales much larger than 2q are no longer important. Renormalization theory shows that the complete equation for $g_{eff}(q)$ in terms of $g_{eff}(2q)$ is finite to all orders in $g_{eff}(2q)$. This is exactly what one expects from the hydrodynamical analogy. When $g_{eff}(q)$ is expressed in terms of $g_{eff}(2q)$, only *one* momentum scale is involved (the range from q to 2q). The effects of much larger momentum scales are already incorporated in the value of $g_{eff}(2q)$. Hence, when $g_{eff}(q)$ is expressed in terms of $g_{eff}(2q)$, high momenta (well above 2q) are unimportant and no divergent sums occur.

It is customary to write a differential equation for $g_{eff}(q)$; to lowest order this becomes

$$g_{eff}(q-\delta q) = g_{eff}(q) + c\ g_{eff}^2(q) \int_{q-\delta q}^{q} \frac{dk}{k} \tag{6}$$

or

$$\frac{dg_{eff}(q)}{d\ \ell n\ q} = -\ c\ g_{eff}^2(q) + \text{order } g_{eff}^3. \tag{7}$$

The solution of this equation (neglecting higher-order terms) is trivial: one finds

$$g_{eff}(q) = \frac{1}{c \ln (q/q_0)} , \quad (8)$$

where q_0 is an arbitrary constant.

The sign of c is a crucial quantity. Asymptotic freedom[4] corresponds to c positive. For positive c, $g_{eff}(q)$ goes to 0 from above as $q \to \infty$. However, this sign of c occurs only for non-Abelian gauge theories;[4] in the example (ϕ^4 theory) c is negative. In the case of gauge theories, the differential equation for g_{eff} actually involves g_{eff}^3 not g_{eff}^2:

$$\frac{dg_{eff}}{d \ln q} = c\, g_{eff}^3 \quad (9)$$

whose solution is

$$g_{eff}^2(q) = \frac{1}{2c \ln(q/q_0)} . \quad (10)$$

In this case, c positive means $g_{eff}^2(q)$ goes to zero from positive values for $q \to \infty$. As one goes to small q one eventually reaches the momentum $q = q_0$, where the logarithm vanishes and $g_{eff}(q)$ goes to ∞. This is an illusion, for once $g_{eff}(q)$ ceases to be small it is no longer permitted to neglect higher-order terms in Eq. (9) for $g_{eff}(q)$. Hence, Eq. (10) is an asymptotic form valid only for sufficiently large q.

The problem of interest for this paper is how to determine $g_{eff}(q)$ for small q for which perturbation theory [Eq. (9)] is not valid because $g_{eff}(q)$ is too large. This is part of the "infrared problem" for non-Abelian gauge theories.

The discussion above suggests that one thinks of a ϕ^4 theory as an infinite set of theories, one for each momentum scale q. The equations of the theory are the same for all q but there is a different coupling constant $g_{eff}(q)$ for each theory. Each theory is used for a single order of magnitude range q.

To prepare for numerical calculations it is necessary to try to minimize the number of degrees of freedom which appear in each of the infinite set of theories making up a field theory. In order to have a discrete degree of freedom we use the lattice gauge theory.[5] The lattice gauge theory has a maximum momentum, that is, a cutoff Λ. For each component of the momentum four-vector the maximum is π/a where a is the lattice spacing. Thus let $\Lambda = \pi/a$ be the cutoff parameter of the lattice theory. The lattice theory only describes physics for momenta less than Λ, in fact only for $q \ll \Lambda$ if cutoff-dependent effects are to be avoided. To make the number of degrees of freedom finite, the lattice theory will be confined to a finite size box. This means that there are a finite number N of lattice sites along each axis, or N^4 lattice sites in the entire box. The effect of the box is that the four-momenta are discrete, with the spacing between momentum values (for a single component of the momentum) being $\Delta q = \pi/Na$. The lattice theory in a box can only describe physics for momenta much larger than the gap momentum Δq, in order that the errors due to the discreteness of momentum are small.

For numerical purposes, therefore, a single continuum field theory is to be described by an infinite set of lattice theories, each with a different lattice spacing. It will be convenient to have the lattice spacing change by a factor of 2 from one theory to the next. It is convenient to plot the range of momentum q covered by each lattice theory versus the cutoff Λ. This is shown in Fig. 3. In this figure, the solid line shows the part of the momentum range where both cutoff-dependent and finite-box effects are negligible.

Each lattice theory has a different value for the bare coupling constant g_0. The coupling $g_0(\Lambda)$ is needed as input in order to perform the lattice calculations. Since the cutoff $\Lambda = \pi/a$ is proportional to the reference momentum q (see Fig. 3) there is no divergence in the relation between $g_{eff}(q)$ and $g_0(\Lambda)$, in fact

$$g_{eff}(q) \simeq g_0(\Lambda) + c\, g_0^{\;3}(\Lambda)\, \ell n\, (\Lambda/q) \tag{11}$$

($g_0^{\;3}$ appears here because the calculations are for the non-Abelian gauge theory).

Since the ratio Λ/q is fixed, this equation in particular gives

$$g_0(\Lambda) = g_{eff}(q) + \text{order } g_{eff}^{\;3}(q)$$

i.e.,

$$g_0^{\;2}(\Lambda) = \frac{1}{2\, c\, \ell n(q/q_0)} + 0\left(\frac{1}{\ell n^2(q/q_0)}\right) \tag{12}$$

i.e., (since Λ/q is fixed)

$$g_0^{\,2}(\Lambda) = \frac{1}{2\ c\ \ell n(\Lambda/\Lambda_0)} + 0\left(\frac{1}{\ell n^2 \Lambda/\Lambda_0}\right) , \tag{13}$$

where Λ_0 is a constant.

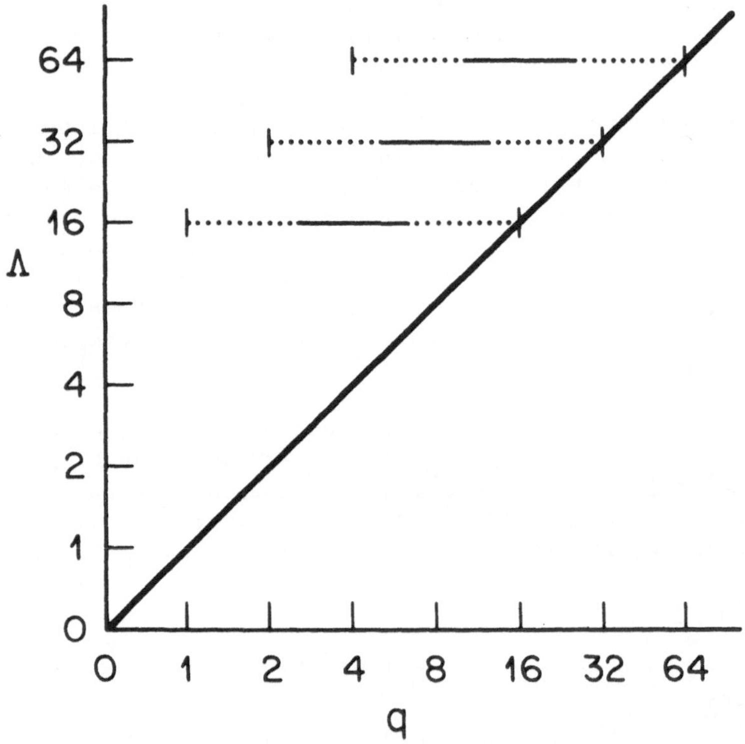

Fig. 3. Range of momentum q encompassed in each of three lattice theories, with cutoffs Λ differing by a factor of 2. The solid line represents the range of q free of infrared or ultraviolet cutoff dependence. The numbers are not to be taken seriously.

Since $g_0(\Lambda)$ has the same dependence on Λ as $g_{eff}(q)$ has on q, to a first approximation $g_0(\Lambda)$ satisfies the *same* differential equation as $g_{eff}(q)$, namely

$$\frac{dg_0(\Lambda)}{d\ \ell n\ \Lambda} = c\ g_0^{\ 3}(\Lambda). \tag{14}$$

However, in higher orders (actually only order $g_0^{\ 7}$ and above) the right-hand side of this equation is different from the equation for $g_{eff}(q)$.

The renormalization problem is therefore reduced to the problem of determining $g_0(\Lambda)$ versus the lattice spacing a where $\Lambda = \pi/a$.

The determination of $g_0(\Lambda)$ is done phenomenologically. The lattice theories for two successive lattice spacings, say a and 2a, must predict the same physics for sufficiently small momentum q. Therefore, all that is required is to compute a specified vacuum expectation value for $q \ll \pi/2a$ from both theories, and adjust $g_0(\pi/2a)$ for the theory with lattice spacing 2a until the vacuum expectation value agrees with that for lattice spacing a.

There are some practical problems and questions that arise when the vacuum expectation values used in the matching calculation are computed by Monte-Carlo numerical methods rather than by perturbation theory. First of all, for a numerical calculation it is necessary that the number of lattice sites N per axis in the box be as small as possible (in the calculations reported later the maximum value of N was 8). This is very difficult to arrange if one has to avoid both errors due to the finite box size (infrared errors) and errors due to the finite cutoff (ultraviolet errors). Fortunately, for the renormalization calculation the finite box size does not introduce errors. As long as one uses periodic boundary conditions (to avoid sharp boundaries which could introduce spurious ultraviolet effects) as well as the *same* physical size box for both lattices, the infrared behavior must be the same for both lattices. The reason for this is that while the infrared behavior is changed by the presence of the box, the infrared behavior still involves only momenta much less than the cutoff, and for these infrared momenta the predictions of the two lattice theories must be the same. Thus for the purpose of renormalization computations, Fig. 3 is replaced by Fig. 4: the only part of the lattice theory that cannot be used for matching is the part with cutoff-dependent errors.

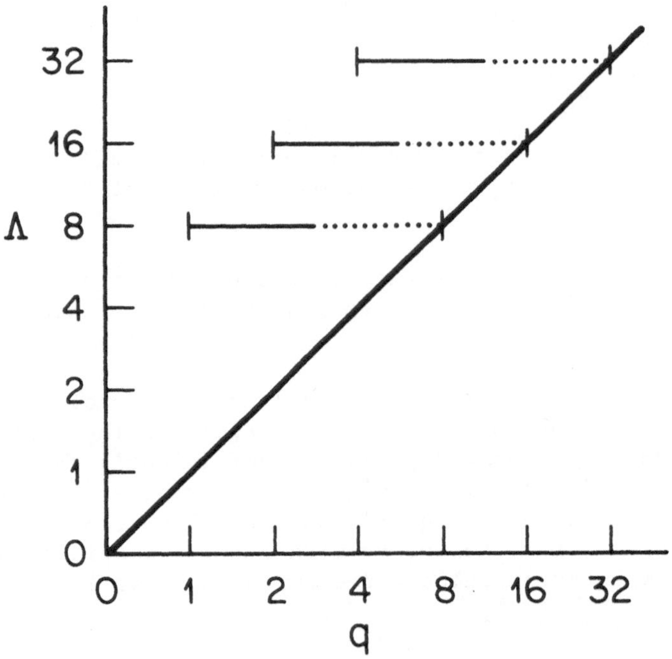

Fig. 4. Same as Fig. 3, except that the solid line represents the range of momentum useful for renormalization calculations, for which only ultraviolet cutoff effects matter.

Since the maximum box size available for the numerical calculations is small, we must discuss the cutoff dependence and how small the external momentum q must be relative to Λ to avoid cutoff dependence. Since the divergent part of the cutoff dependence is removed by the renormalization procedure, the remaining cutoff dependence comes from terms which go to zero as $\Lambda \to \infty$. Consider for example the integral discussed earlier, with a finite cutoff, say

$$\int_{|k|<\Lambda} d^4k \, \frac{1}{k^2+m^2} \, \frac{1}{(k+q)^2+m^2} \, .$$

Apart from the logarithmic integral, which is handled by renormalization, the behavior for large k involves terms such as

$$k\cdot q/k^6,\ q^2/k^6,\ m^2/k^6,\ \text{etc.}$$

The first term integrates to zero due to the symmetry for $k \to -k$; the second and third terms give a cutoff dependence of order $1/\Lambda^2$. This means that for fixed q, doubling Λ (i.e., halving the lattice spacing) reduces the error by a factor of 4. This is a spectacular reduction: one can quite easily imagine that changing q/Λ by three factors of 2 (from 1 to 1/8) cuts the cutoff-dependent error from $\sim 100\%$ to $(1/64) \times 100\%$ or 2%. However, the factor-4 reduction in error for every halving of the lattice spacing is actually only true for asymptotically small lattice spacing; one has to perform numerical studies to investigate the errors when q/Λ is not very small. This will be discussed later.

The next problem is to decide which vacuum expectation value to compute for two consecutive lattic theories (spacing a and spacing 2a) in order to determine whether the two theories have the same infrared behavior. There is a problem here due to the particular features of the gauge theory. The natural candidate, according to perturbation theory, would be a vertex function, namely the vacuum expectation value of three vector potentials. Unfortunately, the vertex function is not gauge invariant. It is expensive to implement the standard relativistic gauges numerically and there are ambiguity problems for these gauges; I have been reluctant to compute non-gauge invariant quantities in the axial gauges which are easy to compute and non-ambiguous. The obvious gauge invariant expectation value is the vacuum expectation value of a gauge loop. Unfortunately, the dominant term in the loop expectation value is a short-distance term behaving as exp(-kL) where L is the length of the loop in units of the lattice spacing and k is a constant of order g_0^2. This term arises due to the loop having no thickness in directions perpendicular to the loop. (When the heavy quark potential is extracted from the loop expectation value, the term proportional to L becomes a cutoff-dependent quark self-energy.)

For very large loops or for large g_0, the term proportional to L is expected to be dominated by a term proportional to the area of the loop, but an area term only makes the numerical problems worse. The problem with the cutoff-dependent term proportional to L is that: 1) it makes the expectation value of the loop small when the product kL is large. In this case, the statistical errors of the Monte-Carlo calculation are larger than the expectation value itself. Even if kL is not so large, it must be removed from the expectation value before one has a low momentum (i.e., large distance) expectation value. In practice, this means taking a difference of logarithms of two different loop expectation values; taking this difference magnifies the effects of statistical errors.

Thus we reject both gauge-dependent vacuum expectation values and loop expectation values. What is left? What is needed is to define a thick loop, one which will avoid a cutoff-dependent length term but is still gauge invariant.

In order to define and understand thick loops, a digression will be made into non-perturbative renormalization-group theory, namely into block spin theory. This digression is necessary also to discuss modified renormalization methods where the cutoff-dependence is reduced compared to just renormalizing g_0.

In block spin renormalization theory the sequence of cutoff lattice theories is defined directly without the need for matching conditions. The general principle is this: the lattice gauge theory has a degree of freedom (group element) $U_{n\mu}$ for each link (from lattice site n in direction μ). On a new lattice with twice the spacing there are fewer degrees of freedom. there is one lattice site of the double-size lattice for every $2 \times 2 \times 2 \times 2$ block of lattice sites of the original lattice. A theory with the reduced number of degrees of freedom can be generated by integrating out (in the lattice form of the Feynman path integral) the excess degrees of freedom. More precisely, what one would like to do is to work in momentum space. In momentum space on the original lattice the possible momentum values, along one axis, are $2\pi\ell/Na$ for $0 \leq \ell < N$. For the lattice with double spacing, but in the same size box, the momenta along the axis are $2\pi\ell/Na$ again but for $0 \leq \ell < N/2$. Thus, one would like to integrate out the high momentum variables of the original lattice which do not occur on the new double-size lattice. Unfortunately, due to the nonlinear constraints (unitarity) on the $U_{n\mu}$ it is not feasible to work with the Fourier transform variables derived from the $U_{n\mu}$. The block spin theory (developed by Kadanoff and Niemeyer and Van Leeuwen[6]) avoids working with the Fourier transform variables but still maintains the spirit of the requirement that the low momentum variables are left unintegrated.

In the Kadanoff form of the block spin theory one simply defines block variables associated with $2 \times 2 \times 2 \times 2$ blocks of lattice sites; then one performs the Feynman integral over the original variables holding the block variables fixed.

Constructing block variables is a somewhat arbitrary process. I will define the block variables used in my calculations; there may be other equally valid ways of defining block variables.

The block gauge field $V_{m\mu}$ is defined for a block m and a direction μ. The variable $V_{m\mu}$ is associated with a link from block m to the next block in the direction μ. The first step in constructing $V_{m\mu}$ is to sum all the link variables $U_{n\mu}$ where the link from n to $n+\hat{\mu}$ joins the block m to the block $m+\hat{\mu}$. This is illustrated in Fig. 5. In four dimensions, one has to sum eight different U's to form one V.

Averaging over all these U's is the first step in building a thick loop, and in so doing emphasizes lower over high momenta. The sum of unitary matrices is not itself unitary. To produce a unitary variable the sum must be normalized. This is easy for SU(2) matrices. A single SU(2) matrix has the form

$$\begin{vmatrix} a & -b^* \\ b & a^* \end{vmatrix}$$

with the only nonlinear constraint being $|a|^2+|b|^2= 1$ (a and b being complex numbers). A sum of SU(2) matrices has the same form except that $|a|^2+|b|^2$ is no longer 1. Such a matrix can be normalized by dividing the matrix by $\sqrt{|a|^2+|b|^2}$.

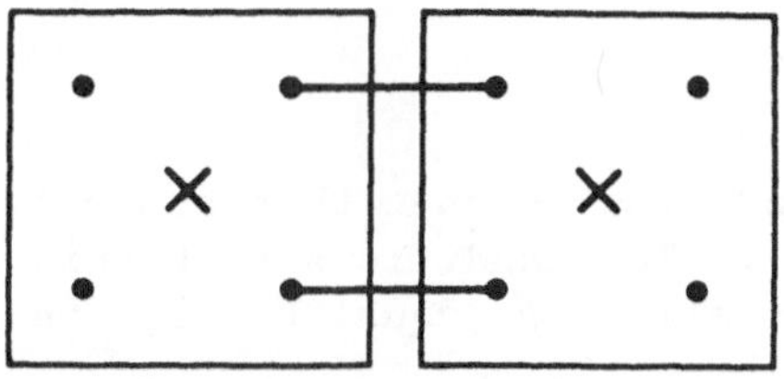

Fig. 5. Construction of block link variables. Two blocks are illustrated; x marks the center of a block. The matrices for the original links shown (solid lines) are added and normalized to form the block link matrix.

Thus the block link variable is

$$V_{m\mu} = W_{m\mu} / |W_{m\mu}|, \tag{15}$$

where $|W_{m\mu}|$ is the norm $(\sqrt{|a|^2+|b|^2})$ of $W_{m\mu}$ and

$$W_{m\mu} = \sum_n U_{n\mu}, \tag{16}$$

where the sum over n is restricted to sites n which are in the block m and for which the site $n+\hat{\mu}$ is in the next block $m+\hat{\mu}$.

The block link variables $V_{m\mu}$ do not have a simple behavior under gauge transformations, since each constituent link variable $U_{n\mu}$ transforms differently under gauge transformations. This situation can be fixed by performing a partial gauge fixing, namely fixing the gauge *within* each $2 \times 2 \times 2 \times 2$ block separately. Once the gauge is fixed within a block, the remaining gauge freedom is to assign a gauge to an entire block, which means every site within the block transforms identically under the gauge. Under this block gauge transformation, the block link variable transforms like an ordinary link variable connecting the blocks. In summary, the block gauge invariance will be preserved while the gauge freedom for different sites within a block is removed by gauge fixing. The gauge fixing used in the calculations is a block version of Landau gauge: one forms the sum

$$\sum_{n\mu} \mathrm{Tr}\, U_{n\mu}$$

for all links n,μ which are within the given $2 \times 2 \times 2 \times 2$ block (i.e., both the sites n and $n+\hat{\mu}$ must be inside the given block). Then one demands that this sum be a maximum with respect to all gauge transformations within the block. A gauge transformation is a set of SU(2) matrices Φ_n for each site n; the gauge transform of the sum is

$$\sum_{n\mu} \mathrm{Tr}\, \Phi_n U_{n\mu} \Phi^{+}_{n+\hat{\mu}};$$

the gauge fixing condition is that this sum has its maximum when all the Φ's are 1. The conditions that this sum has a local stationary point for all Φ's equal to 1, are such that all the sums

$$\sum_{\mu} \{U_{n\mu} - U_{n-\hat{\mu},\mu}\}$$

(where only links $n\mu$ or $n-\hat{\mu},\mu$ which are entirely within the given block are included) must be proportional to the unit matrix.

Using the definition of the block link variables and the gauge fixing condition it is now possible to define an effective action for the block variables. Let $S(U)$ be the action on the original lattice; then

$$e^{S_{eff}(V)} = \Pi_{n\mu} \int_{U_{n\mu}} \Pi_{m\mu} \delta\left(V_{m\mu};\ \frac{W_{m\mu}(U)}{|W_{m\mu}(U)|}\right)$$

$$\times \Pi_{m} \left\{ \Pi_{n\epsilon m} \delta(\text{gauge conditions at site } n) \ \left[\Pi_{n\epsilon m} \int_{\Phi_n}\right.\right.$$

$$\left.\left.\delta(\text{gauge condition transformed by the gauge } \Phi)\right]^{-1}\right\}$$

$$\times\ e^{S(U)}, \tag{17}$$

where the first δ-function is a δ-function on SU(2) group space which sets $V = W/|W|$ and integrates to 1 over Haar measure; $\int U n\mu$ is the SU(2) Haar measure; the next two lines include the gauge fixing δ-functions and the corresponding Faddeev-Popov determinant (I have not written these out in detail because their precise form will not be needed here).

The above formula defines an incomplete integration over the original action. The integration is incomplete due to the δ-functions holding the V's fixed. The result is a functional of all the $Vm\mu$ matrices, which is written in exponential form and defines an action functional on the V's. To complete the integration over the original action, one must integrate $\exp\{S_{eff}(V)\}$ over all the V variables. When the integrals over all the $Vm\mu$ are performed, the δ-functions involving the $Vm\mu$ in Eq.(17) all integrate to 1; the gauge fixing δ-functions and Faddeev-Popov determinants do not change the resulting integral, by the usual arguments, so the result is the complete integral of $\exp\{S(U)\}$ over all the $Un\mu$ variables.

The procedure defined above defines a transformation on a space of actions S. Given an input action $S(U)$ on a lattice of spacing a, the incomplete integration procedure generates a new action $S_{eff}(V)$ on a lattice of spacing 2a. This transformation can be iterated repeatedly to generate a sequence of lattice theories with successively larger lattice spacings, which is exactly what we need to discuss renormalization. Unfortunately, the effective actions generated by the transformation are very complex; they are not the simple plaquette actions with a single coupling constant. Instead the action S_{eff} will contain traces of arbitrarily complicated loops on the V lattice and, in addition, will contain products of traces of loops. This makes these actions hard to compute and hard to work with. The effective action can be calculated to leading order in weak coupling, however, where it reduces to the computation at a discrete sum over a single four-momentum k. In the weak-coupling approximation the effective action can be written as a sum of traces of loops; there is an infinite set of loops, each with a different coefficient.

The transformation Eq.(17) defining S_{eff} from S is an example of a renormalization-group transformation[7] T. When such a transformation is iterated many times, the result is usually a fixed point of the transformation, namely an action $S^*(U)$ which reproduces itself under T. In the present case, after many iterations for weak coupling, the action approaches a fixed point except for an overall normalization, namely after many iterations the effective action has the form

$$S_{eff}(V) = \frac{1}{g_0^2} S^*(V), \qquad (18)$$

where $S^*(V) = c_0^* \sum \mathrm{Tr}$ (simple plaquette loop) + $c_1^* \sum \mathrm{Tr}$ (rectangular loop) + $c_2^* \sum \mathrm{Tr}$ (three-dimensional loop) + ...,(19)

where the plaquette, rectangular and three-dimensional loops are illustrated in Fig. 6, and the sum is over all possible translations and reorientations of these loops.

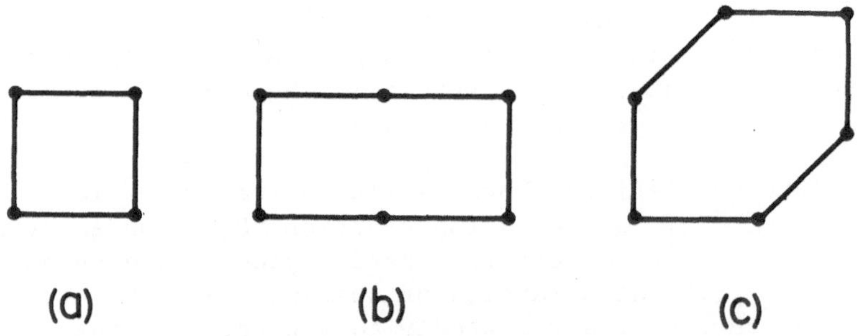

Fig. 6. (a) Simple plaquette loop on a lattice; (b) rectangular loop used in the three-parameter action [Eq.(19)]; (c) three-dimensional loop used in the three-parameter action.

The coefficients c_0^*, c_1^*, c_2^*, etc., are constants which can be computed numerically. A simple plaquette action corresponds to the case $c_0^* = 1$, $0 = c_1^* = c_2^* =$ etc.

Since the original coupling constant g_0 appears in S_{eff}, S^* is not an isolated fixed point of T. Instead there is a fixed line, parametrized by g_0. However, the fixed line exists only to lowest order in g_0. When higher-order terms are computed we expect coupling-constant renormalization to occur and instead of a fixed line, the iteration of T many times should lead to a slowly changing S_{eff} characterized by a changing coupling $g_0(\Lambda)$, where Λ is the cutoff parameter for S_{eff}:

$$S_{eff}(V) = \frac{1}{g_0^2(\Lambda)} S^*(V). \tag{20}$$

This behavior is expected only when g_0 and $g_0(\Lambda)$ are small; once $g_0(\Lambda)$ become large there will be higher-order corrections to $S^*(V)$ itself. The behavior [Eq.(20)] can be justified by formal renormalization-group arguments, but has not been verified directly because the higher-order perturbative calculations on the lattice needed to show the Λ dependence of $g_0(\Lambda)$ are rather complicated to carry out.

The behavior of successive effective actions can be represented by a trajectory in the space of actions. The space of actions is a space whose coordinates are the coupling constants multiplying different traces of loops. To illustrate this space only two coordinates can be shown, one being the simple plaquette coefficient. The trajectory of effective actions is shown in Fig. 7. In the weak-coupling domain the trajectory is a straight line since only the scale factor $1/g_0^2$ changes; as one gets outside the weak-coupling domain the form of the trajectory may be more complicated; its actual behavior is unknown.

It is awkward to work directly with the effective action $S_{eff}(V)$ outside the weak-coupling domain. The problem is that S_{eff} depends on a large number of variables and to explore this dependence numerically would be a very lengthy task. Therefore, we would like to use much simpler actions in the numerical calculation, preferably an action with g_0 as the only free parameter. This is possible, at least in the weak-coupling domain. If one starts, for example, with the simple plaquette action with a particular value of g_0, then iterating the renormalization-group transformation will generate a trajectory such as shown in Fig. 8. The first few iterations remove the cutoff

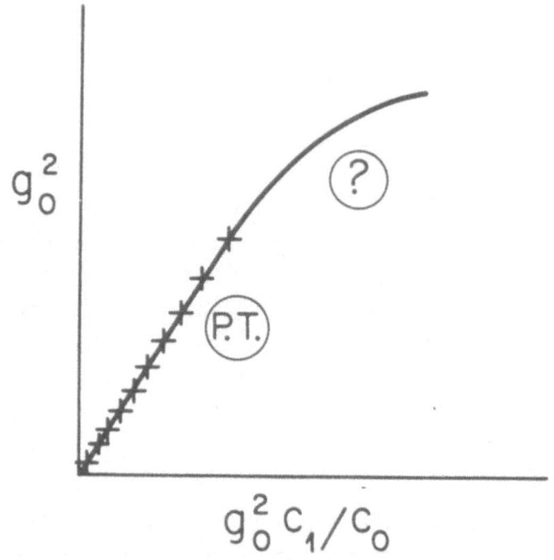

Fig. 7. Trajectory in the space of coupling constants representing successive actions. The crosses represent an actual sequence of actions. In the perturbative region (marked P.T.) the crosses lie on a straight line; beyond this range the behavior of the curve is unknown.

dependence of the original lattice theory; the cutoff dependence is removed when the trajectory, starting from a simple plaquette action with a given g_0 and given lattice spacing, approaches the true renormalized trajectory (which originates from a lattice with g_0 very small and lattice spacing very small). In the weak-coupling domain the two trajectories become very close after about three steps, and continue to approach very rapidly due to the factor-4 reduction in cutoff dependence every step (as was pointed out earlier).

This completes the digression into renormalization-group theory, and we can now consider further the matching problem relating two simple plaquette actions for two different lattice spacings. Consider the simple plaquette action on a lattice of spacing a and coupling constant g_0; we want to find the value g_0' of g_0 to use in the simple plaquette action on a lattice

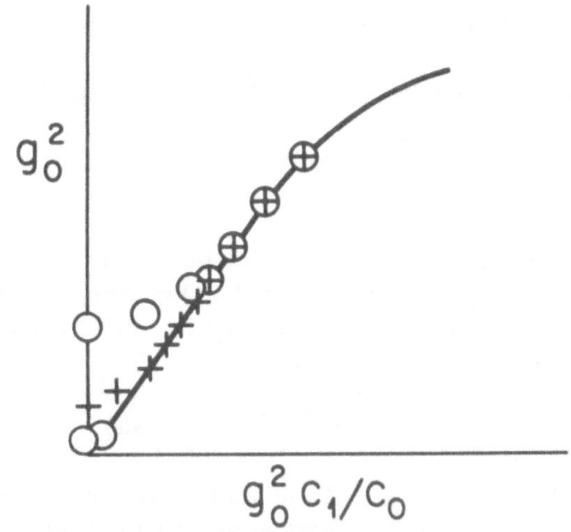

Fig. 8. Trajectories of actions starting from two different initial pure plaquette actions. After two or three steps, the actions lie on the same limiting curve.

of spacing 2a. What we want is for both actions to give the same physics at small q. This means, in the renormalization-group framework, that ideally both actions should lead to the same effective action on lattices of a much larger spacing. For example, we would like both actions to lead to essentially the same effective action on a lattice of spacing 8a.

To avoid computing the effective action directly, I have followed Swendson's[3] approach, namely to compute instead some vacuum expectation values of the block link variables on the lattice of spacing 8a. These block link expectation values must agree for the two theories if the effective actions are the same on the 8a lattice. These block link expectation values can be computed directly as integrals over the original actions, without introducing the effective actions. But to make clear formally the calculation being proposed, suppose the effective actions to be used in the computations; then the computations conceptually are the following. From the simple action with coupling constant g_0 on a lattice of spacing a, one has to iterate the renormalization-group transformation three times:

$$S_{original}(\text{lattice spacing } a) \rightarrow S_{eff}(\text{spacing } 2a) \rightarrow$$
$$S_{eff}(\text{spacing } 4a) \rightarrow S_{eff}(\text{spacing } 8a). \quad (21)$$

From this one computes a loop expectation value for $S_{eff}(8a)$. Next starting from the original action with coupling constant g_0' on a lattice of spacing 2a, one has to apply the renormalization-group transformation twice, giving

$$S_{original}(g_0', 2a) \rightarrow S_{eff}'(4a) \rightarrow S_{eff}'(8a), \quad (22)$$

and after this, one computes a loop expectation value for the effective action $S_{eff}'(8a)$. Then, since we would like to choose g_0' so that $S_{eff}(8a)$ and $S_{eff}'(8a)$ are very close, we demand that loop expectation values be as close as possible for the two actions $S_{eff}(8a)$ and $S_{eff}'(8a)$.

The computation of a block link expectation value is easily reduced to an integration over the original action. This will be illustrated for the first level of block links. Then one has to compute

$$\langle\text{block loop}\rangle = \Pi_{m\mu} \int_{V_{m\mu}} \text{Tr(loop of V's)}\; e^{S_{eff}(V)}/Z, \quad (23)$$

where

$$Z = \Pi_{m\mu} \int_{V_{m\mu}} e^{S_{eff}(V)}, \quad (24)$$

and $S_{eff}(V)$ was defined earlier [Eq.(17)]. Now one uses the definition of S_{eff} to replace $\exp\{S_{eff}(V)\}$ by an integral over U's of the original action. Then the V integrations are cancelled by the δ-functions defining V's in terms of U's; one is left with integrations over U's only, in particular one is left with the expectation value of the loop of V's, with the V's explicitly the block variables built out of U's. There are also the gauge fixing δ-functions defining block Landau gauge, but in this context it is not necessary to use the block Landau gauge. Instead, one can work in an arbitrary gauge, provided the $U_{n\mu}$'s in the arbitrary gauge are gauge transformed to the block Landau gauge before the block variables $V_{m\mu}$ are constructed. For most of my calculations, I have followed the suggestion of Creutz[2] and used no gauge fixing at all, i.e., all $U_{n\mu}$'s are updated without any gauge restrictions. Some early calculations performed in an axial gauge gave identical results.

Because the maximum-size lattice for the numerical calculations is, at present, only $8 \times 8 \times 8 \times 8$, it is desirable to speed up the convergence of the effective actions so that the effective actions for the lattice of size 4a are already quite close. This is not the case if the starting action is the simple plaquette action. Some studies using lowest-order perturbation theory in g_0 showed that an action containing three coupling constants does have this property but the single plaquette action does not. The three terms used are the three traces of plaquettes exhibited explicitly in Eq.(19) and Fig. 6. The coefficients c_0, c_1, c_2 were taken to be

$$c_0 = 4.376 \qquad c_1 = -0.252 \qquad c_2 = -0.17. \tag{25}$$

Note: I use the convention that in any sum over plaquettes or other loops such as Eq.(19), each loop appears twice namely as Tr(loop) and Tr(Hermitian conjugate loop) even though, for SU(2), both traces are identical. As in Eq.(20), the coupling constant g_0 appeared as a factor $1/g_0^2$ times the entire action. The numerical values in Eq.(25) were used because they gave the required rapid convergence in the lowest-order weak-coupling calculations. The coefficients c_0, c_1, and c_2 are normalized such that for long distances the coupling constant is g_0^2 to lowest order with no numerical factor.

Next, the numerical procedures for the calculation will be explained. The basic idea of the Monte-Carlo approach, as originally discussed by Metropolis, Rosenbluth, and Teller,[8] is to think of the exponential $\exp\{S(U)\}$ as a probability distribution in the space of configurations of all the $U_{n\mu}$ variables. To be a probability it must be normalized, that is, the probability distribution is $Z^{-1}\exp\{S(U)\}$ where Z is the integral of $\exp\{S(U)\}$ over all U's. The Monte-Carlo method is one for generating sample configurations of the $U_{n\mu}$'s distributed according to this probability distribution. A single sample configuration of U's consists of a set of 2×2 unitary matrices, one matrix for each nearest-neighbor link of the lattice. For example, on an $8 \times 8 \times 8 \times 8$ lattice with periodic boundary conditions there are 16,384 links. In this case, a single sample involves 16,384 2×2 matrices. A set of K samples involves K times 16,384 matrices. Expectation values of ordinary gauge loops and block gauge loops can be computed by averaging them over the samples.

The procedure for generating the samples is as follows. First an arbitrary set of values for all the matrices $U_{n\mu}$ is constructed and stored in the computer. In the calculations reported below, the initial value for each $U_{n\mu}$ was the unit matrix. In the computer, each $U_{n\mu}$ was represented by the two complex numbers a,b constituting the first column of the matrix. [For an SU(3) calculation the full SU(3) matrix would be stored, namely nine complex numbers.] The Array Processor used for my calculations has 96,000 words of memory, enough for the $8 \times 8 \times 8 \times 8$ calculation but not enough for any larger lattice. Then successive samples of configurations of the $U_{n\mu}$'s are generated by an updating scheme. The updating procedure involves cycling through all the links of the lattice, one by one, updating each $U_{n\mu}$ in turn. The updating procedure for a given link variable $U_{n\mu}$ used in the calculation was as follows. First, a random integer k from 1 to 16 was generated (I used a 48-bit random number generator supplied by M. Kalos: see the Appendix). This integer was used to select the kth SU(2) matrix W_k from a table of 16 matrices set up at the beginning of the calculation. The updated value $U_{n\mu}'$ for the matrix $U_{n\mu}$ is (tentatively), the product $U_{n\mu} W_k$. Next, a decision is made whether to keep the new matrix $U_{n\mu}'$ or to adhere to the old value $U_{n\mu}$. To make this decision, the ratio of the probabilities for $U_{n\mu}'$ and $U_{n\mu}$ is constructed, that is, the quantity $r = \exp\{S(U_{n\mu}') - S(U_{n\mu})\}$. The existing values of all the other U's are used in computing these two actions. Obviously, only the terms in the action containing the link $n\mu$ contribute to this difference. If r is greater than 1, i.e., if substituting $U_{n\mu}'$ for $U_{n\mu}$ increases the action, then the new $U_{n\mu}'$ is accepted as the updated value. (Remember that with the sign convention of this paper, the classical action is a maximum, not a minimum, so increasing the action is desirable.) If r is less than 1, then a random number x between 0 and 1 is generated; if r is greater than x the new $U_{n\mu}'$ is accepted. If $r \leq x$ the new $U_{n\mu}'$ is rejected: the updated value of the link remains $U_{n\mu}$.

The theory of this updating scheme is that a sequence of samples is generated and there is a probability distribution $P_\ell\{U\}$ for the ℓth sample of the sequence. The initial probability $P_0\{U\}$ is a product of δ-functions setting the $U_{n\mu}$ to their initial values. From the updating scheme, there is a linear transformation Q converting the probability P_ℓ to the probability $P_{\ell+1}$.

Ideally, we would like the probability $P_\ell\{U\}$ to approach the probability

$$Z^{-1}\, e^{S(U)}$$

for large ℓ. A necessary condition for this to occur is that the action reproduces itself under the transformation Q. That is, if

$P_\ell\{U\} = Z^{-1}\exp\{S(U)\}$ then $P_{\ell+1}(U)$ should also be $Z^{-1}\exp\{S(U)\}$. The details of the upgrading scheme ensure that this condition is satisfied (see the Appendix).

Once this condition is satisfied one can prove that the limit of $P_\ell(U)$ for $\ell \to \infty$ is indeed $Z^{-1}\exp\{S(U)\}$. The proof requires some minimal assumptions on the matrices W_k [which ensure that the set of all products of W_k matrices spans the entire SU(2) group space]. See the Appendix for details. The Appendix also discusses the updating procedure used by Creutz.[2]

Thus, the idea of the Monte-Carlo calculation is that the configurations generated by the updating scheme become samples of the probability distribution $Z^{-1}\exp\{S(U)\}$ after a number of steps; it is in fact customary to discard say the first 10% of the samples generated and then construct averages over the remaining 90%.

There is one technical comment on the calculations, namely the implementation of the gauge transformation used in defining the block spin variables. The maximization of the sum

$$\sum_{n\mu} \mathrm{Tr}\, \Phi_n U_{n\mu} \Phi_{n+\hat{\mu}}^{-1}$$

over all unitary matrices Φ_n is a highly nonlinear problem in many variables; I knew of no technique guaranteed to obtain the absolute maximum in a reasonable computation time. However, all that I really wanted of this gauge transformation were two requirements: 1) That for weak coupling, the absolute maximum is obtained so that weak-coupling approximations (which assume the absolute maximum is taken) will be correct. Here, there is no trouble: for sufficiently weak coupling the only stationary point is the absolute maximum, and the iteration procedure defined below converges to the absolute maximum. 2) That the effective actions be invariant to interchanges of axes on the lattice. This requirement is needed to ensure that the effective actions are free of terms which violate axis interchange symmetry; in the presence of such terms I am not sure that the effective actions would match properly for two different starting lattices. The reasons why I am not convinced is that simple perturbative renormalization theory breaks down in the presence of such symmetry-breaking terms, namely there is a separate coupling constant for each orientation of the simple plaquette and each of these coupling constants renormalizes differently.

To ensure that requirements 1) and 2) were both met, a numerical iteration procedure was defined to determine the set of 16 gauge transformations Φ_n for each block of 16 sites. Namely, first an ordering of the 16 sites was defined. This ordering was determined separately for each configuration of U's, in such a way that when two U configurations were related by an axis interchange transformation, then the two lattice site orderings were also related by the same transformation. The ordering was constructed by computing the sum of the trace of plaquettes at each site n of the block and rearranging the sites n until these sums were ordered. Then, given the ordering of the sites, a preliminary gauge fixing was performed to an axial gauge inside the block, namely 15 link variables inside the block were gauge transformed to 1. (15 links are required to fix the relative gauges of the 16 sites.) Since there is no *a priori* rotationally invariant way to select 15 links in a block, the location of these 15 links was determined from the site ordering already defined so that a rotation of the U configuration would give a corresponding rotation of the 15 chosen link sites. Finally, an iteration procedure was used to complete the gauge transformation. The iteration procedure used was to cycle through the 16 sites of the block using the ordering constructed earlier; at each site n the expression

$$\sum_{n\mu} \mathrm{Tr}\, \Phi_n U_{n\mu} \Phi^{+}_{n+\hat{\mu}}$$

was maximized with respect to the gauge Φ_n holding the other gauges fixed. This entire cycle was repeated 400 times.

In a typical actual Monte-Carlo run, the link updating cycle through the entire lattice was repeated 8640 times on an $8 \times 8 \times 8 \times 8$ lattice. This required 15 seconds per cycle or 36 hours total. Every eight cycles, a separate cycle through the lattice was made to construct all the block link variables and compute the traces of three different block loops (the block loops constructed included the simple plaquette loop, the third loop of Fig. 6 and the corresponding loop spanning all four dimensions). Due to the periodic boundary conditions, all translations and rotations of the block loops have the same expectation value but the statistical error in the expectation values was reduced by averaging all translations and rotations of the block loops. It was necessary to repeat the cycle of block link construction twice to build the second and third stages of block links, namely the block variables for $4 \times 4 \times 4 \times 4$ and $8 \times 8 \times 8 \times 8$ blocks of the original lattice. The $4 \times 4 \times 4 \times 4$ block link variables are built from the $2 \times 2 \times 2 \times 2$ block link variables by making blocks out of the lattice of $2 \times 2 \times 2 \times 2$ blocks, so a pass is required through this lattice; a second pass is used to build the $8 \times 8 \times 8 \times 8$ block link variables. On an $8 \times 8 \times 8 \times 8$ lattice there is only one $8 \times 8 \times 8 \times 8$ block, namely the entire lattice is a block. But, due to the periodic boundary conditions, there are link variables which connect this block to the block of the next period or, in other words, connect the block to itself. This means there are four link variables V_μ for this single $8 \times 8 \times 8 \times 8$ block, labelled only by their direction μ. A plaquette trace built of these V's has the form

$$\mathrm{Tr}\ V_\mu V_\nu V_\mu^+ V_\nu^+ \ .$$

For an Abelian group this trace would be trivial; for a non-Abelian group the trace is non-trivial and its expectation value provides a measure of the behavior of the lattice system on the scale of $8 \times 8 \times 8 \times 8$ blocks. An additional 18 hours of computing time was required to compute these expectation values through an entire run, i.e., half of the updating time. In addition, the expectation values of ordinary loops were computed; all square and rectangular loops and some moving in three and four dimensions were computed with a maximum length per side of three or four lattice spacings, depending on the run. In computing block loops, all possible shifts of the block locations were used to increase the statistics.

Calculations were performed for a number of choices of g_0 for both the case of the simple plaquette action and the three-loop action and for both $4 \times 4 \times 4 \times 4$ and $8 \times 8 \times 8 \times 8$ size lattices. For the three-loop action only g_0 was varied; the ratio of coefficients of the three different loops was always set to the values given earlier. Two problems arose with the statistical errors. One problem was that for sufficiently large g_0, the expectation values of large-size loops (ordinary or block) were too small to be seen above the statistical errors. For example, at $g_0^2 = 2$ for the single plaquette action, a run of 4800 passes on a $7 \times 7 \times 7 \times 7$ lattice gave an expectation value of 0.0056 for the 3×3 loop with a statistical error of 30%. By comparison, the single plaquette expectation value was 1.002 with a statistical error of 0.3%. The value $g_0^2 = 2$ was found to be near the edge of the strong coupling region (for which simple loop expectation values can be computed from high-order strong-coupling series).

Weak-coupling behavior (qualitatively) occurs for $g_0^2 \lesssim 1.5$. For the three-coupling-constant action the 3×3 loop expectation value goes below the statistical error somewhere between $g_0^2 = 5$ and $g_0^2 = 6$, and at $g_0^2 = 5$ the expectation value of an

$8 \times 8 \times 8 \times 8$ block loop is already below the statistical error (to be precise the strong coupling limit of this expectation value on an $8 \times 8 \times 8 \times 8$ lattice is 0.5 and it is the departure from this strong-coupling limit that is smaller than the statistical error).

The other problem with the statistical errors was the presence of correlations over many iterations of the Monte-Carlo calculation which resulted in a slower decrease of the statistical error than $1/\sqrt{N}$ where N is the number of iterations. This problem arose in particular for the block loop expectation values. To study this problem standard deviations of the expectation values were computed for various sample sizes; first the standard deviation was computed for values of the block loop expectation value from a single pass through the lattice. Then, averages for sets of four consecutive passes of expectation-value calculations (i.e., 32 updating passes with four expectation-value passes) were constructed and standard deviations for these samples were constructed. Then, the standard deviations for averages of 16 expectation-value passes were calculated, etc. For $g_0^2 = 4$, on an $8 \times 8 \times 8 \times 8$ lattice, and for the $4 \times 4 \times 4 \times 4$ block plaquette expectation value, the average was 0.876 and successive standard deviations were 0.104, 0.096, 0.083, 0.061, 0.032 for samples of from 1,4,16,64 and 256 passes, respectively. There were four samples of 256 expectation-value passes. The first four standard deviations change very little, reflecting strong correlations over 64 expectation-value passes (512 updating cycles). The last standard deviation is down by a factor of 2 from the next to the last one, which is to be expected if there are no correlations beyond 64 passes. In the same calculation, the standard deviations for the simple plaquette expectation value dropped a factor 16 from 1 to 256 pass averages, just as expected from the $1/\sqrt{N}$ law, while for 2×2 and 3×3 simple loops, the standard deviations dropped by a factor 7 over the same range. Because of the correlations in the block expectation values, the statistical errors in these values were fairly large even for almost 10,000 updating passes.

Because of the problems with statistical errors, the calculations are far from complete. In addition, a special computer (an Array Processor - the AP-190L made by Floating Point Systems, Inc. in Portland, Oregon) was needed to provide large amounts of cheap computing time in order to carry out these calculations, and this computer was only installed and working a year ago. Since the installation of the Array Processor at Cornell, it took me almost a year to learn the method described here for using the Monte-Carlo technique and prepare the necessary programs. Thus, only a few months of computing were available before this paper was written. The program at present runs eight times faster on the Array Processor than on the IBM 370-168 at Cornell which acts as a host for the Array Processor. The Array Processor is a simple computer which only number crunches, and runs one program at a time (no time-sharing); at present its only input-output capability is to accept a program and data from the 370 from the beginning of a run and return the data at the end of the calculation. It is very cheap (∿ $130,000 for the configuration at Cornell). Its fast speed is partly achieved by having a number of different units: multiply-, add-, address-computation-, branch-, and memory fetch-and-store units, etc., all of which can operate in parallel and all of which can be given new instructions in parallel every machine cycle. In addition, the various units either complete an instruction in one cycle (1/6 microsecond) or can at least accept a new instruction every cycle, to be completed two or three cycles later. To achieve the full speed of the Array Processor some assembly language programming is required, although a Fortran compiler (prepared at Cornell) allowed non-critical parts of the problem to be written in Fortran.

The principal result of my computations so far is a single example of matching between an $8 \times 8 \times 8 \times 8$ lattice and a $4 \times 4 \times 4 \times 4$ lattice with double the lattice spacing. The $8 \times 8 \times 8 \times 8$ lattice calculation involved the three-loop actior of Eq.(19) using the parameters in Eq.(25), with $g_0^2 = 4$, and 8640 updating passes. The $4 \times 4 \times 4 \times 4$ lattice calculation involved the same three-loop action except with $g^2_0 = 4.969$, and 40,000 updating passes. The results are shown in Table I.

The comparison to be made between the two computations is to compare expectation values of loops of $8 \times 8 \times 8 \times 8$ blocks on the $8 \times 8 \times 8 \times 8$ lattice *vs.* $4 \times 4 \times 4 \times 4$ blocks on the $4 \times 4 \times 4 \times 4$ lattice, or compare $4 \times 4 \times 4 \times 4$ blocks on the $8 \times 8 \times 8 \times 8$ lattice *vs.* $2 \times 2 \times 2 \times 2$ blocks on the $4 \times 4 \times 4 \times 4$ lattice. Both comparisons show excellent matching well within statistical errors. The matching becomes much more impressive when one sees how rapidly these expectation values are changing *vs.* g_0 (to illustrate this the results on the $4 \times 4 \times 4 \times 4$ lattice for $g_0^2 = 4.44$ are also shown in Table I). The $2 \times 2 \times 2 \times 2$ block expectation values for the $8 \times 8 \times 8 \times 8$ lattice do not match the simple loop expectation values on the $4 \times 4 \times 4 \times 4$ lattice; but these loops do not match in the weak-coupling approximation either, and in any case these loops are too small in size for agreement to be expected.

To obtain the matching shown in Table I, the first step was to

Table I. Comparison of Block Loop Expectation Values: $8 \times 8 \times 8 \times 8$ lattice *vs.* $4 \times 4 \times 4 \times 4$ lattice with twice the spacing (with a second calculation for the $4 \times 4 \times 4 \times 4$ lattice).

Lattice Size	$8 \times 8 \times 8 \times 8$	$4 \times 4 \times 4 \times 4$	$4 \times 4 \times 4 \times 4$
Coupling g_0^2	4.0	4.969	4.444
No. Passes	8880	41472	9216
No. Subsamples*	4	9	15
Block Size†: Expectation values of simple plaquette of block links			
1×1	1.414 ± 0.001	1.242 ± 0.005	1.347 ± 0.010
2×2	1.308 ± 0.004	0.867 ± 0.010	1.233 ± 0.050
4×4	0.876 ± 0.032	0.850 ± 0.030	1.419 ± 0.140
8×8	0.849 ± 0.051		
Block Size: Expectation values of four-dimensional (eight-link) block loop			
2×2	0.543 ± 0.023	0.129 ± 0.025	0.484 ± 0.080
4×4	0.123 ± 0.056	0.498 ± 0.078	1.000 ± 0.170
8×8	0.501 ± 0.190		
No. Subsamples	36	35	7

*The statistical errors shown in this Table are actual standard deviations for a set of subsample averages. The entire sample (set of configurations) for each calculation was split into the number of subsamples indicated. All calculations in this Table used the three-parameter action of Eq. (19) with coefficients shown in Eq. (25).

†In units of the lattice spacing. The comparison to be made for the renormalization calculation is, for example, 4×4 blocks on the $8 \times 8 \times 8 \times 8$ lattice *vs.* 2×2 blocks on the $4 \times 4 \times 4 \times 4$ lattice.

select an arbitrary value of g_0^2 for the $8 \times 8 \times 8 \times 8$ run, namely $g_0^2 = 4$. Then a number of relatively short runs was made on a $4 \times 4 \times 4 \times 4$ lattice, using different values of g_0^2, until it was found that excellent matching was likely to be obtained for $g_0^2 = 4.969$. Then the long run for $g_0^2 = 4.969$ was made. The error in this value for g_0^2 due to the statistical errors in both the $8 \times 8 \times 8 \times 8$ and $4 \times 4 \times 4 \times 4$ lattice runs was less than 0.1, or 10% of the renormalization difference $\delta(g_0^2) = 4.969 - 4 = 0.969$.

Runs were made for the three-loop action on an $8 \times 8 \times 8 \times 8$ lattice also for $g_0^2 = 4.969$ and $g_0^2 = 2$ (50 hours each). Unfortunately, in both cases the statistical error on the 8×8 block loop expectation values was too great to make an accurate determination of the coupling constant needed for matching. For the run with $g_0^2 = 4.969$, the 8×8 block loop expectation value was too small to be visible above the statistical error. In the case $g_0^2 = 2$ the 8×8 block loop expectation value was larger than the statistical error but unfortunately the matching conditions gave $g_0^2 = 2.22$ on the $4 \times 4 \times 4 \times 4$ lattice, which is very close to 2. As a result the statistical error on the renormalization difference ($\delta g_0^2 = 2.22 - 2 = 0.22$) was large, namely $\sim 100\%$ if the 8×8 block loop was used for matching, or $\sim 30\%$ if the 4×4 block loop was used instead.

Calculations have also been made using the simple plaquette action. No study of renormalization has been made for this case since the lowest-order weak-coupling computations indicated that an $8 \times 8 \times 8 \times 8$ lattice is too small for studying renormalization. However, it may be possible to study the renormalization of the single plaquette action indirectly by determining the value of g_0^2 for the simple plaquette action on an $8 \times 8 \times 8 \times 8$ lattice which matches the large-scale block loops on an $8 \times 8 \times 8 \times 8$ lattice for the three-loop action, and then using the renormalization of the three-loop action to infer the renormalization of the simple plaquette action. The necessary computations have not been carried out. In the meantime it is interesting to see the behavior of simple gauge loops for the simple plaquette action. Creutz has already computed the simple plaquette expectation value as a function of g_0^2, for up to a $10 \times 10 \times 10 \times 10$ lattice, and finds a short and reasonably smooth transition from weak- to strong-coupling behavior for g_0^2 in the range 1.6 to 1.8. My calculations for a variety of lattice sizes ($8 \times 8 \times 8 \times 8$ or $7 \times 7 \times 7 \times 7$ or $8 \times 8 \times 4 \times 4$) are in agreement with Creutz' curve (I do not have his actual numbers). I have calculations for loops of up to size 3×3; some results are shown in Table II.

Table II. Simple Gauge Loop Expectation Values, *vs.* coupling constant (for the simple plaquette action)

Lattice Size	7^4	8^4	8^4	7^4
Coupling	2.0	1.739	1.538	1.25
No. Passes	4800	12096	8640	4800
No. Subsamples	7	20	14	7
Loop Size		Expectation Values		
1 × 1	1.002 ± 0.003	1.204 ± 0.002	1.340 ± 0.002	1.486 ± 0.002
1 × 2	0.517 ± 0.005	0.771 ± 0.003	0.965 ± 0.003	1.177 ± 0.004
1 × 3	0.267 ± 0.004	0.501 ± 0.003	0.706 ± 0.003	0.944 ± 0.007
2 × 2	0.146 ± 0.003	0.359 ± 0.005	0.577 ± 0.004	0.829 ± 0.007
2 × 3	0.042*	0.178 ± 0.004	0.366 ± 0.005	0.611*
3 × 3	0.0056 ± 0.0017	0.070 ± 0.004	0.212*	0.430 ± 0.007

*For stupid reasons the standard deviation was not available in time for this report.

To conclude this paper I will give my view of the status and outlook for the Monte-Carlo approach and also discuss related work. First of all, it is impressive to be able to produce a curve showing, for example, the simple plaquette expectation value as a function of g_0^2 in the intermediate coupling domain, with the statistical accuracy being a fraction of 1% and decreasing according to the $1/\sqrt{N}$ law. For the simple plaquette expectation values my calculations are longer than necessary; the shorter calculations by Creutz[2] are, I believe, quite adequate for most purposes. Although this curve is irrelevant to physics it will be invaluable for testing various weak- and strong-coupling methods to determine their range of validity. Secondly, it is impressive that the ideas of continuum renormalization theory can be transformed into a matching calculation that can be carried out in practice on lattices of size 8 × 8 × 8 × 8 and 4 × 4 × 4 × 4 and give results of the precision obtained here. But many problems and questions remain. Are instantons or other special configurations important in the intermediate coupling domain? If so, are they included in the Monte-Carlo samples? Are the Monte-Carlo results independent of the starting configuration? More important, there is the problem of obtaining real physical results following the renormalization calculations. The problem here is that one must avoid both infrared and ultraviolet errors, which it is hard to do unless the lattice is larger than $8 \times 8 \times 8 \times 8$. The best we can hope for at the moment is that one can use the renormalization calculations to reach a large enough value of $g_0{}^2$ so that high-order strong-coupling series can be used to obtain the string tension and other quantities of physical interest. Even to complete the study of renormalization of the SU(2), pure gauge theory may demand more computing time than is available on the present Array Processor; to study larger lattices or the SU(3) gauge group will be even more demanding of computing time. I know of no way to include quarks in the Monte-Carlo calculations for a reasonable cost. Fortunately, new miraculous developments are expected in the miniaturization and cost reduction of computers, so we can look forward to calculations which would be hopelessly expensive now, becoming feasible some day. In the meantime, the severe restrictions on currently possible calculations are balanced by the even more severe concerns about alternative weak- and strong-coupling methods for studying quark confinement, namely the concern whether any alternative approach (based on instantons or strong-coupling expansions or whatever) has any validity at all in the crucial intermediate coupling domain. Hopefully, the interplay of numerical *vs.* more analytical approaches will lead to more progress than is possible with either type of approach separately.

It has become popular recently to short-circuit the complex renormalization theory used in this paper. Instead Kogut *et al.*[9] and Creutz[10] try to determine how g_0 varies with lattice spacing a in the simple plaquette action, by holding the string tension fixed as the lattice spacing is varied. The string tension is the coefficient of the area in the behavior of large-size gauge loops. This calculation requires that the coefficient of the area be determined for small enough g_0 so that one can make a comparison with the weak-coupling formula for the dependence of g_0 on a. Kogut *et al.* use an extrapolation procedure on the strong-coupling expansion for the string tension in order to determine its behavior for small g_0; Creutz uses a Monte-Carlo calculation similar to the one reported here except that Creutz only calculates the expectation values of simple loops and does not obtain as many passes through the lattice (he uses of order 20-50 passes, I use up to 8000 passes per calculation). To obtain the string tension Creutz fits loop expectation values of sizes 0×0 to 5×5 (or a subset of these) to a formula $\exp\{-(A + Bs + Cs^2)\}$ where s is the width of the loop (0 to 5). The extrapolation procedure of Kogut is subject to all the usual questions about extrapolation of a power series beyond its radius of convergence; studies now under way by Parisi and also by myself of small-size loop expectation values should clarify how serious these questions could be. The procedure of Creutz is subject to the question of how easily one can extract the area law coefficient C from data only for small loops, in particular for small g_0 where C is small. In particular the separation of the C term from the A and B terms requires taking multiple differences of different loop expectation values which may enhance the effects both of statistical errors and of small s corrections other than the A and B terms. For example, for $g_0^2 = 1.74$, I have repeated Creutz' computation of C using the same loop expectation values he uses at this value namely, $\alpha = \ln \langle 1 \times 1 \text{ loop} \rangle$, $\beta = \ln \langle 2 \times 2 \text{ loop} \rangle$ and $\gamma = \ln \langle 3 \times 3 \text{ loop} \rangle$. The formula for C then turns out to be

$$C = \beta - 0.5(\alpha+\gamma).$$

Using a run of 12000 passes through an $8 \times 8 \times 8 \times 8$ lattice, I obtain: $\alpha = 0.186 \pm 0.002$, $\beta = -1.024 \pm 0.014$, $\gamma = -2.66 \pm 0.06$ giving $C = 0.21 \pm 0.04$. Creutz' value (reading from Fig. 4.6 of his preprint) is 0.19 for the same value of g_0^2; he does not quote an error. Clearly there has been almost 90% cancellation between α, β, and γ in the formula for C, which makes the problem of statistical and systematic errors rather acute. However, while my calculations have far greater statistics than Creutz', we are in reasonable agreement at $g_0^2 = 1.74$. The problem of systematic errors due to small s corrections remains.

APPENDIX: VALIDITY OF THE LINK UPDATING PROCEDURE

In this Appendix, it is shown that the probability distribution $Z^1 \exp\{S(U)\}$ is the limiting probability distribution for the configurations of U's generated by the Monte-Carlo updating procedure. I also discuss the alternative updating procedure used by Creutz and its relation to the procedure I actually used, and make some other comments on the Monte-Carlo procedures.

Let the sample of $U_{n\mu}$'s which is used as input to a single updating step have the probability distribution P[U] where U stands for the whole set of $U_{n\mu}$'s. To be precise, this is understood to mean that the probability of obtaining a particular set of values for all the $U_{n\mu}$'s, within a range $dU_{n\mu}$ for each $U_{n\mu}$, is

$$P[U] \, \Pi_{n\mu} \, dU_{n\mu}$$

and the volume element $dU_{n\mu}$, for the considered range of $U_{n\mu}$, is computed using Haar measure [invariant group measure on the SU(2) group space].

The probability $P_1[U]\Pi n\mu\ dUn\mu$ after one updating step for a particular link variable is computed as follows. To arrive at a specific sample $\{Um\nu\}$ for all m and ν, after one step, means first of all that the previous sample had the same values of the $Um\nu$ for all links $m\nu$ except the chosen link $n\mu$. For this link there are only 17 possible values of $Un\mu$ in the previous sample. Call the previous value $Un\mu'$; then $Un\mu'$ can be either $Un\mu$ or one of the 16 products $U_{n\mu} W_k^{-1}$ $(1 \leq k \leq 16)$. Because Haar measure is invariant to group products, the volume element $dUn\mu'$ corresponding to the range $dUn\mu$ is again $dUn\mu$; therefore, the volume elements are the same.

The probability $P_1(U)$ is now easily constructed from the rules given in the text. One simply multiplies the probability that one had $Un\mu'$ before updating along with all the other $Um\nu$ by the probability that the updating replaces $Un\nu'$ by $Un\nu$. The result is as follows. Let U' be the set of link matrices before updating the link $n\mu$ variables from $Un\mu'$ to $Un\mu$, where $Un\mu' = U_{n\mu} W_k^{-1}$. Write $\Theta_{UU'}$ for $\Theta[S(U) - S(U')]$ where $\Theta(x)$ is the Θ function (1 for $x > 0$, 0 for $x < 0$). Then

$$P_1(U) = \frac{1}{16} \sum_{k=1}^{16} \left\{\Theta_{UU'} + \exp[S(U) - S(U')]\ \Theta_{U'U}\right\} P(U')$$
$$+ P(U) \left\{1 - \frac{1}{16} \sum_{k=1}^{16} \Theta_{U''U} - \frac{1}{16} \sum_{k=1}^{16} \exp[S(U'')-S(U)]\Theta_{UU''}\right\}, \quad (A.1)$$

where U'' is a set of link variables which is identical to U except that $U_{n\mu}'' = U_{n\mu} W_k$. In this equation, the first term [involving $\Theta_{UU'}\ P(U)$] arises when U' is updated to U and the action S is increased by the updating. Then the probability that this updating takes place is only the probability that the matrix W_k is used in the updating, namely 1/16. The final term involving $S(U'') - S(U)$ appears in the probability that $Un\mu \rightarrow Un\mu$ without change, because this probability is equal to one minus the probability that $Un\mu$ is changed to some other matrix by updating; the final term gives the probability that $Un\mu$ is updated to $U_{n\mu} W_k = U_{n\mu}''$ when $S(U'')$ is less than $S(U)$.

Once can now verify by substitution that the particular probability distribution $Z^{-1}\exp\{S(U)\}$ is preserved, i.e., $P_1(U) = P(U)$ for this choice of $P(U)$. There is, however, a restriction needed, namely the set of 16 configurations U" must be the same as the 16 configurations U', i.e., the set of matrices W_k must include the inverses W_k^{-1} for all k. This is a standard reciprocity condition present in all Monte-Carlo procedures.

Equation (A.1) defines a linear transformation $Q_{n\mu}$ on the space of functions P[U] on the set of SU(2) group elements U. The full transformation of one complete updating pass is obtained by taking the product of all the transformations $Q_{n\mu}$ over all n and μ. Call the result Q. We now redefine $P_1(U)$ to be the complete transformation Q applied to P(U). Then after ℓ complete passes the probability $P_\ell(U)$ is $Q^\ell P(U)$. The process of applying Q to P many times, defines a Markov process, and because the kernel of Q is positive there are standard limit theorems[11] that describe the limit of $P_\ell(U)$ for large ℓ. Because the space of configurations U is compact, for a finite lattice, [since SU(2) group space is compact] the standard theorems are very powerful. The basic idea is that to study the probability distribution $P_\ell(U)$, for large ℓ, one computes the expectation value of some arbitrary function Ψ(u) with respect to P_ℓ. This is

$$< \Psi P_\ell > \;=\; < \Psi Q^\ell P > \tag{A.2}$$

where the definition of $< \Psi P_\ell >$ is

$$< \Psi P_\ell > \;=\; \Pi_{n\mu} \int dU_{n\mu} \; \Psi(U) \; P_\ell(U). \tag{A.3}$$

Then this can be reinterpreted as $< \Psi_\ell P >$ where

$$\Psi_\ell(U) = (\tilde{Q})^\ell \Psi(U), \tag{A.4}$$

where $\tilde{Q}$ is the operator conjugate to Q. The next step is to determine the limit of $\Psi_\ell(U)$ for large U, namely to show that it is a constant independent of U. The first step is to use the conservation of probability to show that the maximum value of $|\Psi_\ell(U)|$, over all U, is less than or equal to the maximum value of $|\Psi_{\ell-1}(U)|$. The reason for this is as follows: the conservation of probability requires that in the special case that $\Psi_{\ell-1}(U)$ is a constant c, then

$$< \Psi_{\ell-1} \, P_1 > \;=\; c < P_1 > \;= c \tag{A.5}$$

independent of the choice of P, provided only that $< P >$ is 1, but

$$c = < \Psi_{\ell-1} \, P_1 > \;=\; < \Psi_{\ell-1} \, QP > = < \Psi_\ell \; P > . \tag{A.6}$$

Hence, one expects that the function Ψ_ℓ is also c. This means that

$$c = \tilde{Q}c \tag{A.7}$$

(with c being the constant function of U) which can easily be verified explicitly. Now since $\tilde{Q}$ has only positive terms, one has

$$\max \text{ (over all U) } |\Psi_\ell(U)| \leq \tilde{Q}\, |\Psi_{\ell-1}| \leq \tilde{Q} \max[|\Psi_{\ell-1}(U)|] = \max|\Psi_{\ell-1}(U)|, \quad (A.8)$$

where the last equality follows from Eq. (A.7).

The next step is to argue from the standard theory of bounded continuous functions on a closed compact space that there exists a subsequence $\{\ell_k\}$ of the functions $\Psi_\ell(U)$ which approach a continuous limiting function $\Phi_0(U)$ for $\ell_k \to \infty$, and that the further sub-sequences Φ_{ℓ_k+m} for fixed m converge to $\Phi_m(U)$ where $\Phi_m = \tilde{Q}^m \Phi_0$. These convergent sub-sequences exist whether or not the full sequence Ψ_ℓ converges or goes into undamped oscillations for large ℓ. Now, while we do not yet know the convergence of the sequence Ψ_ℓ, we do know that the full sequence of maxima $\max|\Psi_\ell(U)|$ *does* converge because this is a non-increasing sequence of positive numbers. Let the limit of these maximum values be M. Then, because the functions $\Phi_m(U)$ are all limits of sub-sequences of the Ψ_ℓ, these functions all have the same maximum value M. This is a nontrivial fact because there is also the inequality

$$M = \max|\Phi_m(U)| \leq \tilde{Q}^m \max|\Phi_0(U)| = \max|\Phi_0(U)| = M, \quad (A.9)$$

following from Eq. (A.7). Let U be the specific configuration for which $|\Phi_m(U)|$ takes on its maximum value; for convenience we assume $\Phi_m(U)$ is positive. Then at this point the calculation of $\tilde{Q}^m\Phi_0$ gives the same result as when Φ_0 is replaced by its maximum value M. This is possible only if Φ_0 takes on its maximum value M at every configuration U' which contributes to $\Phi_m(U)$. If we were considering $\Phi_1(U)$ this would only be a small sub-set of points in U space, since there would then only be 17 choices of $U_{n\mu}'$ for each link for which $\Phi_0(U')$ has to be equal to M. However, when one convolutes $\tilde{Q}$ many times, to give $\tilde{Q}^m$, the number of configurations U' that contributes, increases rapidly. In particular, it is possible that as $m \to \infty$ the set of configurations U' which contribute to $\Phi_m(U)$ for a particular configuration U becomes everywhere dense in U space. If this is true (as will be discussed below) then $\Phi_0(U)$, by continuity, must be equal to the constant M for all U.

An immediate consequence of this result is that any sub-sequence of the Ψ_ℓ which has a limit, must have the same constant function as its limit. This precludes oscillatory behavior of Ψ_ℓ for large ℓ which would lead to multiple limits; in fact, by general arguments the full sequence $\Psi_\ell(U)$ approaches a limit namely the constant M. In consequence,

$$<\Psi_0 P_\ell> = <\Psi_\ell P_0> \to <M P_0> = M \tag{A.10}$$

for $\ell \to \infty$. The constant M depends on the sequence Ψ_ℓ, that is, on the choice of Ψ_0, but does *not* depend on the choice for the function P. Thus, we can evaluate M by using the particular distribution $Z^{-1}\exp\{S(U)\}$ for P(U). As already discussed above, Q applied to this particular P leaves P unchanged, so $P_\ell = P$ for this case. Hence

$$\lim_{\ell \to \infty} <\Psi_0 P_\ell> = <\Psi_0 Z^{-1} \exp\{S(U)\}> \tag{A.11}$$

for *any* choice of the initial probability distribution $P_0(U)$. This is the desired result.

The one point that has to be cleared up is whether the set of configurations U' for which $\Phi_0(U')$ contributes to $\Phi_m(U)$ becomes everywhere dense in U space as m becomes large, independently of the choice of U. The set of points U' is simply for each link nμ, the set of all products of Unμ with up to m factors chosen from the set of 16 matrices W_k. It is very easy to ensure that this set is everywhere dense in the limit of large m. Let W_1 correspond to a rotation about the z axis through an angle Θ:

$$W_1 = \exp\{\frac{i\Theta}{2} \sigma_z\} . \tag{A.12}$$

Let $\Theta/4\pi$ be an irrational number. Then it is a standard theorem that the set of all products of W_1 is everywhere dense in the subspace of rotations about the z axis. If W_2 is $\exp\{i\Theta\sigma_x/2\}$ with the same Θ, then products of W_2 are everywhere dense in the space of rotations about the x axis. Since an arbitrary rotation can be expressed as a product of three rotations about the x or z axes, the multiple products of W_1 and W_2 combined are everywhere dense in the entire group space. Combined with W_1^{-1} and W_2^{-1}, we see that a set of four matrices is sufficient to provide full coverage of SU(2) space.

Creutz uses a different updating procedure. In his calculations the new value for Unμ is chosen using the probability distribution $Z^{-1}\exp\{S(U)\}$ itself except that this is used as a relative probability for the single matrix Unμ, with the other link variables being held fixed. In this case, it is obvious that the probability distribution $Z^{-1}\exp\{S(U)\}$ is preserved by the updating procedure, and the same arguments show that this is the limit for any initial probability distribution.

In my calculations I have found it useful to update a link variable three times before proceeding onto the next link variable. This requires only a small amount of extra computing time but, in practice, considerably reduces the statistical errors in the calculation. Using more than three updates per link did not improve the statistics much further but began to noticeably increase the computing time. If one were to use a large number of updates at each link before proceeding to the next link, one would expect the final probability distribution for $U_{n\mu}$ after these updates to be independent of the initial value $U_{n\mu}'$ and independent of the number of updates. A distribution in $U_{n\mu}$ alone (the other U's held fixed) which is preserved by the updating scheme is $\exp\{S(U)\}$; as a result, a large number of the Metropolis form updates I use is equivalent to the procedure used by Creutz. The practical observation that there was not much difference between three or more updates suggests that my calculations are not very different from Creutz' as far as the updating is concerned

A cause for concern regarding any Monte-Carlo calculation has to do with the random number generator used. Kalos' random number generator (which I use) is a computer program that generates a sequence of numbers; the $\ell + 1$st random number $R_{\ell+1}$ is given by

$$R_{\ell+1} = 11^{13} R_{\ell} \text{ (modulo } 2^{48}). \tag{A.13}$$

No such formula can generate a truly random sequence of integers; for one reason, the sequence necessarily repeats itself after at most 2^{48} steps. There are various tests[12] one can make of a random number generator such as this one, to try to check that the effects of non-randomness are small; these tests are complex and I have not tried to perform any of them.

Instead of formal proofs of the validity of the random number generator, I relied on comparisons of Monte-Carlo runs with weak- and strong-coupling expansions in the weak- and strong-coupling domain. The results agree to within the statistical errors of the Monte-Carlo calculations. (I use a reasonably conservative estimate for these errors, say the standard deviation of eight subsamples *without* dividing by $\sqrt{8}$.) However, I have been unable to construct strong-coupling expansions for the block spin expectation values, so the detailed strong-coupling comparisons apply only to the simple loop expectation values. (I do however know the strong-coupling *limit* for block expectation values, and the Monte-Carlo results are equal to the correct limit for very strong coupling.) In weak coupling there is a complication due to the presence of an exactly zero momentum mode of the gauge field (due to the periodic boundary conditions, which I use). Integrals over this zero momentum mode diverge in a conventional perturbation treatment; in the lattice theory the true integrals converge but one must consider regions of U space where the $U_{n\mu}$'s are far from weak. It turns out that the leading weak-coupling behavior can still be computed although the details are complicated and involve some multi-dimensional integrals which I have again computed by Monte-Carlo methods. The zero momentum mode is unimportant for small loops but is important for the expectation values of large-size block loops. Also in weak coupling, the statistical errors on large block loop expectation values were a fairly large percentage of the perturbative term, so the tests were not very precise.

REFERENCES

1. M. Creutz, L. Jacobs and C. Rebbi, Phys. Rev. Lett. 42, 1390 (1979); Phys. Rev. D (to be published).
2. M. Creutz, Phys. Rev. Lett. 43, 553 (1979).
3. R.H. Swendson, Phys. Rev. B 20, 2080 (1979); H.W.J. Blöte and R.H. Swendson, Phys. Rev. B 20, 2077 (1979); S. Ma, Phys. Rev. Lett. 37, 461 (1976).
4. See, e.g., A. Peterman, Phys. Reports 53, 159 (1979).
5. K. Wilson, Phys. Rev. D 10, 2445 (1974); K. Wilson, in: *New Developments in Quantum Field Theory and Statistical Mechanics*, (Plenum Press, New York, 1977).
6. See, e.g., Th. Niemeijer and J.M.J. Van Leeuwen, in: *Phase Transitions and Critical Phenomena*, eds. C. Domb and M.S. Green, (Academic Press, N.Y., 1976), Vol.6, p.425.
7. See, e.g., K. Wilson, Rev. Mod. Phys. 47, 773 (1975).
8. J.M. Hammersley and D.C. Handscomb, *Monte Carlo Methods*, (Wiley, New York, 1964).
9. J.B. Kogut, R.B. Pearson and J. Shigemitsu, Phys. Rev. Lett. 43, 484 (1979).
10. M. Creutz, *Solving Quantized SU(2) Gauge Theory*, Brookhaven preprint (1979).
11. See, e.g., W. Feller, *An Introduction to Probability Theory and its Applications*, (Wiley, New York, 1966) Vol.II, p.26488.
12. D.E. Knuth, *The Art of Computer Programming*, (Addison-Wesley, Reading, Mass, 1969), Vol.II, Chap.3.

THE 1/N EXPANSION IN ATOMIC AND PARTICLE PHYSICS

Edward Witten

Lyman Laboratory of Physics
Harvard University
Cambridge, Massachusetts 02138

ABSTRACT

The 1/N expansion in atomic physics is explained in order to motivate consideration of the 1/N expansion in QCD. Some comments are made concerning possible future development of the 1/N expansion in QCD.

I. INTRODUCTION

Almost as old as QCD itself is 't Hooft's suggestion of the 1/N expansion.[1] The 1/N expansion has provided us with some very interesting qualitative insights concerning QCD, including explanations of some qualitative phenomena whose theoretical status is otherwise mysterious.

The 1/N expansion also offers a hope--really only the glimmer of a hope, since the difficulties involved are very great--of a possible systematic expansion which might unravel the secrets of this model.

Since various reviews of the 1/N expansion are available,[2,3,4,5,6] these lecture notes do not aim at a systematic review of the subject. Instead I will discuss just a few aspects of the 1/N expansion, the purpose being to explain the motivation and discuss the prospects.

II. ATOMIC PHYSICS

Let us consider the familiar Hamiltonian of the hydrogen atom:

$$H = \frac{p^2}{2m} - \frac{e^2}{r} \qquad (1)$$

Since for $e^2 = 0$ we can solve this problem exactly--it is simply the problem of the motion of a free particle--one's first hope might be to try to understand the hydrogen atom in the small e^2 regime by treating the interaction term, $-e^2/r$, as a perturbation.

This hope is frustrated because in the hydrogen atom e^2 is not really a relevant parameter--it can be eliminated from the problem by redefining the scale of distances. After a rescaling $r \to tr$, $p \to p/t$, with $t = 1/me^2$, the Hamiltonian becomes

$$H = (me^4)\left(\frac{p^2}{2} - \frac{1}{r}\right) \qquad (2)$$

and one sees that the "coupling constant" e^2 appears only in the overall factor me^4 which serves merely to define the overall scale of energies (and which could be absorbed in a rescaling of the time coordinate). Therefore, except for the overall scale of lengths and times, the physics of the hydrogen atom--and atomic and molecular physics in general--is independent of e^2, and perturbation theory in e^2 is meaningless.

Otherwise, atomic and molecular physics would be completely different! The literature would be full of, for instance, weak coupling calculations of the structure of the iron atom.

Atoms and molecules can be described by reduced Hamiltonians with e^2 scaled out. For hydrogen the reduced Hamiltonian is

$$\hat{H} = \frac{p^2}{2} - \frac{1}{r} \qquad (3)$$

The reduced Hamiltonian contains no evident free parameter.

Without a free parameter there is no perturbation expansion. Without a perturbation expansion what can we do?

Apparently, there is nothing to do except to try to diagonalize the reduced Hamiltonian exactly. This is not so bad, since we can actually diagonalize the hydrogen Hamiltonian exactly, and since more complicated atoms can be analyzed on a computer.

But suppose--since this is the case in QCD and in many other analogous problems--that we were unable to diagonalize the Hamiltonian exactly, and that even a computer solution were impractical. How then might we proceed?

To make progress, we must make an expansion of some kind. Since there is no obvious expansion parameter we must find a hidden one.

To find a hidden expansion parameter, we may treat as a free, variable parameter a quantity that one usually regards as given and fixed.

For instance, we may take a cue from the spectacular developments of this decade in critical phenomena. After decades in which the study of critical phenomena was frustrated by the absence of an expansion parameter, Wilson and Fisher suggested[7] that to introduce an expansion parameter, one should regard the number of spatial dimensions not as a fixed number, three, but as a variable parameter. They showed that critical phenomena are simple in four dimensions and that in $4-\varepsilon$ dimensions critical phenomena can be understood by perturbation theory in ε. Even at $\varepsilon = 1$--the original three dimensional problem--this perturbation theory is quite successful. (That latter fact--a successful perturbation expansion at $\varepsilon = 1$--may come as a surprise, but shouldn't. It simply illustrates the important rule that approximations that are correct qualitatively tend to be successful also quantitatively, although sometimes for reasons that are understood only in hindsight.)

How, by analogy, can we create an expansion parameter "from thin air" for atomic physics?

Instead of studying atomic physics in three dimensions, where it possesses an O(3) rotation symmetry, let us consider atomic physics in N dimensions, so that the symmetry is O(N). We will see that atomic physics simplifies as $N \to \infty$ and can be solved for large N by an expansion in powers of 1/N.

I should explain in advance that the 1/N expansion in atomic physics is not very useful at $N = 3$ (N must apparently be at least six or seven for the 1/N expansion to give good results). The 1/N expansion for atomic physics is discussed here only in order to explain the ideas.

Also, by "atomic physics" I mean the 1/r potential, irrespective of the number of dimensions. It is very important in atomic physics that e^2 has dimensions of energy times length, and in generalizing atomic physics to N dimensions we should preserve this property, if we wish the generalization to be smooth.

Now, how would we carry out the 1/N expansion for atoms? Considering first hydrogen, we would like to solve the Schrodinger equation

$$\left(-\frac{\nabla^2}{2m} - \frac{e^2}{r}\right)\Psi = E\Psi \tag{4}$$

For simplicity, let us consider the s-wave states only, although it is not difficult to include orbital angular momentum in this procedure. For s-wave states, Ψ is a function of r only. The Laplacian is conveniently rewritten in terms of radial derivatives, and the Schrodinger equation becomes

$$\left(-\frac{1}{2m}\left(\frac{d^2}{dr^2}+\frac{N}{r}\frac{d}{dr}\right)-\frac{e^2}{r}\right)\Psi = E\Psi \tag{5}$$

(Actually, the "N" in (5) should be N - 1. For large N, the difference between N and N - 1 is negligible. Similar approximations will be made in some of the equations that follow.)

To eliminate the first derivative term from the Hamiltonian, we make the transformation

$$H \to r^{\frac{1}{2}N}\, H\, r^{-\frac{1}{2}N}$$

(this is equivalent to redefining the wave function by

$$\Psi = r^{-\frac{1}{2}N}\,\tilde{\Psi}),$$

whereupon the Hamiltonian becomes

$$H = -\frac{1}{2m}\frac{d^2}{dr^2}+\frac{N^2}{8mr^2}-\frac{e^2}{r} \tag{6}$$

If now we rescale the radial coordinate, defining $r = N^2R$, then in terms of R, H becomes

$$H = \frac{1}{N^2}\left(-\frac{1}{2mN^2}\frac{d^2}{dR^2}+\frac{1}{8R^2m}-\frac{e^2}{R}\right) \tag{7}$$

Apart from the overall factor of $1/N^2$, which only determines the overall scale of energy or time, the only N in this Hamiltonian is the N^2 that appears with the mass in the kinetic energy term. The Hamiltonian (7) describes a particle with an effective mass $M_{eff} = mN^2$, moving in an effective potential

$$V_{eff} = \frac{1}{8R^2m} - \frac{e^2}{R} \tag{8}$$

Since for large N the effective mass is very large, the particle for large N simply sits in the bottom of the potential V_{eff}--the quantum fluctuations are negligible. The ground state energy is simply the absolute minimum of V_{eff}.

To calculate the excitation spectrum, one may, for large N, simply make a quadratic approximation to V_{eff} near its minimum, since the large effective mass ensures that the particle stays very close to the minimum of V_{eff}. The anharmonic terms in the expansion of V_{eff} around its minimum can be included as perturbations; this leads to an expansion in powers of 1/N.

Of course, the hydrogen atom is exactly soluble, so it is perhaps not startling that it can be solved by an expansion in powers of 1/N. But the 1/N expansion can also be carried out for more complex atoms, such as helium.

If x_i and y_i are the positions of the two electrons, then the Hamiltonian of the helium atom is

$$H = -\frac{1}{2m}\sum_i\left(\frac{d^2}{dx_i^{\,2}} + \frac{d^2}{dy_i^{\,2}}\right) - 2e^2\left(\frac{1}{|x|} + \frac{1}{|y|}\right) + \frac{e^2}{|x-y|} \tag{9}$$

Let us once again consider s-waves, or states of zero angular momentum. The wave function of such a state is a function only of the rotationally invariant quantities, which (figure (1)) are the distances $x = \sqrt{x^2}$ and $y = \sqrt{y^2}$ of the two electrons from the nucleus, and also the angle θ between the two electrons. Acting on such a wave-function, the kinetic energy becomes

$$H_{Kin} = -\frac{1}{2m}\left(\frac{d^2}{dx^2} + \frac{d^2}{dy^2} + \frac{N}{x}\frac{d}{dx} + \frac{N}{y}\frac{d}{dy} + \left(\frac{1}{x^2} + \frac{1}{y^2}\right)\frac{d^2}{d\theta^2}\right.$$
$$\left. + N\left(\frac{1}{x^2} + \frac{1}{y^2}\right)\frac{\cos\theta}{\sin\theta}\frac{d}{d\theta}\right) \tag{10}$$

This expression does not look very promising, but it simplifies after a transformation

$$H \to (xy\sin\theta)^{\frac{1}{2}N}\, H (xy\sin\theta)^{-\frac{1}{2}N}.$$

The Hamiltonian of the system becomes

$$H = -\frac{1}{2m}\left(\frac{d^2}{dx^2} + \frac{d^2}{dy^2} + \left(\frac{1}{x^2} + \frac{1}{y^2}\right)\frac{d^2}{d\theta^2}\right)$$
$$+ \frac{N^2}{8m}\left(\frac{1}{x^2} + \frac{1}{y^2}\right)\frac{1}{\sin^2\theta} - \frac{2e^2}{x} - \frac{2e^2}{y} + \frac{e^2}{\sqrt{x^2+y^2-2xy\cos\theta}} \tag{11}$$

If, again, we rescale the coordinates, $x = N^2X$, $y = N^2Y$, we find

$$H = \frac{1}{N^2}\left[\frac{1}{2mN^2}\left(-\frac{d^2}{dx^2} - \frac{d^2}{dY^2} - \left(\frac{1}{X^2} + \frac{1}{Y^2}\right)\frac{d^2}{d\theta^2}\right)\right.$$
$$\left. + V_{eff}(X,Y,\theta)\right] \tag{12}$$

where

$$V_{eff}(X,Y,\theta) = \frac{1}{8m}\left(\frac{1}{X^2}+\frac{1}{Y^2}\right)\frac{1}{\sin^2\theta} - \frac{2e^2}{X} - \frac{2e^2}{Y} + \frac{e^2}{\sqrt{X^2+Y^2-2XY\cos\theta}} \tag{13}$$

Once again, for large N, the kinetic energy is suppressed, and the particle simply sits in the minimum of the effective potential. The ground state energy is the absolute minimum of V_{eff}, and the excitation energies could be computed by making a quadratic approximation to V_{eff} near its minimum.

Notice that for large N the helium atom is very much like a molecule. There are well-defined "bond lengths," the equilibrium values of x and y, and a well defined "bond angle," the equilibrium value of θ. These bond lengths and angles are well defined, with negligible quantum fluctuations, because the effective mass, $M_{eff} = N^2m$, is large, just as in a molecule the bond parameters are well defined and the quantum fluctuations small because the nuclear masses are large.

Unfortunately, the 1/N expansion is not extremely successful in atomic physics at N = 3. For instance, for hydrogen, the ground state binding energy, to lowest order in 1/N, is $(2/N^2)me^4$. At N = 3, this is $(2/9)me^4$, while the actual value in three dimensions is $\frac{1}{2}me^4$.

III. QCD

Let us now turn our attention to quantum chromodynamics, or QCD.

In QCD, as in atomic physics, the coupling constant can be scaled out of the problem. In atomic physics this is obvious on dimensional grounds. In QCD, it is not obvious, but it is still true. It is a consequence of the renormalization group and the nonzero β function that in QCD a change in the coupling constant can be absorbed in a change in the scale of distances, so that the coupling constant can be scaled out of the problem.

Because of this, in QCD, perturbation theory does not illuminate the central unsolved problems of confinement, dynamical mass generation, and chiral symmetry breaking. Changing the coupling constant in QCD would change the energy at which these phenomena occur, but would not change the phenomena themselves.

(Atomic physics has a natural energy scale, the Rydberg, $1Ry = \frac{1}{2}me^4$, but has no characteristic dimensionless parameter. If one scatters a high energy external probe from an atom, there is a small expansion parameter--the ratio of the Rydberg to the energy of the probe. A perturbation expansion in this parameter exists; it is called the Born expansion. Likewise, in QCD, although there is no characteristic coupling constant, there is a characteristic mass scale. In the response to an external probe which has an energy much larger than this mass scale, there is a small parameter--the "effective coupling constant", $\bar{g}(Q^2)$. The perturbation expansion in this parameter is, of course, very important--it is the basis of most of what we know about QCD. But in hadronic physics itself, as opposed to the response of a hadron to a probe carrying very high energy, the coupling constant is not a relevant parameter and can be scaled out.)

Thus, in QCD the basic difficulty is the same as in atomic physics--the seeming absence of an expansion parameter. Moreover, QCD is so complicated an exact solution is out of the question. And the obstacles to solving the problem numerically are enormous. (For efforts in this direction, see the lectures of K. Wilson at this school.) Faced with these difficulties, one naturally asks whether in QCD there is some way to find a hidden expansion parameter.

The 1/N expansion, originally suggested by 't Hooft in 1974,[1] is an attempt to find such a parameter. In this approach, one generalizes QCD from three colors and an SU(3) gauge group to N colors and an SU(N) gauge group, and one tries to solve the theory for large N.

The basic idea of the 1/N expansion in QCD is very simple. For large N, the gluon field $A_\mu{}^i{}_j$ is an N x N matrix, with N^2 components. (Actually, it is a traceless matrix, with only $N^2 - 1$ components, but the difference between N^2 and $N^2 - 1$ is unimportant if N is large.) There are thus N^2 (or $N^2 - 1$) gluon particle states which might appear as intermediate states in Feynman diagrams.

For large N, the Feynman diagrams contain large combinatoric factors, arising from the large number of possible intermediate states. Only the diagrams with the largest possible combinatoric factors need to be included, if N is large. So for large N the theory simplifies, only a subclass of diagrams being relevant.

To see how this works in more detail, let us consider (figure (2)) the lowest order contribution to the gluon vacuum polarization. It is not hard to see that for any choice of initial and final states, there are N possibilities for the intermediate state in this diagram. (If the initial state gluon is of type $i\bar{j}$, the intermediate state gluons may be $i\bar{k}$ and $k\bar{j}$, where N values of k are possible, corresponding to N intermediate states.) Therefore this diagram has a combinatoric factor of N, from the sum over N possible intermediate states.

At first sight it seems that this diagram diverges without limit for large N, in proportion to N, and therefore Yang-Mills theory simply does not have a smooth large N limit. However, we must remember that the one loop diagram contains factors of coupling constant at each of the two vertices. If the coupling constant is chosen to be $g/\sqrt{N}$, where g is to be kept fixed as $N\to\infty$, then the N dependence cancels out of the one-loop diagram, since $N(g/\sqrt{N})^2 = g^2$ is independent of N. So the choice of the coupling as $g/\sqrt{N}$ gives a smooth limit to the one-loop diagram.

However, in supposing that the coupling constant is $g/\sqrt{N}$, we have made a fateful choice. Complicated diagrams will have factors of $g/\sqrt{N}$ at each vertex, and so will vanish for large N unless, like the simple one-loop diagram, they have combinatoric factors large enough to cancel the factors at the vertex. It turns out that a certain class of diagrams with the largest combinatoric factors survives for large N, while all other diagrams vanish as $N\to\infty$.

For example, the three-loop diagrams in figure (3) each have factors of $(g/\sqrt{N})^6$ from the six vertices. The first turns out to have a combinatoric factor of N^3, and so survives for large N ($(g/\sqrt{N})^6 N^3 = g^6$), while the second has only a combinatoric factor of N, and vanishes for large N as $1/N^2$.

The class of diagrams that survives for large N was originally determined by 't Hooft. The diagrams that survive are the "planar" diagrams. A planar diagram is a diagram that can be drawn in the plane with no two lines crossing. The first diagram of figure (3) is a planar diagram, and survives as $N\to\infty$. But the second diagram in figure (3) is not a planar diagram, since two gluon lines cross at the center of the diagram, and it vanishes for large N.

The planar diagrams are a vast class of diagrams; a typical planar diagram with many loops is indicated in figure (4). Since 1974, when the 1/N expansion was originally suggested, there has been very little progress toward actually summing the planar diagrams. And this--or something equivalent--would have to be done, in order for the 1/N expansion to become really important in particle physics.

Nevertheless, the 1/N expansion has already given us a fair amount of insight into the phenomenology of QCD. The reason for this is that there are certain "selection rules" or qualitative properties that are preserved, diagram by diagram, by each of the planar diagrams, but not by the nonplanar diagrams.

These selection rules, which can be regarded as predictions or tests of the 1/N expansion, are observed to be rather well satisfied in nature. If one assumes that the 1/N expansion is a good approximation to the real world, which has $1/N = 1/3$, then we obtain by means of the 1/N expansion an understanding of some qualitative properties of the strong interactions that are not otherwise well understood. Conversely, the success of the selection rules is a good reason--probably our best one--to believe that the 1/N expansion really is relevant to nature.

I will not go into detail here, since there are a variety of reviews,[2,3,4,5] and I have discussed the subject elsewhere.[6] Some of the most interesting qualitative predictions of the 1/N expansion are the following:

(a) For given J^{PC} and flavor quantum numbers, there are (for large N) an infinite number of resonances. These resonances are narrow, their widths being of order 1/N.

(b) Zweig's rule is satisfied to lowest order in 1/N, so that, for instance, ρ and ω are degenerate at $N = \infty$. Also, the quark-antiquark sea is absent at $N = \infty$, and mesons are pure $q\bar{q}$ (not $q\bar{q}q\bar{q}$).

(c) Exotics, or meson-meson bound states, are absent at $N = \infty$.

(d) Multi-body decays of unstable mesons are dominated by resonant two body final states, whenever these are available. (A quick glance at the particle data tables will convince one that this rule is well satisfied. For instance, the observed decay $B\to\pi\pi\pi\pi$ is observed to proceed mainly as $B\to\omega\pi$, followed by $\omega\to\pi\pi\pi$.) A non-resonant meson decay with a k body final state has a partial width in the 1/N expansion of $1/N^{K-1}$.

(e) The meson-meson scattering amplitude is of order 1/N, and is given, as in the Regge pole theory, by an (infinite) sum of one meson exchange diagrams.

It will be noted that many of the above properties are also properties of dual models. Since 't Hooft,[1] many writers have suspected a connection between the 1/N expansion and dual models. No solid argument for such a connection is known. If a connection were solidly established, many of the leading mysteries of QCD would be solved. The dual model has built into it confinement, with quarks at the end of a string (let's not forget that some of these concepts were discovered via the dual model!). Also, a clear connection between QCD and the dual model would mean that the problem of dynamical mass generation had been solved, since the dual model certainly has a mass scale (the Regge slope). A solidly established connection between the 1/N expansion to QCD and dual models might also mean at least a partial solution of the problem of understanding chiral symmetry breaking, since the dual model has at least an indication of a Goldstone pion.

(1)

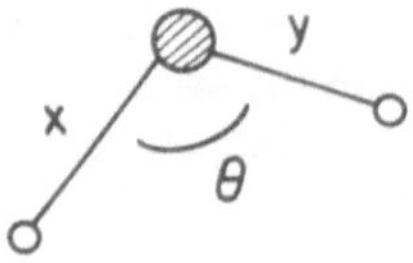

(2)

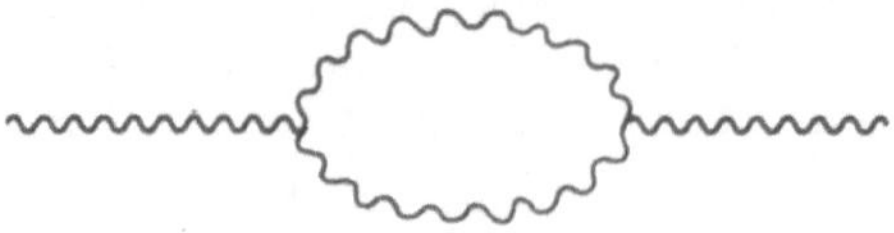

(3)

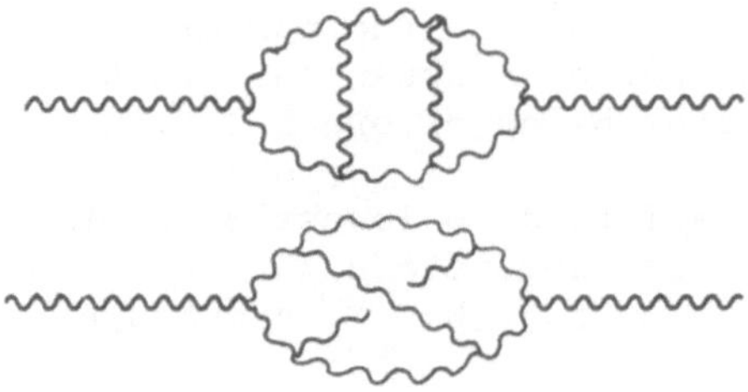

(4)

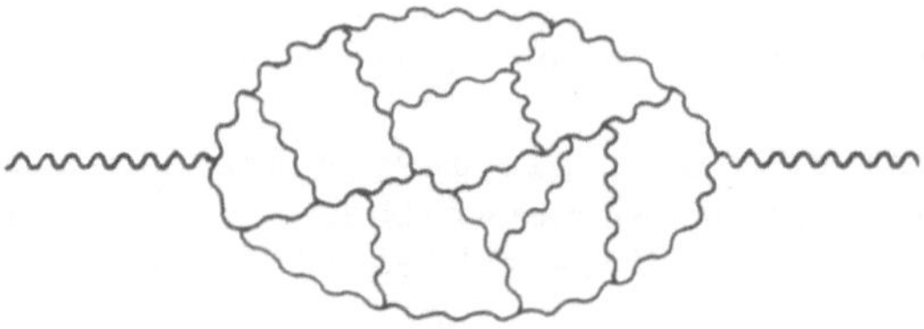

(5)

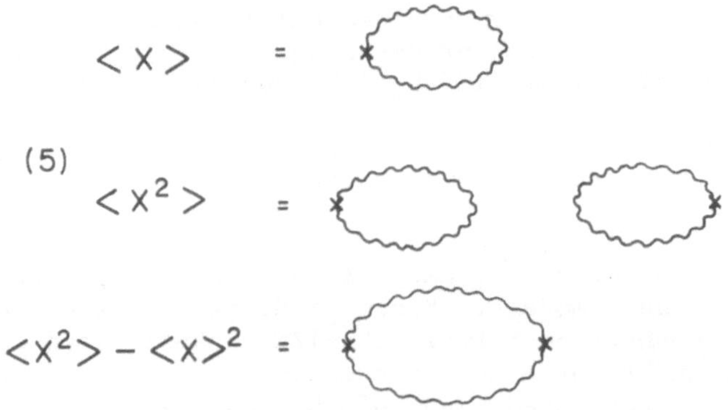

Figures

1. The helium atom, with a nucleus (shaded circle) and two electrons (small circles). For large N the atom has well-defined "bond angles" and "bond lengths".
2. The lowest order contribution to the gluon vacuum polarization.
3. Two three-loop diagrams.
4. A typical large planar diagram.
5. Illustration of the rule $\langle X^2\rangle = \langle X\rangle^2$.

It is suspected by many physicists that even at $N = \infty$, QCD is only approximately related to a dual model. The connection probably applies only asymptoticaly, to the highly excited hadron states of large angular momentum.

Incidentally, there is a very interesting prediction of the 1/N expansion in QCD whose validity is so far not proved or disproved. This concerns glueball states. For large N, QCD definitely contains glueball states. In fact (as for mesons), there are infinitely many glueball states for each J^{PC}. Moreover, the glueball states should be even narrower than mesons (widths or order $1/N^2$). They interact weakly with mesons (a glueball-meson-meson coupling is of order 1/N), and they mix with mesons only weakly (the amplitude for mixing between a quarkless glueball state and a $q\bar{q}$ meson is of order $1/\sqrt{N}$). If the glueballs are found experimentally, it will be interesting to see how well they fit these expectations.

IV. PROSPECTS

To sum the planar diagrams--or otherwise solve QCD for $N\to\infty$--is clearly a very difficult problem. Very possibly, a conceptual breakthrough would be needed before this problem could be solved.

This problem has been the subject of interesting recent work following several different approaches.[8,9,10,11] Here I would like to make a few remarks along lines suggested by the recent work of Breyin, Parisi, Itzykson, and Zuber[12] and Casartelli, Marchesini, and Onofri[13] (see also a recent paper by Itzykson and Zuber[14]). The remarks that follow do not seem to lead to a solution of the problem, at least not without some very non-trivial additional input, which remains to be identified.

Let X be any gauge-invariant operator in QCD, such as a local operator like Tr $F_{\mu\nu}F_{\mu\nu}(x)$, or a non-local operator, perhaps a Wilson loop or something even more exotic like det(iD - m), the determinant of the Dirac operator. Then, to lowest order in 1/N there is a simple factorization rule

$$\langle X^2\rangle = \langle X\rangle^2 \tag{14}$$

which states that the expectation value of X squared is the square of the expectation value of X. (This fact has recently been discussed in another context by Migdal.[9])

The factorization (14) can easily be seen in terms of the Feynman diagrams. For instance, let X be a local operator like Tr $F_{\mu\nu}F_{\mu\nu}$, and consider the free field theory diagrams of figure (5). We see that $\langle X\rangle$ is of order N^2, since there are N^2 gluon species that can run around the loop. So $\langle X\rangle^2$ is of order N^4; it is given by two disconnected one-loop diagrams, each of order N^2. But the connected part of $<X^2>$ is only of order N^2, since it comes from a diagram with only one loop (and two insertions of X); any of the N^2 gluon species can run around the loop, so the loop is of order N^2. So, the connected part of $<X^2>$, of order N^2, is negligible for large N compared to the disconnected part, which is of order N^4. This establishes the factorization (14).

(14) has a simple generalization. Let X, Y, Z, etc., be any gauge invariant operators (local or not). Then, to lowest order in 1/N

$$<XYZ> = <X><Y> <Z> \tag{15}$$

The argument is just as before.

(14) can also be rewritten to say that any gauge invariant operator X has a vansishing mean square fluctuation at $N = \infty$:

$$<(X - <X>)^2> = 0 \text{ at } N = \infty \tag{16}$$

The factorization (15) or absence of fluctuation (16) are characteristic of c-numbers rather than of quantum mechanical operators. Indeed, as $N \to \infty$, the gauge invariant "operators" become c-numbers.

What are the implications of this fact? Let us think in terms of the usual Euclidean space path integral formulation of QCD, defined by the "partition function"

$$Z = \int dA_\mu \exp - \frac{1}{4} \int d^4 x \operatorname{Tr} F_{\mu\nu}{}^2 \tag{17}$$

and for any operator X,

$$<X> = \frac{1}{Z} \int dA_\mu \, X \exp - \frac{1}{4} \int d^4 x \operatorname{Tr} F_{\mu\nu}{}^2 \tag{18}$$

From the point of view of the path integral, how does it come about that, ordinarily, $<X^2> \neq <X>^2$?

This arises from the basic property of the path integral of averaging over different classical fields. The various gauge fields over which we integrate in doing the path integral have various different values of any given operator X, and upon doing the path integral we obtain not just an average value, but a whole spectrum of possible values. In other words, we obtain not just the vacuum expectation value $<X>$, but also a mean square fluctuation $<X^2> - <X>^2$.

Conversely, what does it mean that for large N the operators do not have nonzero square fluctuations?

It means--and this is the main point--that at $N = \infty$ it is not really necessary to do a path integral, or to average over different gauge fields. There must be a single classical gauge field that gives the whole answer!

Indeed, suppose that even at $N = \infty$ we are still doing an averaging over different gauge fields. Suppose that at least two distinct gauge fields, $A_\mu(1)$ and $A_\mu(2)$, are still being considered in the path integral. If $A_\mu(1)$ and $A_\mu(2)$ are really different (and are not related by a gauge transformation), then there will be some gauge invariant operator X which distinguishes them, in the sense that its value in the classical field $A_\mu(1)$ is different from its value in the classical field $A_\mu(2)$. The operator X would then have a certain probability to assume its value at $A_\mu(1)$, and a certain probability to assume its value at $A_\mu(2)$. With nonzero probabilities to assume these two different values, the operator X would inevitably have a nonzero mean square fluctuation.

So the vanishing of the mean square fluctuations of the gauge invariant operators means that at $N = \infty$ we are not averaging over two or more different gauge fields. There must be (up to a gauge transformation) a single classical gauge field $A_\mu^{cl}(x)$ which gives the whole answer of the path integral. A_μ^{cl} must have the property that, given any gauge invariant operator X, its vacuum expectation value is equal to its value in the classical field A_μ^{cl},

$$\langle X\rangle = X(A_\mu^{cl}) \tag{19}$$

This leads to the required factorization, since for any two operators X and Y, the value of XY in A_μ^{cl} is, of course, the product of the value of X and of the value of Y.

If we could determine this one classical gauge field A_μ^{cl}, we would have the solution of QCD at $N = \infty$! Unfortunately, the above argument only shows the existence of A_μ^{cl} and gives no indication of how to determine it. However, there is a general argument that gives a surprisingly strong restriction on A_μ^{cl}.

Since vacuum expectation values are Poincaré invariant, the value $X(A_\mu^{cl})$ must be invariant under Poincaré transformations, for any gauge invariant operator X. This means that A_μ^{cl} must itself be Poincaré invariant, up to a gauge transformation--the change in A_μ^{cl} under a rotation or translation of space must be equivalent to a gauge transformation.

It is easy to show that if A_μ^{cl} is Poincaré invariant up to a gauge transformation, there must be a gauge in which it is actually translation invariant and thus space-time independent:

$$A_\mu^{cl}(x) = A_\mu^{cl}(0), \text{ for all } x \tag{20}$$

(In other words, if A_μ^{cl} is translation invariant up to a gauge transformation, it is possible to make a gauge transformation which undoes the gauge transformation that would have had to accompany the translation. The ability to do this is related to the fact that the translation group acts transitively on space-time.)

Moreover, in this gauge, a rotation or Lorentz transformation must be equivalent to a global gauge transformation, since only a global gauge transformation would preserve the fact that A_μ^{cl} is space-time independent. That a rotation is equivalent to a global gauge transformation means that $A_\mu^{cl}(0)$ transforms as a vector operator under some representation of the Lorentz group.

So a knowledge of four matrices $A_\mu^{cl}(0)$, which moreover are Lorentz vector operators, would lead to a solution of QCD in the large N limit. Unfortunately, the four matrices $A_\mu^{cl}(0)$ are very large N x N matrices, and since all of this is true only at $N = \infty$, we should really think of $A_\mu^{cl}(x)$ as infinite matrices, or operators on Hilbert space.

How could we proceed if we knew $A_\mu^{cl}(x)$? We would like, for example, to calculate the meson mass spectrum. To do this, we must include quarks in the path integral, and study

$$Z(J) = \int d\Psi d\bar{\Psi}\, dA_\mu \, \exp - \frac{1}{4}\int d^4 x \, \mathrm{Tr}\, F_{\mu\nu}F_{\mu\nu}$$

$$\exp \bar{\Psi}(i\not{D} - m + J)\Psi \tag{21}$$

where the quarks--assumed to be in the fundamental representation of SU(N)--have an arbitrary bare mass m, and where a source J has been introduced which has arbitrary space-time dependence and gamma matrix structure.

At least formally, we can integrate out the quarks and write

$$Z(J) = \int dA_\mu \, \exp - \frac{1}{4}\int d^4 x \, \mathrm{Tr}\, F_{\mu\nu}F_{\mu\nu}$$

$$\exp \mathrm{Tr}\, \ell n(i\not{D} - m + J) \tag{22}$$

This can also be written

$$Z(J) = Z(0)\left(\frac{1}{Z(0)}\int dA_\mu \, \exp - \frac{1}{4}\int d^4 x \, \mathrm{Tr}\, F_{\mu\nu}F_{\mu\nu} \right.$$
$$\left. \exp \mathrm{Tr}\, \ell n(i\not{D} - m + J)\right)$$

$$= Z(0)\langle \exp \mathrm{Tr}\, \ell n(i\not{D} - m + J)\rangle \tag{23}$$

where the expectation value in the last part of (23) is evaluated in the world without quarks, by a gluon path integral only.

Now for given J, exp Tr $\ell n(i\not{D} - m + J)$ is a gauge invariant operator, although a non-local one. We can apply to it our basic formula $\langle X\rangle = X(A_\mu^{cl})$, whose validity is not limited to local operators. From (23) we thus get

$$Z(J) = Z(0)\ \exp\ \mathrm{Tr}\ \ell n(i\not{D}^{cl} - m + J) \tag{24}$$

where $\not{D}^{cl} = \not{\partial} + i\not{A}^{cl}$ is the Dirac operator in the classical field A_μ^{cl}.

Now, to determine the meson mass spectrum, we must study the two-point function of a quark bilinear operator. It is obtained by differentiating ℓn Z twice with respect to J at J = 0:

$$\left(\frac{\delta}{\delta J(x_1)}\ \frac{\delta}{\delta J(x_2)}\right)_{J=0} \ell n\ Z(J) = \left(\frac{\delta}{\delta J(x_1)}\ \frac{\delta}{\delta J(x_2)}\right)_{J=0} \mathrm{Tr}\ \ell n(i\not{D}^{cl} - m + j) \tag{25}$$

More generally, a connected scattering amplitude involving k mesons is given by differentiating ℓn Z k times with respect to J at J = 0:

$$\left(\frac{\delta}{\delta J(x_1)}\ \frac{\delta}{\delta J(x_2)} \cdots \frac{\delta}{\delta J(x_k)}\right)_{J=0} \ell n\ Z(J)$$
$$= \left(\frac{\delta}{\delta J(x_1)}\ \frac{\delta}{\delta J(x_2)} \cdots \frac{\delta}{\delta J(x_k)}\right)_{J=0} \mathrm{Tr}\ \ell n(i\not{D}^{cl} - m + J) \tag{26}$$

But the derivatives of Tr $\ell n(i\not{D}^{cl} - m + J)$ with respect to J are simply one-loop diagrams for the propagation of quarks in the background field A_μ^{cl}! So, if we knew A_μ^{cl}, we could calculate the meson masses and scattering amplitudes simply (figure (6)) by evaluating one-loop diagrams in the background field.

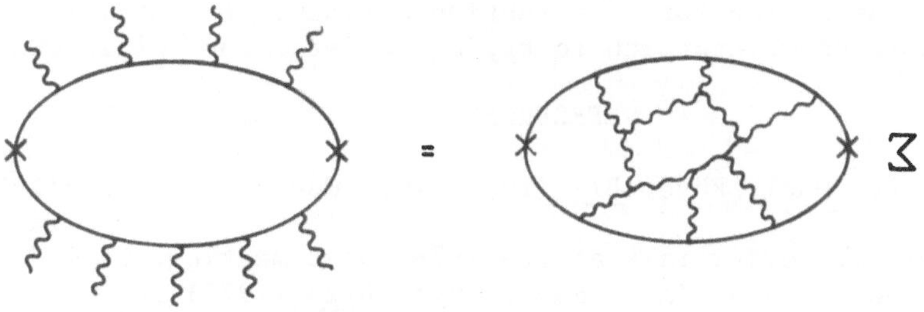

(9)

Figure 6. The sum of the planar diagrams, which gives the meson spectrum, is equal to the one-loop diagram for propagation of quarks in the background field $A_\mu cl$.

There is a similar but more intricate argument which indicates that the glueball spectrum and scattering amplitudes could be calculated from, roughly speaking, the one-loop diagram for gluon propagation in the background field A_μ^{cl}. The situation for baryons is more complicated, however.[6] A baryon, with its N quarks, perturbs the large N vacuum so strongly that the field A_μ^{cl} would have to be recalculated to take into account the presence of the baryon. The baryon spectrum could then be calculated by studying quark propagation in the modified A_μ^{cl}. I will not discuss here in detail the status of glueball states and baryons from this point of view.

The foregoing is rather tantalizing, but I do not believe that it will lead much farther unless a very deep and far-reaching idea can be supplied. I do not have any advice to offer about how to determine A_μ^{cl}, and it is simply not obvious that it is any easier to determine A_μ^{cl} than it would be, for instance, to diagonalize the Hamiltonian exactly. Some of the difficulties can be seen from the recent paper of Itzykson and Zuber.[14]

It would be very interesting if one could find a dynamical principle which even in principle offered a way to determine A_μ^{cl}. For instance, one might try to find an effective action whose minimum is A_μ^{cl}. This effective action could not be the classical action, because the minimum of the classical action is $A_\mu = 0$, and $A_\mu = 0$ obviously does not satisfy $\langle X \rangle = X(A_\mu)$. (Also, it is possible to show on general grounds, by considering the Ward identities satisfied by the objects $\langle X \rangle$, that A_μ^{cl} is <u>not</u> a solution of the classical Yang-Mills equations, because the Ward identities contain commutator terms that are not reflected in the classical equations. This again shows that if A_μ^{cl} is an extremum or minimum of some effective action, this action is not the classical action.)

Itzykson and Zuber[14] have determined a suitable effective action in the n-vector model. They have also shown some of the difficulties in trying to find an effective action for QCD. While I am not extremely optimistic that this approach will bear fruit, I think that the search for an effective action is probably the most reasonable known approach to trying to determine A_μ^{cl} in QCD.

REFERENCES

1. G. 't Hooft, Nucl. Phys. <u>B72</u> (1974) 461, Nucl. Phys. <u>B75</u> (1974) 461.
2. G. Veneziano, review talk at the 1978 Tokyo meeting; G. C. Rossi and G. Veneziano, Nucl. Phys. <u>B123</u> (1977) 507.
3. G. G. Chew and C. Rosenzweig, Physics Reports <u>41C</u> (1978) No. 5.
4. A. De Rújula, talk at 1979 European Physical Society Meeting (CERN preprint).
5. S. Coleman, Erice lectures, 1979.

6. E. Witten, "Baryons in the 1/N Expansion," (Harvard preprint HUTP-79/A007), to appear in Nuclear Physics.
7. K. Wilson and M. Fisher, Phys. Rev. Lett. 28 (1972) 240.
8. C. Thorne, Phys. Rev. D 17 (1978) 1073; R. Giles, L. McLerran, and C. Thorne, Phys. Rev. D 17 (1977) 2058; R. Brower, R. Giles, and C. Thorne, Phys. Rev. D 18 (1978) 484.
9. A. Polyakov, Phys. Lett. 77B (1978) 211, 82B (1979) 247; Y. Nambu, Phys. Lett. 80B (1979) 372; I. Ya. Araf'eva, Lett. Math. Phys. 3 (1979) 241; D. Foerster, Nucl. Phys. 119B (1977) 141, Phys. Lett. 87B (1979) 87; A. Neveu and J. L. Gervais, Phys. Lett. 81B (1979) 225; A. Migdal, preprint, 1979.
10. E. Corrigan and B. Hasslacher, Phys. Lett. 80B (1979) 225; T. Eguchi, Phys. Lett. 87B (1979) 91; T. Eguchi and S. Wadia, U. of Chicago preprint EFI 78165; K. Kikkawa and M. Sato, Phys. Rev. Lett. 38 (1977) 1309; D. Weingarten, Phys. Lett. 87B (1979) 97; L. Durand and E. Mendel, Univ. of Wisconsin preprint COO-881-88 (1979); N. V. Borisov, M. I. Eides, and M. V. Ioffe, Phys. Lett. 81B (1979) 101, Yad. Fiz. 29½ (1979) 1421.
11. A. Jericki and B. Sakita, "The Quantum Collective Field Method," preprint, Brown University, 1979; A. Jevicki and G. Papanicolaou, to be published.
12. E. Brezin, C. Itzykson, G. Parisi, and J. B. Zuber, Comm. Math. Phys. 59 (1978) 35; D. Bessis, Saclay preprint DPh-T 23/79, to appear in Comm. Math. Phys.; Y. Goldschmidt, Saclay preprint DPh-T 153/79; D. J. Gross and E. Witten, Princeton preprint, 1979.
13. M. Casartelli, G. Marchesini, and E. Onofri, Univ. of Parma preprint (February, 1979).
14. C. Itzykson and J. B. Zuber, Saclay preprint DPh-T 82/179; M. L. Mehta, Saclay preprint DPh-T 124/79.

GENERALIZED NON-LINEAR σ-MODELS WITH GAUGE INVARIANCE

Jean Zinn-Justin

Service de Physique Théorique
CEN Saclay
BP n°2 - 91190 Gif-sur-Yvette, France

INTRODUCTION

In these lectures we shall give a brief description of a family of models which generalize the non-linear σ-model, and possess in addition a local gauge invariance without containing explicitly a gauge field.

For more details we shall refer to various published articles.

In the first section we shall review the well known properties of the non-linear σ-model. In section II we characterize the largest class of boson field theories we know which have classically an infinite number of non local conserved currents.

In section III we describe some properties of these theories as asymptotic freedom and non perturbative spectrum in two dimensions.

I. THE NON LINEAR σ-MODEL[1]

Definition

The correlation functions of the non-linear σ-model are given by the functional integral :

$$Z(\underset{\sim}{J},H) = \int \prod_{x} d\underset{\sim}{\pi}(x)\,[1-\underset{\sim}{\pi}^2(x)]^{-1/2} \exp -\mathcal{A}(\pi) + \frac{1}{g}\int dx\,(\underset{\sim}{J}(x).\underset{\sim}{\pi}(x) + H(x)\,\sigma(x)) \tag{1}$$

with

$$\mathcal{A}(\pi) = \frac{1}{2g} \int dx \; [\partial\mu \; \underset{\sim}{\pi}(x)^2 + \partial\mu \; \sigma(x)^2] \tag{2}$$

and $\sigma(x) = [1-\underset{\sim}{\pi}^2(x)]^{1/2}$. (3)

The $\pi(x)$ field is an n-1 component field which transforms under a nõn linear representation of the O(n) group :

$$\delta\underset{\sim}{\pi}(x) = \underset{\sim}{\omega} \; \sigma(x) \tag{4}$$

and the action $\mathcal{A}(\pi)$ is O(n) invariant. The model can be considered as defined on a coset space O(n)/O(n-1). The coupling constant g plays the role of the temperature in Statistical Mechanics. The perturbative expansion in powers of g is thus equivalent to a low temperature expansion.

We shall now describe without proof a few properties of this model.

Perturbative expansion and renormalization

Power counting suggests that the model might be renormalizable in two dimensions. If we expand $\mathcal{A}(\pi)$ in powers of the $\underset{\sim}{\pi}(x)$ field:

$$\mathcal{A}(\pi) = \frac{1}{2g} \int dx \; [(\partial\mu \; \underset{\sim}{\pi})^2 + (\underset{\sim}{\pi}.\partial\mu \; \underset{\sim}{\pi})^2 + O(\pi^6)] \; . \tag{5}$$

We see that the classical spectrum consists in (n-1) massless particles corresponding to the π field. In two dimensions the correlation functions of a massless field are in general infrared divergent as integrals of the form $\int d^2p/p^2$ are encountered. One has to provide an infrared cut-off. This can most easily be done by expanding Z(J,H) around H(x)=h, with h constant and non vanishing. By expanding $h\int\sigma(x)\,dx$ in powers of π, we see that this term provides a mass for the π field.

It should nevertheless be pointed out that a conjecture[2] verified explicitly for the three first terms[3] of the perturbation series, states that the O(n) invariant correlation functions are not infrared divergent. An example is :

$$<\underset{\sim}{\pi}(o).\underset{\sim}{\pi}(x) + \sigma(o)\; \sigma(x)> \; . \tag{6}$$

An ultraviolet regularization which respects the O(n) symmetry has also to be found to be able to renormalize the theory. As the symmetry is non linearly realized, one cannot use a simple Pauli-Villar regularization. The most detailed discussion is allowed by the lattice regularization which transforms the model in a model of Statistical Mechanics : the classical Heisenberg ferromagnet. For practical purpose it is however convenient to use the dimensional regularization. It is then possible to prove that only two renormalization constants are needed to render the theory finite in two dimensions : A coupling constant and a wave function renormalization.

$$\underset{\sim}{\pi}(x) \rightarrow \sqrt{Z}\, \underset{\sim}{\pi}(x) \tag{7}$$

with now : $$\sigma^2(x) + \underset{\sim}{\pi}^2(x) = 1/Z \ . \tag{8}$$

Asymptotic freedom

It is easy to calculate the renormalization group functions. The coupling constant renormalization group function $\beta(g)$ is at leading order, in two dimensions :

$$\beta(g) = -(n-2)g^2 + O(g^3) \ . \tag{9}$$

We see that for $n > 2$ the theory is asymptotically free, as gauge theories are in 4 dimensions. This analogy is one of the most interesting features of the model.

Another consequence of this result is that for d larger than 2, and d-2 infinitesimal we have :

$$\beta(g) = (d-2)g - (n-2)g^2 + O(g^3, g^2(d-2)) \ . \tag{10}$$

Therefore the theory possesses an ultra-violet stable fixed point g^* :

$$g^* = \frac{d-2}{n-2} + O((d-2)^2) \ . \tag{11}$$

In the language of phase transitions g^* is a critical temperature which separates two different phases of the model.

Phase structure and spectrum (for $n > 2$)

For d larger than 2 one can calculate the vacuum expectation value of the field $\sigma(x)$ in zero field from renormalization group arguments :

$$\langle\sigma\rangle \sim \left(1 - \frac{g}{g^*}\right)^{\beta} \qquad g \lesssim g^* \ (h=0) \tag{12}$$

with

$$\beta = \frac{n-1}{2(n-2)} + O((d-2)) \ . \tag{13}$$

In the small coupling (low temperature) phase one component of the vector (π,σ) has a non-zero expectation value and the other components (the $\underset{\sim}{\pi}$ field) correspond to (n-1) Goldstone bosons of the O(n) spontaneously broken symmetry.

For $g = g^*$, the expectation value of $\sigma(x)$ vanishes and the symmetry is restored. For $g > g^*$ presumably the π and σ fields remain degenerate and become massive. This conjecture receives a direct confirmation from the 1/n expansion and from the numerical analysis of high temperature series expansions.

In two dimensions g^* vanishes, and therefore the true spectrum, consisting of n massive and degenerate states, is different for all values of the coupling constant from the perturbative spectrum. This again is a similarity with gauge theories.

The large n expansion

Using simple algebraic transformations it is possible to generate a systematic 1/n expansion. The leading term corresponds to the continuous version of the spherical model. The procedure is the following : we first introduce a Lagrange multiplier $\alpha(x)$ to enforce the condition

$$\tilde{\pi}^2(x) + \sigma^2(x) = 1 \tag{14}$$

and rewrite $Z(\tilde{J},H)$ as :

$$\begin{aligned} Z(J,H) = \int[d\pi\, d\sigma\, d\alpha] \exp\Big[& -\mathcal{H}(\pi,\sigma) \\ & + \frac{1}{2g}\int dx\, \alpha(x)\,(1-\tilde{\pi}^2(x)-\sigma^2(x)) + \frac{1}{g}\int\Big(\tilde{J}(x).\tilde{\pi}(x) \\ & + H(x)\,\sigma(x)\Big)dx\Big] \ . \end{aligned} \tag{15}$$

In this form the functional integral is gaussian in $\tilde{\pi}(x)$. We can thus explicit the dependence in n by integrating over the field $\pi(x)$. Restricting ourselves to the σ correlation functions for simplicity we get :

$$\begin{aligned} Z(H) = \int[d\sigma d\alpha] \exp & -\frac{1}{g}\int dx\left\{\frac{1}{2}\,\partial\mu\,\sigma(x)^2\right. \\ & \left. + \frac{1}{2}\,\alpha(x)\,\sigma^2(x) - \frac{1}{2}\alpha\,(x) - H(x)\sigma(x)\right\} \\ & - \frac{(n-1)}{2}\,\mathrm{tr}\,\ln\,(-\Delta+\alpha(x)) \ . \end{aligned} \tag{16}$$

If we now take g of order 1/n we can generate a 1/n expansion by a steepest descent calculation.

The results are the following :

The theory can be defined even in zero field for all dimensions between 2 and 4.

It can be shown that the non-linear σ-model is identical to the ϕ^4 theory with O(n) symmetry when the ϕ^4 coupling takes it infrared fixed point value.

In two dimensions the infrared instability leads to a restoration of the O(n) symmetry and a spectrum of n massive particles.

In higher dimensions there is a phase transition at a finite value of the coupling constant.

Conservation laws and S-matrix in two dimensions

We shall derive in section II that the model has classically and infinite number of non local conserved currents. As Lüscher[4] has shown, if one assumes the form of the spectrum, and that these currents are also conserved quantum mechanically, then one can calculate explicitly the S-matrix of the model. The S matrix one obtains in this way is the S-matrix previously found by Zamalodchikov and Zamalodchikov[5] by postulating the absence of production of particles and the property of factorization These properties can also be derived from local conservation laws. Such conservation laws have been postulated by Polyakov[6] and seem to hold in the framework of the 1/n expansion[7].

Such an S-matrix depends on the coupling constant only through a scale of momenta. There is therefore no possible perturbative check. On the other hand it agrees with the results obtained from the 1/n expansion up to order $1/n^2$[8].

All these known features of the non-linear σ-model are by themselves sufficiently interesting to motivate a study of possible generalizations of the model. Our starting point will be the characterization of models which possess classically an infinite number of non local conservation laws.

II. NON-LOCAL CONSERVATION LAWS IN TWO DIMENSIONS

Following reference [9] we shall show how one can construct an infinite number of non local currents in two dimensions and then characterize a family of models which have the necessary properties.

A set of conserved currents

Let us assume that we have found a set of matrices $A_\mu^{\alpha\beta}(x)$ with the following properties :

(i) $A_\mu(x)$ is a pure gauge, i.e. there exists a non singular matrix g such that :

$$A_\mu(x) = g^{-1}(x)\, \partial\mu\, g(x) \tag{17}$$

(ii) $A_\mu(x)$ is a conserved current :

$$\partial\mu\, A_\mu(x) = 0 \quad . \tag{18}$$

We can then construct an infinite number of non local conserved currents by the following inductive procedure. We introduce first a covariant derivative D_μ :

$$D_\mu^{\alpha\beta} = \partial\mu\, \delta_{\alpha\beta} + A_\mu^{\alpha\beta} \quad . \tag{19}$$

As A_μ is a pure gauge D_μ satisfies :

$$[D_\mu, D_\nu] = F_{\mu\nu}(A) = 0 \quad . \tag{20}$$

The conservation of the current implies :

$$[\partial\mu, D_\mu] = \partial\mu\, A_\mu = 0 \quad . \tag{21}$$

We now assume that we have constructed the nth conserved current $J_\mu^{(n)}$. There exists therefore a matrix $\chi^{(n)}(x)$ such that :

$$J_\mu^{(n)} = \varepsilon_{\mu\nu}\, \partial_\nu\, \chi^{(n)} \qquad n \geq 1 \quad . \tag{22}$$

We define the (n+1)th current by :

$$J_\mu^{(n+1)} = D_\mu\, \chi^{(n)} \qquad n \geq 0 \quad . \tag{23}$$

We start the induction with :

$$\chi^{(o)} = \mathbb{I} \quad , \tag{24}$$

and thus :

$$J_\mu^{(1)} = A_\mu \quad . \tag{25}$$

We have now to show that $J_\mu^{(n+1)}$ is conserved :

$$\partial\mu\, J_\mu^{(n+1)} = \partial\mu\, D_\mu\, \chi^{(n)} = D_\mu\, \partial\mu\, \chi^{(n)} \quad . \tag{26}$$

We express then $\partial\mu\, \chi^{(n)}$ in terms of $J_\mu^{(n)}$:

$$\partial\mu\, J_\mu^{(n+1)} = -\, \varepsilon_{\mu\nu}\, D_\mu\, J_\nu^{(n)} \quad . \tag{27}$$

Introducing now $\chi^{(n-1)}$ we can write

$$\partial\mu \, J_\mu^{(n+1)} = -\varepsilon_{\mu\nu} \, D_\mu \, D_\nu \, \chi^{(n-1)}$$

$$= -\frac{1}{2} \varepsilon_{\mu\nu} \, [D_\mu, D_\nu] \, \chi^{(n-1)} = 0 \quad . \tag{28}$$

This completes the induction[10]. We shall present all models known to us which possess a curvature free conserved current. As shown by Eichenherr and Forger[11] they are associated to a geometrical structure called symmetric space.

All models are defined in the following way. Their classical euclidean action is :

$$\mathfrak{R} = \frac{1}{2} \int dx \ \mathrm{tr} \ [\partial\mu \ g(x) \ \partial\mu \ g^{-1}(x)]$$

$$= -\frac{1}{2} \int dx \ \mathrm{tr} \ A_\mu^2(x) \tag{29}$$

with

$$A_\mu(x) = g^{-1}(x) \ \partial\mu \ g(x) \tag{30}$$

in which $g(x)$ belongs to a matrix representation of a compact simple Lie group G. Their differ only by the domain of variation of $g(x)$ in the group space.

A) Chiral models[12]

The matrix **g** varies over the whole group G. The action is invariant under the global transformations of G×G .

$$g(x) \rightarrow g_1^{-1} \ g(x) \ g_2 \qquad g_1, g_2 \in G \quad . \tag{31}$$

The classical equations of motion are obtained by varying $g(x)$:

$$g(x) \rightarrow g(x) \ (1+\varepsilon(x)) \tag{32}$$

$$A_\mu(x) = g^{-1}(x) \ \partial\mu \ g(x) \rightarrow A_\mu + \partial\mu\varepsilon + [A_\mu \ \varepsilon] \quad . \tag{33}$$

The variation of the action is then

$$\delta\mathfrak{R} = \int dx \ \mathrm{tr} \ \{A_\mu(\partial\mu\varepsilon+[A_\mu,\varepsilon])\} \quad . \tag{34}$$

So that the equation of motion is :

$$\partial\mu \ A_\mu = 0 \quad . \tag{35}$$

The equation of motion states just the conservation of a curvature free current $A_\mu(x)$.

An infinite number of conservation laws therefore follows.

B) A second family of models[13]

The matrices g(x) vary now over a subset S of the group space defined in the following way.

We consider an involutive automorphism of the group

$$g \rightarrow g^*$$

with :

$$\begin{aligned}(g_1 g_2)^* &= g_1^* g_2^* \\ (g^*)^* &= g .\end{aligned} \qquad (36)$$

This automorphism is allowed to be trivial $g^* \equiv g$. We introduce now a fixed element of G which satisfies the condition :

$$g_o\, g_o^* = \lambda \mathbb{1} \qquad (37)$$

in which λ is a real or complex number[13].

The subset S is then defined by the elements g(x) of G of the form :

$$g(x) = u^{-1^*}(x)\; g_o\; u(x) \qquad (38)$$

in which u(x) is an arbitrary element of G. It follows immediately from the definition that

$$g(x)\; g(x)^* = \lambda \mathbb{1} \quad . \qquad (39)$$

Let H be the subgroup of G leaving g_o invariant

$$h \in H \Leftrightarrow g_o = h^{-1^*}\; g_o\; h \quad . \qquad (40)$$

The elements g(x) are invariant under a left multiplication of u(x) by an element of H. This defines a coset space G/H. The equation defining H can also be written :

$$g_o^{-1}\; h^*\; g_o = h \quad . \qquad (41)$$

This shows that H is the subgroup of G invariant under the involutive automorphism :

$$u \rightarrow \bar{u} = g_o^{-1}\; u^*\; g_o \quad . \qquad (42)$$

The coset space G/H is therefore a symmetric space corresponding to the automorphism $u \rightarrow \bar{u}$.

It is easy to verify that again the equation of motion is as in the case of chiral models :

$$\partial\mu\, A_\mu \;=\; 0 \quad , \tag{43}$$

and an infinite number of conservation laws follows.

Examples

The classification of symmetric spaces can be found in text books and for the orthogonal and unitary group in references[13,15].

A) $G \equiv O(N)\ ;\ g^* \equiv g.$

1) $\lambda = +1$

Without loss of generality g_o can be chosen to be diagonal. It has p +1 eigenvalues and N-p -1 eigenvalues. The coset space is then $O(N)/O(p)\times O(N-p)$. For p=1 it reduces to the usual O(N) non linear σ-model

2) $\lambda = -1$

Then N is of the form

$N = 2M$.

The matrix g_o can be chosen to be :

$$g_o \;=\; \begin{bmatrix} 0 & 1\!\!1 \\ -1 & 0 \end{bmatrix} \tag{44}$$

and the symmetric space is O(N)/U(M).

B) $G \equiv U(N)$

1) $g \equiv g^*$

The matrix g_o can be chosen diagonal with p+1 and N-p -1 eigenvalues. The symmetric space is $U(N)/U(p)\times U(N-p)$. For p=1 we recover the CP^{N-1} models introduced by Eichenherr [16].

2) The star automorphism is non trivial, it is just the complex conjugation.

Depending on the choice $\lambda = +1$ or $\lambda = -1$ one obtains respectively the symmetric spaces U(N)/O(N) or U(N)/Sp(N) respectively, Sp(N) being the symplectic subgroup of U(N).

III. THE QUANTUM THEORY[13]

Models defined on homogeneous spaces G/H can naturally be written as models possessing a gauge invariance associated with the group H without an explicit gauge field. Actually the corresponding gauge field is in these models a composite field function of the group element g(x). A simple way to see it is to start from the chiral models and to make them explicitly gauge invariant under a subgroup H of G. We can write the action as :

$$\mathcal{A} = -\frac{1}{2}\int dx \ \mathrm{tr}\ (g^{-1}D_\mu g)^2 \tag{45}$$

with $D_\mu = \partial\mu + g\,A_\mu$

and A_μ is a dynamical field belonging to the Lie algebra of H. The action can also be written :

$$\mathcal{A} = -\frac{1}{2}\int dx\ \mathrm{tr}\ (\partial\mu\, g\, g^{-1} + A_\mu)^2 \tag{46}$$

This shows that the integration on A_μ projects $\partial\mu\, g\, g^{-1}$ onto the subspace spanned by the generators of the Lie algebra of G orthogonal to those of the Lie algebra of H. In the parametrization of the models used in the previous section the gauge field A_μ will be therefore given by the projection of $g^{-1}\partial\mu\, g$ on the Lie algebra of H.

We shall consider from now on only the models which have been studied the most extensively the $O(n)/O(n-p)\times O(p)$ and the $U(n)/U(n-p)\times U(p)$ models[17]. Furthermore as these two models have very similar properties we shall present only the unitary model. We shall give some results for both models and indicate the properties which can be generalized to all symmetric spaces.

As the group elements g(x) have only +1 and -1 eigenvalues it is natural to reparametrize g(x) by introducing a projector of rank p and set :

$$g(x) = 2P(x) - 1 \ , \tag{47}$$

so that the action becomes :

$$\mathcal{A} = \frac{1}{2}\int dx\ \mathrm{tr}\ [\partial\mu\, P(x)\ \partial\mu\, P(x)] \tag{48}$$

with

$$P = P^+ \ ,\ P^2 = P \ ,\ \mathrm{tr}\ P = p \ . \tag{49}$$

It is then convenient (specially in view of the large n limit) to consider an orthonormed basis in the eigenspace on which P(x) projects :

$$P_{\alpha\beta}(x) = \varphi^i_\alpha(x)^*\, \varphi^i_\beta(x) \qquad \begin{matrix} 1 \le i \le p \\ 1 \le \alpha,\beta \le n \end{matrix} \ . \tag{49}$$

The p vectors satisfy then :

$$\varphi_{\alpha}^{*i}\,\varphi_{\alpha}^{j} = \delta_{ij} \quad . \tag{50}$$

At each point x the projector P(x) defines the vectors $\varphi^{i}(x)$ only up to a U(p) unitary transformation. As a result the action written in terms of the fields φ (x) will posses a U(p) gauge invariance, part of the original U(p)×U(n-p) gauge invariance which has survived our parametrization.

Of course this parametrization breaks the p → n-p symmetry of the problem. Let us write the action in terms of the new variables :

$$\mathfrak{S} = \int dx \left\{ \partial\mu\, \varphi_{\alpha}^{*i}\, \partial\mu\, \varphi_{\alpha}^{i} + \frac{1}{2} [(\varphi_{\alpha}^{j}\, \partial\mu\, \varphi_{\alpha}^{*i}\, \varphi_{\beta}^{i}\, \partial\mu\, \varphi_{\beta}^{*j}) + c.c.] \right\} . \tag{51}$$

We have now still to fix the gauge. The simplest choice is to write φ_{α}^{i} as :

$$\varphi_{\alpha}^{i} = \begin{pmatrix} u_{k}^{i} \\ v_{\alpha}^{i} \end{pmatrix} \qquad \begin{matrix} 1 \leq k \leq p \\ p+1 \leq \alpha \leq n \end{matrix} \tag{52}$$

The gauge freedom corresponds to a p×p hermitian matrix. We can therefore impose the condition :

$$U = U^{+} \quad . \tag{53}$$

The orthogonality condition allows us then to calculate U in terms of V :

$$U_{ij} = [1 - V_{\alpha}^{*} V_{\alpha}]_{ij}^{1/2} = 1 + O(V^{2}) \quad , \tag{54}$$

which are the p×(n-p) independent fields.

In terms of the fields V_{α}^{i} , the action is an infinite power series :

$$\mathfrak{S} = \frac{1}{t_{o}} \int dx\, [\partial\mu\, V_{\alpha}^{*i}\, \partial\mu\, V_{\alpha}^{i} + O(V^{4})] \quad . \tag{55}$$

The theory is renormalizable by power counting in two dimensions. To calculate the correlation functions of the gauge invariant operators $P_{\alpha\beta}(x)$, we need only two renormalization constants:

The coupling constant renormalization

$$t_{o} = t\, Z_{1} \tag{56}$$

and the multiplicative renormalization constant Z_2 of the traceless operator $P_{\alpha\beta} - p/n\, \delta_{\alpha\beta}$.

Infrared regularization

At the classical level the theory contains 2p(n-p) massless states, Goldstone bosons of the spontaneously broken U(n) symmetry. As in the case of the non linear σ-model, we have to introduce a mass term for the v-field to make the theory infrared finite. A gauge invariant mass term is for example provided by :

$$\sum_{\alpha=p+1}^{n} (P_{\alpha_\alpha} - \frac{p}{n}) = \sum_{i=1}^{p} v_\alpha^{*i} v_\alpha^{i} - \frac{p(n-p)}{n} \quad . \tag{57}$$

The renormalized action is then :

$$= \frac{1}{tZ_1} \left\{ \int d^2x \, \{\partial\mu \, v_\alpha^{*i} \, \partial\mu \, v_\alpha^{i} + O(v^4) - h\, Z_2 \, [\frac{p(n-p)}{n} - v_\alpha^{*i} v_\alpha^{i}] \right\} \tag{58}$$

Renormalization group functions[13]

The calculation of the renormalization group functions is straightforward. In 2+ε dimensions we obtain :

$$\beta(t) = \varepsilon t - nt^2 - 2\,[p(n-p)+1]\; t^3 + O(t^4) \tag{59}$$

$$\gamma(t) = -nt + O(t^3) \tag{60}$$

for the U(n)/U(p)×U(n-p) model

and :

$$\beta(t) = \varepsilon t - (n-2)t^2 - (2p(n-p)-n)\; t^3 + O(t^4) \tag{61}$$

$$\gamma(t) = -nt + O(t^3) \tag{62}$$

for O(n)/O(p)×O(n-p) .

These results contain as special cases the O(n) non linear σ-model and the CP^{n-1} model.

The first remark is that these models are all asymptotically free in two dimensions.

Actually it is expected from the Mermin-Wagner theorem[19], that all scalar models should have a phase transition at zero temperature.

For a model with only one coupling constant this means that it should be asymptotically free, while a model with many coupling constants should have at least one direction of infrared instability.

In reference [3] the β function has been calculated for

chiral models :

$$G \equiv SU(n) \quad \beta(t) = -\frac{1}{4}(n^2-1)t^2 - \frac{1}{32}(n^2-1)^2 t^3 + 0(t^4) \qquad (63)$$

$$G \equiv O(n) \quad \beta(t) = -\frac{n}{4}(n-2)t^2 - \frac{1}{32}(n(n-2))^2 t^3 + 0(t^4) \ . \qquad (64)$$

In reference (13) asymptotic freedom was shown for all models on symmetric spaces. (This includes chiral models which can be considered as G×G/G symmetric spaces), which are the most general scalar model with only one coupling constant we known in two dimensions.

In 2+ε dimension, all these A.F. models have a phase transition at a finite temperature t^* :

$$\beta(t) = \varepsilon t - \beta_2 t^2 + 0(t^3) \qquad (65)$$

$$t^* = \frac{\varepsilon}{\beta_2} + 0(\varepsilon^2) \ . \qquad (66)$$

Below t^*, the G symmetry is spontaneously broken and the spectrum consists in massless Goldstone bosons. Above t^*, the symmetry is restored and the particles form massive degenerate multiplets. When the models possess like the U(n)/U(p)×U(n-p) a gauge invariance, it is expected in addition that the non gauge invariant fields like the φ here correspond to confined particles.

In the case of the non linear σ model the true spectrum of the model for $t > t^*$ can be obtained from the 1/n expansion. For the general models the situation is not so simple as we shall see. Notice that we are able to find the large n limit only of U(n)/U(p)×U(n-p) and O(n)/O(p)×O(n-p) models, and then only in the limit p/n << 1 which breaks the p ↔ n-p symmetry.

The large n limit

We shall now use the remark at the beginning of this section, i.e. that a gauge field could have been introduced explicitly Also all these models can be considered as the formal limit of usual gauge invariant non linear model, when the gauge coupling constant becomes infinite :

$$\mathfrak{H} = \int dx \ \mathrm{tr} \left\{ \frac{1}{4e_o^2} F_{\mu\nu}(A)^2 - \frac{1}{2t_o} (g^{-1} D_\mu g)^2 \right\} \qquad (67)$$

$$= \int dx \ \mathrm{tr} \left\{ \frac{1}{4e_o^2} F_{\mu\nu}(A)^2 + \frac{1}{2t_o} D_\mu \, g \, D_\mu^* \, g^{-1} \right\} \ . \qquad (68)$$

If we could calculate the integral over the field g(x) in the large n limit, we could solve all the models corresponding to symmetric spaces in this limit. Unfortunately this involves the summation of planar diagrams. Instead we shall use part of the gauge invariance to eliminate some degrees of freedom of g(x). This can only be done in the cases in which the gauge group H is factorized, i.e.

$$U(n)/U(p)\times U(n-p) \quad \text{and} \quad O(n)/O(p)\times O(n-p) \quad .$$

By eliminating the U(n-p) factor, we are left with a number of degrees of freedom proportional to n rather to n^2 for n large.

Taking the orthogonal model as an example we can write the action

$$\mathcal{R} = \int dx \sum_{\alpha} (\partial\mu\, \varphi^i_\alpha + A^{ij}_\mu\, \varphi^j_\alpha)^2 \tag{69}$$

with the constraints :

$$\varphi^i_\alpha \varphi^j_\alpha = \delta_{ij} \qquad i,j = 1 \ldots p \quad . \tag{70}$$

It is easy to verify that the integration over the A_μ fields yields the action of the orthogonal model. Instead now we can enforce the constraints (70) through Lagrange multipliers[20]

$$\mathcal{R} = \int dx \left\{ \frac{1}{2t} [\partial\mu\, \varphi^i_\alpha\, \partial\mu\, \varphi^i_\alpha + 2A^{ij}_\mu\, \varphi^j_\alpha\, \partial\mu\, \varphi^j_\alpha + A^{ij}_\mu A^{ij}_\mu] \right.$$
$$\left. + \lambda(\varphi^i_\alpha\varphi^i_\alpha - p) + \mu_{ij}\, \varphi^i_\alpha\, \varphi^j_\alpha \right\} \tag{71}$$

with $\mathrm{tr}\mu = 0$.

The action is now quadratic in φ and the integral over the φ fields can be performed.

The result is an effective action[21], function only of A_μ, λ and μ , in which the n dependence is explicit exactly as in the case of the non linear σ model. If we take t of order 1/n, the complete action is proportional to n, and a steepest descent calculation of the path integral generates a 1/n expansion.

The essential difference with the σ model comes from the presence of the gauge field A_μ . Due to the gauge invariance, the field A_μ is massless, and so the massless Goldstone bosons have been traded against massless gauge fields generating long range forces and a different kind of infrared instability. The leading term in the 1/n expansion does not reveal the true nature of the spectrum. One has to use non relativistic approximations, as in the bound state problem of Q.E.D., to get some ideas about its structure. The result in two dimensions is that charged particles with respect to the O(p) gauge symmetry are confined. Physical states correspond in the original model to bilinears fields $\varphi^i_\alpha\varphi^i_\beta$, i.e. to the projector. Unfortunately it is difficult to obtain more quantitative results, and to understand more generally what happens when the dimension varies from two to four.

REFERENCES

[1] J.Zinn-Justin, Cargèse Summer School lectures 1976 M. Levy and P. Mitter ed.(Plenum Press) and references therein contained.
E. Brézin and J. Zinn-Justin P.R.L. 36, 691 (1976) and Phys. Rev. B14, 3110 (1976).
E. Brézin, J.C. Le Guillou and J. Zinn-Justin, Phys. Rev. D14, 2615 and 4976 (1976).
W.A. Bardeen, B.W. Lee and R.E. Schrock, Phys. Rev. D14, 985 (1976).

[2] S. Elitzur, I.A.S. preprint (1979).

[3] A. Mc Kane and M. Stone, Cambridge Univ. preprint 1979.

[4] M. Lüscher, Nucl. Phys. B135, 1 (1978).

[5] A.B. Zamolodchikov and A.B. Zamolodchikov, Nucl. Phys. B133, 525 (1978).

[6] A.M. Polyakov, Phys. Lett. B72, 224 (1977).

[7] I.Y. Arefeva, P.P. Kulish, E.R. Nissinov and S.J. Pacheva, LOMI preprint, E-I (1978).

[8] B. Berg, M. Karowski, V. Kurak and P. Weisz, Phys. Lett. 76B, 502 (1978).

[9] E. Brézin, C. Itzykson, J. Zinn-Justin and J.B. Zuber, Phys. Lett. 82B, 442 (1979).

[10] These non local conservation laws can also be derived from compatibility conditions of a linear system, see for example H. de Vega, Phys. Lett. 87B, 233 (1979).

[11] H. Eichenherr and M. Forger, Nucl. Phys. B155, 381 (1979).

[12] Non linear realizations of symmetry groups have been extensively discussed in the literature, see for example S. Coleman, J. Wess and B. Zumino, Phys. Lett. 177, 2239 (1969).

Chiral models have also a long history. An early reference is A.A. Slavnov and L.D. Faddeev, Teor. Mat. Fiz. 8, 297 (1971). English translation 8, 843 (1971).

[13] E. Brézin, S. Hikami and J. Zinn-Justin, Saclay preprint DPh-T/79-135, Nuclear Physics to be published.

[14] It is sufficient more generally that $g_o g_o^*$ commutes with all elements of the group.

[15] R. Pisarski, Princeton Univ. preprint (1979).

[16] H. Eichenherr, Nucl. Phys. B146, 215 (1978).

[17] V.L. Golo and A.M. Perolomov, Phys. Lett. 79B, 112 (1978).
A.V. Mikhailov and V.E. Zakharov, Pis'ma Zh. Eksp. Teor. Fiz. 27 (1978) 47 [JETP Lett. 27, 42 (1978)], Zh. Eksp. Teor. Fiz. 74 (1978) 1953, C.W. Misner, Phys. Rev. D18, 45, 10 (1978).
J. Fröhlich, IHES preprint (1979).
M. Dubois-Violette and Y. Georgelin, Phys. Lett. 82B, 251 (1979).

[18] The renormalization group functions for these models have also been obtained by S. Duane, Oxford preprint (1979).
The CP^{n-1} case had beentreated by S. Hikami, Prog. Theor. Phys. 62, 226 (1979).

[19] N. Mermin and H. Wegner, Phys. Rev. Lett. 17, 1133 (1966).
P. C. Hohenberg, Phys. Rev. 158, 383 (1973)
S. Coleman, Comm. Math. Phys. 31, 295 (1973).

[20] The method and the results are very similar to those of the CP^{n-1} model
A. D'Adda, P. Di Vecchia and M. Lüscher, Nucl. Phys. B146, 63 (1978).
E. Witten, Nucl. Phys. B149, 285 (1979).

[21] I.Y. Aref'eva, Steklov Institute Moscow preprint, and E.R. Nissinov and S.J. Pacheva, Sofia preprints, have studied the renormalizability of the 1/n expansion of the CP^n models.

INDEX